Cardiovascular Molecular Morphogenesis

Series Editor

Books in the Series

Myofibrillogenesis

Dipak K. Dube
Editor

Foreword by Roger R. Markwald

With 105 Figures, Including 11 Color Figures

Birkhäuser
Boston • Basel • Berlin

Dipak K. Dube
Department of Medicine and Department of Cell and Developmental Biology
SUNY Upstate Medical University
Syracuse, NY 13210
USA

Cover illustration: Adapted from Figure 1.7 (see chapter 1, "Myofibrillogenesis in Cardiac Muscle," by J.W. and J.M. Sanger for additional details) and from Figure 12.2A (see chapter 12, "Cellular, Molecular, and Developmental Studies on Heart Development in Axolotls," by L.F. Lemanski et al.).

Library of Congress Cataloging-in-Publication Data
Myofibrillogenesis / edited by Dipak K. Dube.
   p. cm.—(Cardiovascular molecular morphogenesis; 6)
   Includes bibliographical references and index.
   ISBN 0-8176-4226-9 (alk. paper)
   1. Myocardium. 2. Heart cells. I. Dube, Dipak K. II. Series.
QP113.2 .M965 2001
612.1'7—dc21   00-069765

Printed on acid-free paper.   ***Birkhäuser*** ®
© 2002 Birkhäuser Boston

ISBN 0-8176-4226-9
ISBN 3-7643-4226-9   SPIN 10834011

Production managed by Louise Farkas; manufacturing supervised by Jerome Basma.
Typeset by Best-set Typesetter Ltd., Hong Kong.
Printed and bound by Maple-Vail Book Manufacturing Group, York, PA.
Printed in the United States of America.

9 8 7 6 5 4 3 2 1

Contents

Contributors

Paul J.R. Barton, National Heart and Lung Institute, Imperial College of Medicine, SW3 6LY, London, England

Narasimhaswamy S. Belaguli, Department of Molecular and Cellular Biology, Department of Surgery, Baylor College of Medicine, Houston, TX 77030, USA

Pankaj K. Bhavsar, National Heart and Lung Institute, Imperial College of Medicine, SW3 6LY, London, England

Neil E. Bowles, Department of Pediatrics Cardiology, Baylor College of Medicine, Houston, TX 77030, USA

Nigel J. Brand, National Heart and Lung Institute, Imperial College of Medicine, SW3 6LY, London, England

Stephanie Burge, Department of Molecular Genetics, Biochemistry, and Microbiology, University of Cincinnati Medical Center, Cincinnati, OH 45267-0524, USA

Martin E. Cullen, National Heart and Lung Institute, Imperial College of Medicine, SW3 6LY, London, England

Kimberley A. Dellow, National Heart and Lung Institute, Imperial College of Medicine, SW3 6LY, London, England

Christopher R. Denz, Department of Medicine, SUNY Upstate Medical University, Syracuse, NY 13210, USA

Dipak K. Dube, Department of Medicine and Department of Cell and Developmental Biology, SUNY Upstate Medical University, Syracuse, NY 13210, USA

Syamalima Dube, Department of Medicine, SUNY Upstate Medical University, Syracuse, NY 13210, USA

Elizabeth Ehler, Institute of Cell Biology, ETH-Honggerberg, CH-8093 Zurich, Switzerland

Carol A. Eisenberg, Department of Cell Biology and Anatomy, Medical University of South Carolina, Charlston, SC 29425, USA

Leonard M. Eisenberg, Department of Cell Biology and Anatomy, Medical University of South Carolina, Charlston, SC 29425, USA

Dalton Foster, Department of Medical Physiology, Texas A & M University System Health Science Center, College Station, TX 77843, USA

Carol C. Gregorio, Department of Molecular and Cellular Biology and Department of Cell Biology and Anatomy, The University of Arizona, Tucson, AZ 85724, USA

Fukuko Hasebe-Kishi, Department of Anatomy and Cell Biology, Chiba University School of Medicine, Chiba 260-8670, Japan

Xupei Huang, Department of Medical Physiology, Texas A & M University System Health Science Center, College Station, TX 77843, USA

Ganapathy Jagatheesan, Department of Molecular Genetics, Biochemistry, and Microbiology, University of Cincinnati Medical Center, Cincinnati, OH 45267-0524, USA

Larry F. Lemanski, Associate Vice President for Research and Professor of Medical Physiology, Texas A & M University System, College Station, TX 77843; Vice President for Research, Florida Atlantic University, Boca Raton, FL 33431-0991, USA

Sharon L. Lemanski, Department of Medical Physiology, Texas A & M University System, Health Science Center, College Station, TX 77843, USA

Quing Li, Department of Medical Physiology, Texas A & M University System Health Science Center, College Station, TX 77843, USA

Eduardo Mascareno, Department of Anatomy and Cell Biology, State University of New York Health Science Center at Brooklyn, Brooklyn, NY 11203, USA

Matthew D. McLean, Department of Cell and Developmental Biology, SUNY Upstate Medical University, Syracuse, NY 13210, USA

Catherine McLellan, Department of Molecular and Cellular Biology, The University of Arizona, Tucson, AZ 85724, USA

Fanyin Meng, Department of Medical Physiology, Texas A & M University System Health Science Center, College Station, TX 77843, USA

Joseph M. Metzger, Department of Physiology, University of Michigan, Ann Arbor, MI 48109, USA

Daniel E. Michele, Department of Physiology, University of Michigan, Ann Arbor, MI 48109, USA

Takashi Mikawa, Department of Cell Biology, Cornell University Medical College, New York, NY 10021, USA

Antony J. Mullen, National Heart and Lung Institute, Imperial College of Medicine, SW3 6LY, London, England

Tin Moe Nwe, Department of Anatomy and Cell Biology, Chiba University School of Medicine, Chiba 260-8670, Japan

Jean-Claude Perriard, Ph.D., Institute of Cell Biology, ETH-Honggerberg, CH-8093 Zurich, Switzerland

Kathy Pieples, Department of Molecular Genetics, Biochemistry, and Microbiology, University of Cincinnati Medical Center, Cincinnati, OH 45267-0524, USA

Rethinasamy Prabhakar, Department of Molecular Genetics, Biochemistry, and Microbiology, University of Cincinnati Medical Center, Cincinnati, OH 45267-0524, USA

Jean M. Sanger, Department of Cell and Developmental Biology, University of Pennsylvania, School of Medicine, Philadelphia, PA 19104-6058, USA

Joseph W. Sanger, Department of Cell and Developmental Biology, University of Pennsylvania, School of Medicine, Philadelphia, PA 19104-6058, USA

Robert J. Schwartz, Department of Molecular and Cellular Biology, Baylor College of Medicine, Houston, TX 77030, USA

Jorge Sepulveda, Department of Pathology, Baylor College of Medicine, Houston, TX 77030, USA

Saiyid Shafiq, Department of Anatomy and Cell Biology, State University of New York Health Science Center at Brooklyn, Brooklyn, NY 11203, USA

Yutaka Shimada, Professor Emeritus, Department of Anatomy and Cell Biology, Chiba University School of Medicine, Chiba 260-8670, Japan

M.A.Q. Siddiqui, Department of Anatomy and Cell Biology, State University of New York Health Science Center at Brooklyn, Brooklyn, NY 11203, USA

Homare Suzuki, Department of Anatomy and Cell Biology, Chiba University School of Medicine, Chiba 260-8670, Japan

Jeffrey A. Towbin, M.D., Department of Pediatrics Cardiology, Molecular & Human Genetics, and Cardiovascular Sciences, Baylor College of Medicine, Houston, TX 77030, USA

Michael Wagner, Department of Anatomy and Cell Biology, State University of New York Health Science Center at Brooklyn, Brooklyn, NY 11203, USA

Robert E. Welikson, Department of Cell Biology, Cornell University Medical College, New York, NY 10021, USA

David F. Wieczorek, Department of Molecular Genetics, Biochemistry, and Microbiology, University of Cincinnati Medical Center, Cincinnati, OH 45267-0524, USA

Robert W. Zajdel, Department of Cell and Developmental Biology, SUNY Upstate Medical University, Syracuse, NY, 13210, USA

Chi Zhang, Department of Medical Physiology, Texas A & M University System Health Science Center, College Station, TX 77843, USA

Series Preface

The overall scope of this new series will be to evolve an understanding of the genetic basis of (1) how early mesoderm commits to cells of a heart lineage that progressively and irreversibly assemble into a segmented, primary heart tube that can be remodeled into a four-chambered organ, and (2) how blood vessels are derived and assembled both in the heart and in the body. Our central aim is to establish a four-dimensional, spatiotemporal foundation for the heart and blood vessels that can be genetically dissected for function and mechanism.

Since Robert DeHaan's seminal chapter "Morphogenesis of the Vertebrate Heart" published in *Organogenesis* (Holt Rinehart & Winston, NY) in 1965, there have been surprisingly few books devoted to the subject of cardiovascular morphogenesis, despite the enormous growth of interest that occurred nationally and internationally. Most writings on the subject have been scholarly compilations of the proceedings of major national or international symposia or multiauthored volumes, often without a specific theme. What is missing are the unifying concepts that can make sense out of a burgeoning database of facts. The Editorial Board of this new series believes the time has come for a book series dedicated to cardiovascular morphogenesis that will serve not only as an important archival and didactic reference source for those who have recently come into the field but also as a guide to the evolution of a field that is clearly coming of age. The advances in heart and vessel morphogenesis are not only serving to reveal general basic mechanisms of embryogenesis but are also now influencing clinical thinking in pediatric and adult cardiology.

Undoubtedly, the Human Genome Project and other genetic approaches will continue to reveal new genes or groups of genes that may be involved in heart development. A central goal of this series will be to extend the identification of these and other genes into their functional role at the molecular, cellular, and organ levels. The major issues in morphogenesis highlighted in the series will be the local (heart or vessel) regulation of cell growth and death, cell adhesion and migration, and gene expression responsible for the cardiovascular cellular phenotypes.

Specific topics will include the following:

- The roles of extracardiac populations of cells in heart development.
- Coronary angiogenesis.
- Vasculogenesis.
- Breaking symmetry, laterality genes, and patterning.
- Formation and integration of the conduction cell phenotypes.

- Growth factors and inductive interactions in cardiogenesis and vasculogenesis.
- Morphogenetic role of the extracellular matrix.
- Genetic regulation of heart development.
- Application of developmental principles to cardiovascular tissue engineering.

Cardiovascular Developmental
  Biology Center
Medical University of South Carolina
Charleston, South Carolina

Roger R. Markwald

Foreword

Cardiovascular disease is the single most common cause of death worldwide and the leading cause of disability/morbidity in the Western world. Most specifically, heart failure is the predominant long-term complication of virtually all cardiac disease. In turn, congenital heart disease is the most frequent human birth defect. In *Myofibrillogenesis*, Dr. Dipak Dube has assembled a stellar team of international authors who have combined their multidisciplinary expertise in a unique "cradle to grave" overview of the underlying developmental mechanisms that can conceptually result in fetal, postnatal, and adult disorders of the myocardial contracture apparatus. The result is something special, a first: *Myofibrillogenesis* is novel in that it is the first book to be dedicated entirely to the developing myocardium and the processes by which presumptive myocardial cells acquire and assemble their highly structured contractile apparatus. Dr. Dube has laid out a foundation of fundamental underpinnings for one of nature's most dynamic and complex phenomena that precede most others in development but yet must be repeated, in part, over and over again throughout prenatal and adult life. Integrated throughout the fourteen chapters of this book are the molecular and morphological nuances of myofibrillogenesis and how they can be used to fashion a normal, mature four-chambered heart, but, if developmentally perturbated, can produce cardiomyopathies, dystrophies, and structural birth defects or if abnormally "reawakened," reprogrammed, or remodeled may lead to heart failure.

Specifically, a comprehensive body of information is progressively and sequentially presented to reveal how totipotent cells of a fertilized ovum proliferate, the mechanisms that signal their restriction into a mesodermal then a myocardial lineage, how such cells, once committed to a cardiac fate, acquire the potential to encode cardiac-specific structural proteins and, finally, how these and related proteins are assembled into membrane-associated, nascent prototypes or templates for the future assembly of organized myofibrils. Using real-time imaging, we are also given a rare view of myofibrillogenesis as it occurs in living cells and tissues. Answers are given or suggested for the many gaps that exist in our understanding of sarcomerogenesis, including its link to nonmuscle, cytoskeletal proteins that can have direct relevance to transducing mechanical forces or to the differentiation of myocardial cells into electrical propagating cells of the central and peripheral cardiac conduction system. In one chapter, the challenging question of how sarcomeric phenotype is stabilized is considered. In normal cardiogenesis, controlled destabilization of myofibrils ("reverse myofibrillogenesis") is emerging as a major morphogenetic mechanism required for the septation of the cardiac chambers, the creation and alignment of ventricular-arterial connections, and the formation of valves—all examples of how the process of myofibrillogenesis can directly relate to the major forms of congenital heart disease.

I have enjoyed reading *Myofibrillogenesis* and have gained much by doing so. It is to the authors' credit that they have not just produced another book on heart development but rather a finely crafted synthesis of information that lies at the core of heart development today and is, surely, a cornerstone for future translation of a genetic code into the developmental bases of cardiovascular diseases.

Roger R. Markwald
Cardiovascular Developmental
Biology Center
Department of
Cell Biology and Anatomy
Medical University of South Carolina
Charleston, South/Carolina

Preface

Myofibrillogenesis has been studied extensively over the last 100 years. Until recently, we have not had a comprehensive understanding of this fundamental process. The emergence of new technologies in molecular and cellular biology combined with classical embryology has started to unravel some of the complexities of myofibril assembly in striated muscles.

The myofibril is the contractile apparatus of striated muscle cells, including cardiac and skeletal. muscles. In vertebrates, the ultrastructure and molecular composition of the sarcomeres are remarkably similar in various types of muscles. In striated muscles, the contractile proteins are arranged in a highly ordered three-dimensional lattice known as the sarcomere. Thus, the assembly of a myofibril involves the precise ordering of several proteins into a linear array of sarcomeres. The process is complicated further because most of these proteins may have multiple isoforms. Hence, it is difficult to define the precise role of each component in the process. This volume has been compiled as a comprehensive reference on myofibrillogenesis. It focuses on myofibrillogenesis in the developing heart because the heart is the first functional organ in vertebrate embryos.

Emphasis is given also to the maintenance of the fully differentiated muscle sarcomere, which is a less well understood process. Historically, sarcomeres were perceived to be stable and static. We now know that the sarcomeric structure is dynamic: new proteins are synthesized and incorporated to replace older and damaged contractile proteins. This repair process has been coined "sarcomere maintenance."

In addition, we have incorporated a section on the conduction system in the heart. The cardiac excitation–contraction systems initiate, perpetuate, and coordinate rhythmic contraction in the mature heart. The whole system is composed of a specialized subset of heart cells that form a network of fibers.

The book also contains reviews on myofibrillar disarray under various pathological conditions. For example, abnormal myofibrillar organization has been found to be associated with familial hypertrophic cardiomyopathy (FHC) in humans. FHC has been characterized as a disease of the sarcomere with morphological changes, including myocyte hypertrophy, disarray, and fibrosis. To date, at least seven genes of sarcomere proteins have been implicated in FHC in humans. Various mutations in these genes can result in similar pathophysiological changes.

It is now well established that myofibrillar organization is remarkably similar among vertebrates. The overall similarities among different species certainly imply the universality of the process. However, to explore potential variations in the myofibrillar process among different species, we have presented studies using different animal models—for example, mammalian, avian, and amphibian. Comparative analyses will help researchers better understand the strengths and weaknesses of various models.

I would like to express my sincere gratitude to Dr. Roger R. Markwald, the editor for the cardiovascular morphogenesis series, whose encouragement and support have been invaluable. I am also grateful to Dr. Robin Smith for his encouragement. Finally, I extend my appreciation to all the contributors for presenting their excellent work without which this compilation would not have been possible.

SUNY Upstate Medical University   *Dipak K. Dube*
Syracuse, New York

PART I

Assembly of Myofibrillar Proteins in Striated Muscle

CHAPTER 1

Myofibrillogenesis in Cardiac Muscle

Joseph W. Sanger and Jean M. Sanger

INTRODUCTION

Cardiomyocytes are among the first cells in an embryo to become organized into a functioning organ. During heart formation, these cells form myofibrils, contract, and undergo cell division. In this chapter, we will discuss the process of myofibril assembly in cardiomyocytes, based mainly on experiments and observations of primary cultures of embryonic chick cardiomyocytes. On the basis of the arrangement of specific cytoskeletal proteins in these cells, there appears to be a hierarchical progression of fibril formation leading to myofibrillogenesis, beginning with premyofibrils, followed by nascent myofibrils and mature myofibrils (Figure 1.1).

PREMYOFIBRIL MODEL FOR THE ASSEMBLY OF MATURE MYOFIBRILS

In the embryonic chick model system, hearts can be removed from 5- to 8-day-old embryos and treated with enzymes to isolate cardiomyocytes (reviewed by Dabiri et al., 1999a, 1999b). The freshly isolated heart cells, when placed in tissue culture, attach to the substrate, spread, and assemble myofibrils (Figure 1.2). The cells are capable of cell division, repeating in culture processes detected in the intact heart, namely the disassembly of myofibrils, chromosome separation, assembly of a cleavage furrow, cytokinesis, and the subsequent reformation of myofibrils in the two contractile daughter myocytes (Chacko, 1973; Kaneko et al., 1984; Dabiri et al., 1999a).

Z-Bands

The premyofibril model for the stepwise assembly of myofibrils in cardiac muscle cells illustrated in Figure 1.1 was deduced from immunofluorescent images of cardiomyocytes (Figure 1.3) that had been fixed and stained during different stages of spreading in tissue culture (Rhee et al., 1994). The α-actinin antibody delineates the boundaries of the sarcomeres, the smallest repeating units of a myofibril (Figure 1.1). When an anti-α-actinin antibody was used to stain the spreading cells, different populations of striated fibrils were evident (Figure 1.3). At the peripheral spreading edges of the cells, linear arrays of closely spaced bodies of α-actinin could be detected.

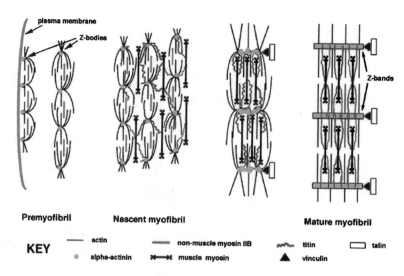

FIGURE 1.1. Model for steps in the assembly of a mature myofibril. The first fibrils (premyofibrils) form along the cell membrane with bands of nonmuscle myosin IIB lying between periodic deposits of α-actinin in minisarcomeric arrays. Actin filaments overlap along the premyofibrils, resulting in a continuous pattern of staining with phalloidin or actin antibody. Nascent myofibrils are characterized by the lateral association of α-actinin-rich Z-bodies and the appearance of titin and a second myosin isoform, muscle myosin II. The muscle myosin II filaments overlap along the fibrils and presumably bind to titin. In mature myofibrils, Z-bodies have fused to form mature Z-bands; nonmuscle myosin IIB is absent and muscle myosin filaments are now aligned into A-bands. The costameric proteins, talin and vinculin, first appear in mature myofibrils.

These localized concentrations of α-actinin we called Z-bodies. The Z-bodies mark the repeating units of minisarcomeres that compose the premyofibrils (Figures 1.1 and 1.3). Probes that specifically stained F-actin demonstrated that actin filaments were present between these Z-bodies in a pattern suggesting that the filaments overlapped (Figure 1.1). Toward the central region of the cell, Z-bodies were spaced further apart and aligned laterally across the fibrils. Adjacent to these fibrils in the central region of the spreading cells, myofibrils with mature solid-staining Z-bands, spaced about two microns apart, were readily visible (Figure 1.3). The continuous α-actinin staining of Z-bands composed of fused Z-bodies could be detected in immunofluorescently stained cells (Figure 1.3). It appeared from the patterns of α-actinin staining in the cardiomyocytes that these mature Z-bands resulted from fusion of Z-bodies (Figures 1.1 and 1.3).

Myosin II

Double labeling of these spreading cardiomyocytes with the muscle-specific α-actinin antibodies and two different isoforms of myosin II antibodies (nonmuscle myosin IIB or muscle myosin II antibodies) revealed distinctive patterns of myosin II localization in the spreading cardiomyocytes (Figures 1.1, 1.4, and 1.5). Premyofibrils stained only with the nonmuscle myosin IIB (Figure 1.4), whereas mature myofibrils exhibited only muscle myosin II staining (Figure 1.5). These myosins were localized in banded patterns, with their fluorescence concentrated between the concentrations

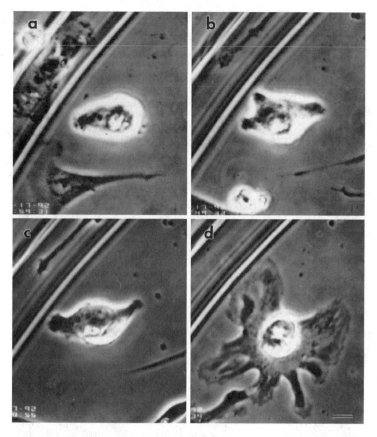

FIGURE 1.2. A live embryonic chick cardiomyocyte spreading in culture. The cell was beating as it spread. The expired times after the first image in (a) are: (b) 6 h 50 m; (c) 9 h 40 m; (d) 24 h. Bar = 10 μm. (Reproduced by permission of Cell Motility and the Cytoskeleton.)

of α-actinin. Thus, the nonmuscle myosin IIB was localized between the Z-bodies of the premyofibrils, and the muscle myosin II was concentrated in A-bands between the Z-bands (Figure 1.1). There was also a population of fibrils in an area between the premyofibrils and the mature myofibrils that stained with both antibodies (Figures 1.1 and 1.5). These fibrils with two isoforms of myosin II we called nascent myofibrils (Figure 1.1). The nonmuscle myosin IIB was present in striated bands between the Z-bodies, but the muscle myosin II was present in a continuous pattern. We interpreted the continuous staining pattern to be due to myosin thick filaments overlapping along the nascent myofibrils (Figure 1.1). The restricted distribution of the two myosins within the cytoplasm of single myocytes could occur if there were a transition from premyofibrils to nascent myofibrils to mature myofibrils with the nonmuscle isoform of myosin replaced by the muscle isoform. Future experiments are needed to determine how the localized assembly of the two different types of myosin filaments is determined.

There are two different isoforms of nonmuscle myosin II in vertebrate cells: IIA and IIB. The IIA isoform of nonmuscle myosin is concentrated in blood cells and in the intestinal lining (Murakami et al., 1993). Nonmuscle myosin IIB is found in cardiomyocytes and in the brain (Murakami et al., 1993). In cardiomyocytes,

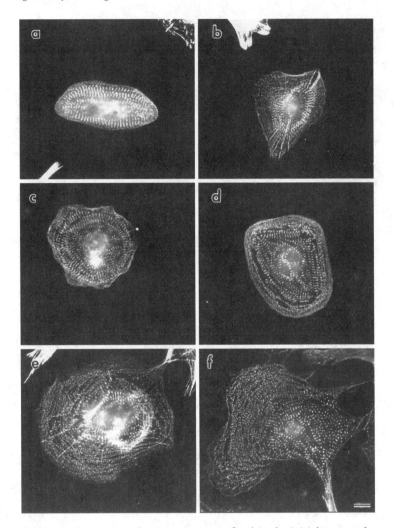

FIGURE 1.3. Chick embryonic cardiomyocytes were fixed in the initial stages of spreading in culture and then stained with muscle-specific α-actinin antibodies. (a to f) Note the central location of mature myofibrils and the peripheral location of premyofibrils. Bar = 10 μm. (Reproduced by permission of Cell Motility and the Cytoskeleton.)

nonmuscle myosin IIB is concentrated in the cleavage furrows of dividing cells (Conrad et al., 1991). Several cell types contain both isoforms (e.g., cardiac fibroblasts, Rhee et al., 1994; skeletal muscle cells; Fallon & Nachmias, 1980), but only the IIB form is in cardiomyocytes (Rhee et al., 1994). The reported absence of nonmuscle myosin II in cardiac muscle cells (Lu et al., 1992) was found to be due to the use of the nonmuscle myosin IIA antibodies to stain these cardiomyocytes (Rhee et al., 1994; Turnacioglu et al., 1997a). It is of interest that when nonmuscle myosin IIB was knocked out in mice, normal heart and brain development was inhibited in the few animals that survived to birth (Tullio et al., 1997). Since there were no descriptions of unusual cytokineses in these animals, we assume that these few animals that were able to survive did so by the induction of the synthesis of the IIA isoform of nonmuscle myosin in their cardiomyocytes and brain cells (Sanger et al., 2000).

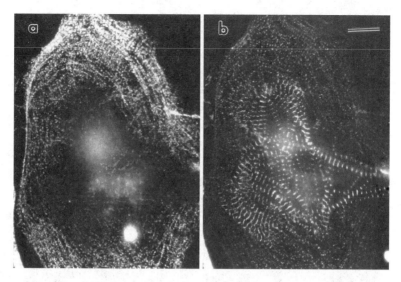

FIGURE 1.4. Spreading embryonic chick cardiomyocyte doubly stained with antibodies against nonmuscle myosin IIB (a) and muscle-specific α-actinin (b). The nonmuscle myosin II is concentrated in the peripheral area of the cell in a banded pattern in the premyofibrils. Alpha-actinin appears in small beads along these premyofibrils and in Z-bands in the mature myofibrils in the central region of the cell, where nonmuscle myosin IIB is absent. Bar = 10 μm. (Reproduced by permission of Cell Motility and the Cytoskeleton.)

PRECARDIAC MESODERM AND THE TRANSITION TO CARDIOMYOCYTES

The precardiac mesoderm tissue in vertebrate embryos gives rise to the cells that form the future heart (reviewed in Imanaka-Yoshida et al., 1998). Freshly isolated chick precardiac mesoderm tissue from stage 6 chick embryos (eggs 23 to 25 hours old) does not stain with muscle-specific α-actinin antibodies but does stain with a pan-α-actinin antibody that reacts with both nonmuscle and muscle isoforms of α-actinin (Imanaka-Yoshida et al., 1998). The staining is periodic along thin fibers similar in appearance to stress fibers (Sanger et al., 1983). After 10 hours of culture, the explants stained with muscle-specific α-actinin antibodies along fibrils that resemble the premyofibrils in cardiomyocyte cultures from 5- to 8-day chick embryos (Imanaka-Yoshida et al., 1998). The spacings between bands of α-actinin increased gradually from 0.83 μm to 1.86 μm over 68 hours in culture. The staining patterns suggested that the Z-bodies of these premyofibrils aligned and fused with one another to form the Z-bands of mature myofibrils (Imanaka-Yoshida et al., 1998). These authors also concluded that ". . . the assembly of Z-lines in precardiac explant cells is quite similar to those in cardiomyocytes cultured from older embryos (Rhee et al., 1994; LoRusso et al., 1997)." Reactivity of these same explant cells with a muscle-specific myosin II antibody was seen first at 10 hours in overlapping patterns reminiscent of nascent myofibrils (see Figure 1.1). The initial alignment of the myosin II staining into A-bands (mature myofibrils) was first detected at 20 hours. Future work will be needed to determine the presence of nonmuscle myosin IIB and the appearance of titin in the transition of precardiac mesoderm to embryonic cardiomyocytes.

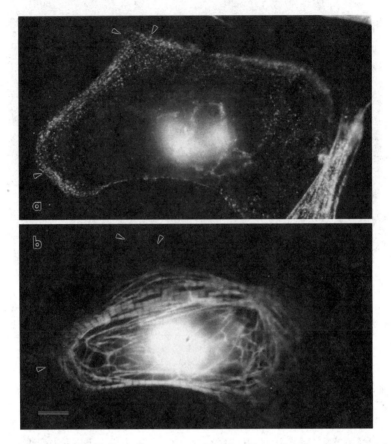

FIGURE 1.5. Embryonic chick cardiomyocyte stained with antibodies directed against non-muscle myosin IIB (a) and muscle-specific muscle myosin II (b). Arrowheads mark the margin of the cell. Muscle-specific myosin II is absent from the premyofibrils along the periphery of the cell. Some regions, especially on the right-hand side of the cell, contain both types of myosin in nascent myofibrils, and the area with clear A-bands of muscle myosin (mature myofibrils) is largely devoid of the nonmuscle myosin IIB. Bar = 10 μm. (Reproduced by permission of Cell Motility and the Cytoskeleton.)

DETECTION OF PREMYOFIBRILS IN NEONATAL AND ADULT CARDIAC MUSCLE CELLS

Evidence for a premyofibril model for the formation of mature myofibrils also exists in cardiomyocytes isolated from neonatal and adult mammalian hearts (LoRusso et al., 1997; Aoki et al., 1998, 2000). Figure 1.6 shows an isolated rat neonatal car-diomyocyte that was fixed after two days in culture and then stained with a muscle-specific α-actinin antibody. Short minisarcomeres, marked by the short spacings of α-actinin, were detected at the spreading edges of the neonatal cardiomyocyte. When neonatal rat cardiomyocytes were serum-starved for several days and the serum was then replaced or the hypertrophic agent angiotensin II added, the cells assembled pre-myofibrils within three hours (Aoki et al., 1998). We have repeated this regimen with cultured cardiomyocytes from 8-day chick embryos and obtained similar results.

Angiotensin II was shown to activate Rho A in the neonatal rat cardiomyocytes and inhibition of Rho with C3 inhibited premyofibril formation (Aoki et al., 1998). The

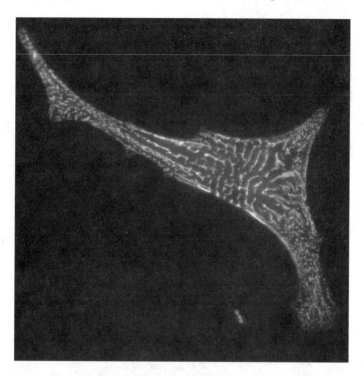

FIGURE 1.6. Neonatal rat cardiomyocyte fixed and stained with a muscle-specific α-actinin antibody. Note the beaded arrays of α-actinin typically found in premyofibrils (arrow) and the Z-bands of the mature myofibrils. Bar = 10 μm.

Rho family of small G proteins—Rho, Rac, and CDC42—exerts important control of the organization of the actin cytoskeleton in nonmuscle cells (Tapon and Hall, 1997). In broad outline, each of the three small G proteins links extracellular signals to the assembly of stress fibers (Rho), lamellipodia (Rac), and filopodia (CDC42), and all are involved in formation of focal adhesion complexes. The analogous structure and composition of myofibrils and stress fibers (Sanger et al., 1983, 1994a, 1994b) suggest that Rho might also promote an increase in myofibril assembly.

Freshly isolated cardiomyocytes from adult cat and rat hearts exhibit only mature myofibrils with no fibrils in the cells staining with nonmuscle myosin IIB antibodies (Figure 1.7). After a day in culture, however, these cells start to hypertrophy and extend lamellipodia. Fibrils formed in these newly spread areas with short spacings of α-actinin (Figures 1.8 and 1.9) and a banded pattern of nonmuscle myosin IIB (Figure 1.10) (LoRusso et al., 1997), suggesting a premyofibril model for the assembly of mature myofibrils in hypertrophying cardiomyocytes.

PROGRESSION OF PREMYOFIBRILS TO MATURE MYOFIBRILS IN TIME-LAPSE STUDY

The assembly of myofibrils in living muscle cells was initially carried out with fluorescently labeled proteins that were microinjected into the cells (Sanger et al., 1984a, 1984b, 1986a, 1986b; Dome et al., 1988; Sanger et al., 1990, 2000). The seminal

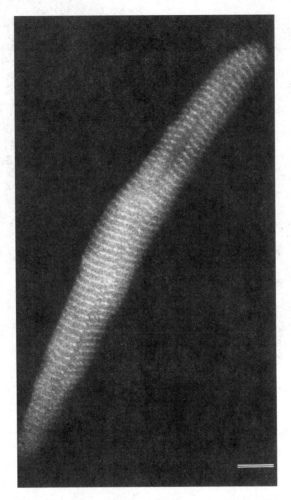

FIGURE 1.7. Freshly isolated adult cat cardiomyocyte fixed and stained with an antibody directed against α-actinin. The staining is concentrated in the Z-bands, which mark the boundaries of the sarcomeres. The average size of the sarcomeres in the mature myofibrils in this image is about 2 μm. Note that there are no signs of premyofibrils in this recently isolated cell. Bar = 10 μm. (Reproduced by permission of Cell Motility and the Cytoskeleton.)

discovery that the 28-kd bioluminescent green fluorescent protein (GFP) could be expressed fused to another protein, allowing that GFP to be followed in living cells, has provided a catalyst for the study of the dynamics of molecules in living cells (Chalfie et al., 1994; reviewed in Dabiri et al., 1999a). Proteins in low abundance or that are too insoluble to microinject into living cells can be studied. Fragments of large proteins such as titin can also be tested for their ability to target to different areas of a cell (Ayoob et al., 2000). Another advantage of transfecting cells with GFP plasmids encoding different sarcomeric proteins is the ability to follow the same cell over several days on a heated microscopic stage (Dabiri et al., 1999a,b). The transfected cells are able to continually synthesize the GFP fusion protein as the cell grows or spreads and thus the signal remains strong. Figure 1.11 illustrates one time point of a living embryonic chick cardiomyocyte transfected with a plasmid encoding the

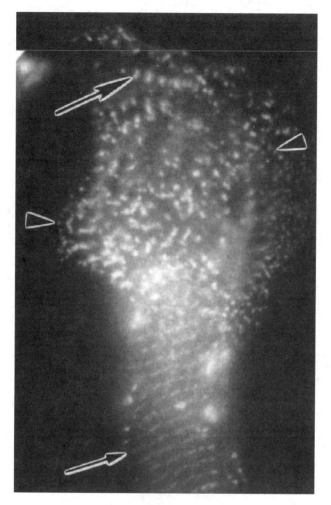

FIGURE 1.8. Live adult cat cardiomyocyte that was microinjected with rhodamine-labeled α-actinin. The labeled α-actinin is located in premyofibrils at the spreading edges of this cell (arrowheads). Note the presence of two fused Z-bodies (at the edge of the large arrow). Mature myofibrils are present in the lower part of the figure (small arrow). (Reproduced by permission of Cell Motility and the Cytoskeleton.)

DNA for α-actinin linked to GFP. The fluorescent probe is localized to the Z-bands and the intercalated disk. The transfection process and the localization of the fluorescent probe to the Z-bands did not inhibit the normal beatings of this cell. Fixation of similar transfected cells and subsequent staining with muscle-specific α-actinin antibodies revealed a colocalization of native α-actinin with the GFP-actinin construct.

Time points of the same transfected spreading muscle cell taken over several days in culture allowed a direct observation of myofibrillogenesis in culture (Figure 1.12). Premyofibrils with GFP-α-actinin formed in the cell in a spreading cytoplasmic margin that initially was devoid of fibrils. With time, the number of fibrils in the margin increased and the GFP-positive Z-bodies aligned laterally. Wider fibrils with beaded Z-bands formed and, subsequently, mature myofibrils were seen in the same

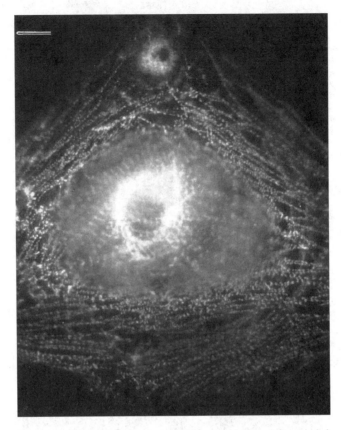

FIGURE 1.9. Adult rat cardiomyocyte that was in culture for several days and then stained with a muscle-specific α-actinin antibody. The Z-bodies seen in the premyofibrils in the outer region of the cell are spaced from 0.4 to 1.0 μm apart, in marked contrast to the 2-μm spacings of the Z-bands in the mature myofibrils in the center of the cell. Bar = 10 μm. (Reproduced by permission of Cell Motility and the Cytoskeleton.)

areas previously occupied by premyofibrils. Our results are consistent with the hypothesis, illustrated in Figure 1.1, that myofibril formation begins at the spreading edges of the cardiomyocytes with the formation of premyofibrils that subsequently fuse at the level of the Z-bodies to form mature myofibrils (Dabiri et al., 1997; Sanger et al., 2000).

ROLE OF TITIN IN THE FORMATION OF MATURE MYOFIBRILS

Part of what is now known to be the N-terminal region of titin (Labeit and Kolmerer, 1995; Turnacioglu et al., 1996) was originally isolated as a Z-band protein called zeugmatin (Maher et al., 1985). Localization studies with antibodies to zeugmatin indicated that this protein appeared when the Z-bodies were fusing (nascent myofibrils) to one another to form the mature myofibrils (Rhee et al., 1994). This was the same pattern observed with titin antibodies; see Figure 1.1 for the appearance of titin in nascent myofibrils (Rhee et al., 1994). We determined, through sequencing, that zeugmatin is actually part of the Z-band region of titin, that it contains the targeting site

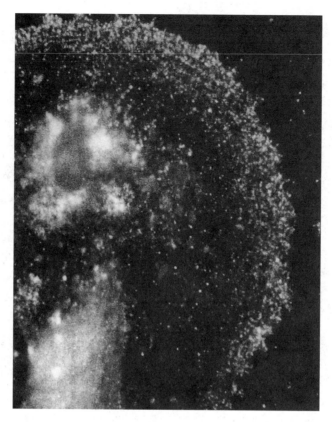

FIGURE 1.10. Adult cat cardiomyocyte in culture when stained with a nonmuscle myosin IIB antibody. Note the concentration of small bands of this myosin at the spreading edge of the cell. (Reproduced by permission of Cell Motility and the Cytoskeleton.)

for titin's localization to the Z-band, and that overexpression of a fragment containing the targeting site leads to the inhibition of new myofibrils and the loss of existing myofibrils (Turnacioglu et al., 1996, 1997a,b). The expressed fragment of this protein bound α-actinin and vinculin in an immunoprecipitation assay, suggesting that titin may be involved in the costameric localization of proteins in muscle (Turnacioglu et al., 1997b; Sanger et al., 2000).

A 1.1-kb cDNA fragment from the Z-band region of titin (Figure 1.13) was linked to the cDNA for green fluorescent protein (GFP) in an enhanced GFP (EGFP) expression plasmid. Transfection of muscle cells resulted in incorporation of the GFP fusion protein into the Z-bands of contracting myofibrils (Turnacioglu et al., 1997b). A dominant negative phenotype was also observed in living cells expressing high levels of this fusion protein, with myofibril disassembly occurring as titin-GFP fragments accumulated. A muscle-specific myosin II antibody was used to stain the A-bands in cardiac myocytes expressing different levels of the GFP fusion protein. When the level of the fusion protein was low, discrete 1.9-μm A-bands were localized between the fluorescent Z-bands that were spaced along the myofibrils. However, when a cardiomyocyte expressed high levels of GFP-titin fragment and therefore became intensely fluorescent, the A-bands were disrupted and scattered in these cells (Turnacioglu et al., 1997a). These data indicate that the Z-band region of

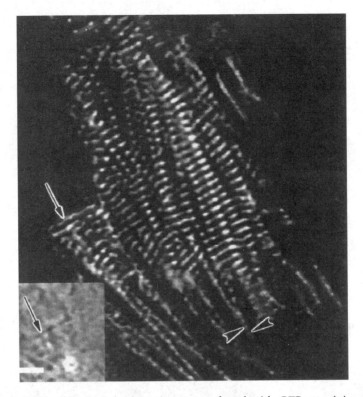

FIGURE 1.11. Embryonic chick cardiomyocyte transfected with GFP-α-actinin was beating in culture. Note the localization of the fluorescent α-actinin in the Z-bands of the mature myofibrils, in Z-bodies (arrowheads), and in the intercalated disk (arrow). Bar = 5 μm. (Images reproduced with permission from the Proceedings of the National Academy of Sciences.)

titin plays an important role in maintaining and organizing the structure of the myofibril. (Turnacioglu et al., 1997a).

Comparison of the complete cDNA for the Z-band region of titin from chicken cardiac titin (Ayoob et al., 2000) with an 11.5-kb cDNA derived from chicken skeletal muscle DNA (Yajima et al., 1996) revealed isoform differences (Figure 1.13). The Z-band region of cardiac titin contained six z-repeat motifs (Gautel et al., 1996) of approximately 45 amino acids, whereas the homologous region of skeletal titin contained only two or four z-repeats (Figure 1.13).

To determine which regions of this piece of titin were capable of targeting to the Z-bands in live cardiomyocytes, we assembled GFP-linked constructs for smaller fragments of the cardiac Z-band titin (Figure 1.13). GFP fusion peptides coding for either the entire Z-band region (Figure 1.14) or the six z-repeats (Figure 1.13) localized to the Z-bands, whereas the expression of a construct lacking the z-repeats resulted predominately in diffuse fluorescence in the cells (Figures 1.13 and 1.15). Single z-repeats also targeted to the Z-band as long as a suitable linker separated the expressed fragments from the EGFP (Ayoob et al., 2000). It is of interest that a z-repeat construct containing z-repeats not found in avian skeletal muscles cells (i.e., z-repeats 1 and 2) (Figure 1.13) are also able to target to the Z-bands of skeletal muscle cells (Ayoob et al., 2000). The z-repeats may be able to bind to the Z-bands of both cardiac and skeletal muscles due to their ability to bind α-actinin (Turnacioglu et al., 1996; Ohtsuka et al., 1997; Young et al., 1998).

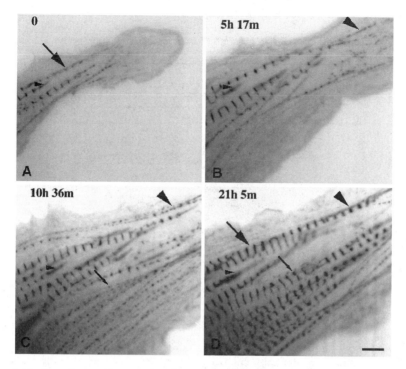

FIGURE 1.12. Four images from a time-lapse sequence of a spreading embryonic chick cardiomyocyte transfected with GFP-α-actinin. This cell was contracting as it spread in culture. Premyofibrils appeared in the lower right spreading margin of cytoplasm, and at 21h 5m myofibrils with solid Z-bands had formed. The image is shown in reverse contrast to render the Z-bands and Z-bodies more visible. The same prominent adhesion plaque is marked in each section of the figure as a reference mark (A to D, horizontal arrowhead). Note the amount of spreading with respect to this adhesion plaque. As the cell spread toward the lower right of the image, punctate Z-bodies appeared and assembled into linear arrays that fused laterally into Z-bands of myofibrils. A myofibril indicated by a large arrow in (A) doubles in length over the next 21 hours (D). This appears to happen by the lateral fusion of myofibrils (B to D) (large oblique arrowheads). Scale = 5 μm. (Images reproduced with permission from the Proceedings of the National Academy of Sciences.)

TITIN AND COSTAMERES

The Z-bands of mature myofibrils are attached to the cell membrane at sites called costameres, where proteins such as vinculin, tensin, and talin are concentrated; see Figure 1.1 for the distribution of vinculin and talin (Pardo et al., 1983; Danowski et al., 1992; Mondello et al., 1996). Whereas integrin, a transmembrane protein located at the Z-band sites, is known to bind a number of costameric proteins (e.g., α-actinin, talin, and tensin), it does not bind the classic costameric protein vinculin (reviewed in Mondello et al., 1996). Vinculin's association with integrin could be mediated through a number of proteins that bind both integrin and vinculin (α-actinin, talin, and/or tensin). There is also evidence that vinculin binds to the Z-band region of titin (Turnacioglu et al., 1996, 1997a, 1997b). Titin also binds α-actinin and a 19-kd protein, telethonin (Gregorio et al., 1998; Mues et al., 1998). Of the proteins that might link Z-bodies and Z-bands to integrin in the cell membrane, α-actinin is the only candidate in premyofibrils, as most proteins associated with costameres are first detected

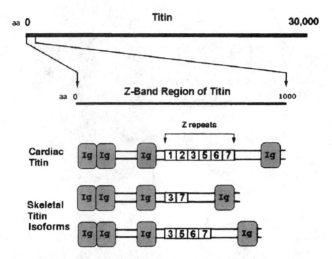

FIGURE 1.13. A diagram comparing the N-terminal region of titin from chicken cardiac muscle (Ayoob et al., 2000) with that region from chicken skeletal muscle (Ohtsuka et al., 1997). The oval boxes represent Ig-like domains, and the numbered boxes represent the z-repeat motifs. The six cardiac repeats have been numbered 1–3 and 5–7 to correspond to the numbering of the homologous sequences of mammalian cardiac titin (Sorimachi et al., 1997). The chicken skeletal titin has two variants with either two or four repeats. Chicken cardiac repeats 3 and 7 are identical in sequence to skeletal 2 and 5; likewise, cardiac 5 and 6 are identical in sequence to X and Y. (Reproduced by permission of Cell Motility and the Cytoskeleton.)

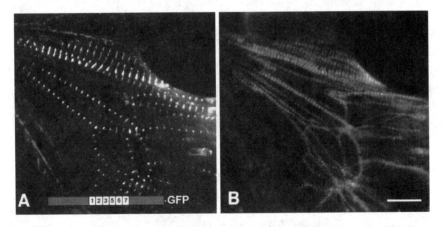

FIGURE 1.14. Chick embryonic cardiomyocyte transfected after one day in culture with (A) the full-length Z-band titin-GFP probe and counterstained with (B) a muscle-specific myosin II antibody. The full-length titin Z-band probe showed targeting to the Z-bands of all the muscle. Bar = 10 μm. (Reproduced by permission of Cell Motility and the Cytoskeleton.)

in the mature myofibrils (Sanger et al., 1997). Titin's appearance at the nascent myofibril stage may reflect a role in linking premyofibrils in their transition to mature myofibrils (Figure 1.1). In addition to membrane linkages, the Z-bands of adjacent mature myofibrils are linked to each other by intermediate filaments, composed of desmin (Granger and Lazarides, 1979), but binding partners of desmin that might effect this linking are not known. Future work is needed to determine if any other

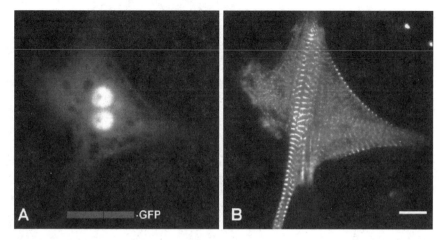

FIGURE 1.15. Chick embryonic cardiomyocyte transfected after one day in culture with a plasmid encoding the full Z-band titin-GFP construct from which the six z-repeats were deleted. (A) The transfected cell displayed a diffuse GFP fluorescence in the cytoplasm and fluorescent nuclei. (B) The Z-bands in the cell are shown counterstained with the α-actinin antibody. Bar = 10 μm. (Reproduced by permission of Cell Motility and the Cytoskeleton.)

Z-band proteins bind to titin to form costameric attachments of the mature myofibril to the cell surface or attachments to the intermediate filament system.

CELL DIVISION OF CARDIOMYOCYTES

Embryonic and neonatal cardiomyocytes undergo mitosis and cytokinesis to increase the number of cells in the developing heart (Chacko, 1973; Sanger, 1977; Clubb and Bishop, 1984; Kaneko et al., 1984; Conrad et al., 1991). The entry of cardiomyocytes into mitosis leads to a disassembly of mature myofibrils in the region of the mitotic spindle (Kaneko et al., 1984; Dabiri et al., 1997a). A contractile furrow of actin, α-actinin, and nonmuscle myosin IIB forms in the area between the two separating sets of chromosomes (Conrad et al., 1991; Sanger et al., 1993, 1994b; Dabiri et al., 1999a). Muscle myosin II (Conrad et al., 1991) and muscle-specific titin (Sanger et al., 2000) are not detected in the cleavage furrows. It appears that only proteins present in the premyofibrils are used for the assembly and contraction of cleavage furrows in dividing cardiomyocytes. As in nonmuscle cells, these cleavage furrow structures disassemble during the terminal stages of cytokinesis (Sanger and Sanger, 1980, 2000; Sanger et al., 1994a). At the end of cytokinesis, the two daughter cardiomyocytes reform their myofibrils. Future studies on live cells will have to determine whether the same premyofibril pathway as illustrated in Figure 1.1 is used after cytokinesis.

CONCLUSIONS

A number of approaches have indicated that the formation of mature myofibrils is mediated by the initial deposition of premyofibrils in close association with the cell's surface (Figure 1.1). Future work is needed to discover how the macromolecular properties of the sarcomeric proteins give rise to three different stages of myofibril assembly: premyofibrils, nascent myofibrils, and mature myofibrils. The first fibrils appear at the cell membrane, indicating that the cell surface may provide positional information to initiate this assembly. The premyofibril model of myofibrillogenesis also appears to be applicable to the formation of mature myofibrils in skeletal muscle cells (Sanger et al., 1984a, 1984b, 1989a; Dome et al., 1988; Ferrari et al., 1998; Sanger et al., 1999).

ACKNOWLEDGMENTS

Our work reported and discussed in this chapter has been supported by grants from the American Heart Association, the Muscular Dystrophy Association, the National Institutes of Health, and the National Science Foundation.

REFERENCES

Aoki, H., Izumo, S., and Sadoshima, J. 1998. Angiotensin II activates RhoA in cardiac myocytes. A critical role of RhoA in angiotensinII-induced premyofibril formation. *Circ. Res.* 82:666–676.

Aoki, H., Sadoshima, J., and Izumo, S. 2000. Myosin light chain kinase mediates sarcomere organization during cardiac hypertrophy in vitro. *Nat. Med.* 6:183–188.

Ayoob, J.C., Turnacioglu, K.K., Mittal, B., Sanger, J.M., and Sanger, J.W. 2000. Targeting of cardiac titin fragments to the Z-bands and dense bodies of muscle and non-muscle cells. *Cell Motil. Cytoskel.* 45:67–82.

Chacko, S. 1973. DNA synthesis, mitosis and differentiation in cardiac myogenesis. *Dev. Biol.* 35:1–18.

Chalfie, M., Tu, Y., Euskirchen, G., Ward, W.W., and Prasher, D.C. 1994. Green fluorescent protein as a marker for gene expression. *Science* 263:802–805.

Clubb, F.J. and Bishop, S.P. 1984. Formation of binucleated myocardial cells in the Neomatal rat. Lab. *Invest.* 50:571–577.

Conrad, A.H., Clark, W.A., and Conrad, G.W. 1991. Subcellular compartmentalization of myosin isoforms in embryonic chick ventricle myocytes during cytokinesis. *Cell Motil. Cytoskel.* 19:189–206.

Dabiri, G.A., Turnacioglu, K.K., Sanger, J.M., and Sanger, J.W. 1997. Myofibrillogenesis visualized in living embryonic cardiomyocytes. *Proc. Natl. Acad. Sci. USA* 94:9493–9498.

Dabiri, G.A., Turnacioglu, K.K., Ayoob, J.C., Sanger, J.M., and Sanger, J.W. 1999a. Transfections of primary muscle cell cultures with plasmids coding for GFP linked to full length and truncated muscle proteins. In: K.F. Sullivan and S.A. Kay, Eds., *Green Fluorescent Proteins*, Methods in Cell Biology, Vol. 58, Academic Press, New York, pp. 239–260.

Dabiri, G.A., Ayoob, J.C., Turnacioglu, K.K., Sanger, J.M., and Sanger, J.W. 1999b. Use of Green Fluorescent Proteins linked to cytoskeletal proteins to analyze myofibrillogenesis in living cells. In: P.M. Conn, Ed., Methods in Enzymology, Vol. 302, *Optical Imaging and Green Fluorescent Proteins*, Academic Press, New York, pp. 171–186.

Danowski, B.A., Inimaka-Yoshida, K., Sanger, J.M., and Sanger, J.W. 1992. Costameres are sites of force transmission to the substratum in adult rat cardiomyocytes. *J. Cell Biol.* 118:1411–1420.

Dome, J.S., Mittal, B., Pochapin, M.B., Sanger, J.M., and Sanger, J.W. 1988. Incorporation of fluorescently labeled actin and tropomyosin into muscle cells. *Cell Differ.* 23:37–52.

Fallon, J.R. and Nachmias, V.T. 1980. Localization of cytoplasmic and skeletal myosins in developing muscle cells by double immunofluorescence. *J. Cell Biol.* 87:237–247.

Ferrari, M.B., Ribbeck, K., Hagler, D.J., and Spitzer, N.C. 1998. A calcium signaling cascade essential for myosin thick filament assembly in *Xenopus* myocytes. *J. Cell Biol.* 141:1349–1356.

Gautel, M., Goulding, D., Bullard, B., Weber, K., and Furst, D.O. 1996. The central Z-disk region of titin is assembled from a novel repeat in variable copy numbers. *J. Cell Sci.* 109:2747–2754.

Granger, B.L. and Lazarides, E. 1979. Desmin and vimentin coexist at the periphery of the myofibril Z disc. *Cell* 18:1059–1063.

Gregorio, C.C., Trombitas, K., Center, T., Kolmerer, B., Stier, G., Kunke, K., Sizuki, K., Obermayr, F., Herrmann, B., Granzier, H., Sorimachi, H., and Labeit, S. 1998. The NH$_2$ terminus of titin spans the Z-disc: its interaction with a novel 19-kD ligand (T-cap) is required for sarcomeric integrity. *J. Cell Biol.* 143:1013–1027.

Imanaka-Yoshida, K., Knudsen, K.A., and Linask, K.K. 1998. N-cadherin is required for the differentiation and initial myofibrillogenesis of chick cardiomyocytes. *Cell Motil. Cytoskel.* 39:52–62.

Kaneko, H., Okamoto, M., and Goshima, K. 1984. Structural changes of myofibrils during mitosis of newt embryonic myocardial cells in culture. *Exp. Cell Res.* 153:483–498.

Labeit, S. and Kolmerer, B. 1995. Titins: giant proteins in charge of muscle ultrastructure and elasticity. *Science* 270:293–296.

Li, F., Wang, X., Bunger, P.C., and Gerdes, A.M. 1997. Formation of binucleated cardiac myocytes in rat heart: I. Role of actin-myosin contractile ring. *J. Mol. Cell. Cardiol.* 29:1541–1551.

LoRusso, S.M., Rhee, D., Sanger, J.M., and Sanger, J.W. 1997. Premyofibrils in spreading adult cardiomyocytes in tissue culture: evidence for reexpression of the embryonic program for myofibrillogenesis in adult cells. *Cell Motil. Cytoskel.* 37:363–377.

Lu, M.-H., Dilullo, C., Schultheiss, T., Holtzer, S., Murray, J.M., Choi, J., Fischman, D.A., and Holtzer, H. 1992. The vinculin/sarcomeric-alpha-actinin/alpha actin nexus in cultured cardiac myocytes. *J. Cell Biol.* 117:1007–1022.

Maher, P.A., Cox, G.F., and Singer, S.J. 1985. Zeugmatin: a high molecular weight protein associated with Z lines in adult and early embryonic striated muscle. *J. Cell Biol.* 101:1871–1883.

Mondello, M.R., Bramanti, P., Cutroneo, G., Santoro, G., DiMauro, D., and Anastasi, G. 1996. Immunolocalization of the costameres in human skeletal muscle: confocal scanning laser microscope investigations. *Anat. Rec.* 245:481–487.

Mues, A., van der Ven, P.F.M., Young, P., Furst, D.O., and Gautel, M. 1998. Two immunoglobulin-like domains of the Z-disc portion of titin interact in a conformational-dependent way with telethonin. *FEBS Lett.* 428:111–114.

Murakami, N., Trenkner, E., and Elzinka, M. 1993. Changes in expression of nonmuscle myosin heavy chain isoforms during muscle and nonmuscle tissue development. *Dev. Biol.* 157:19–27.

Ohtsuka, H., Yajima, H., Maruyama, K., and Kimura, S. 1997. The N-terminal z repeat of connectin/titin binds to the C-terminal region of alpha-actinin. *Biochem. Biophys. Res. Commun.* 235:1–3.

Pardo, J.V., Siliciano, J.D., and Craig, S.W. 1983. Vinculin is a component of an extensive myofibril-sarcolemma attachment region in cardiac muscle fibers. *J. Cell Biol.* 97:1081–1088.

Rhee, D., Sanger, J.M., and Sanger, J.W. 1994. The premyofibril: evidence for its role in myofibrillogenesis. *Cell Motil. Cytoskel.* 28:1–24.

Sanger, J.W. 1977. Mitosis in beating cardiac myoblasts treated with cytochalasin-B. *J. Exp. Zool.* 201:463–469.

Sanger, J.M. and Sanger, J.W. 1980. Banding and polarity of actin filaments in interphase and cleaving cells. *J. Cell Biol.* 86:568–575.

Sanger, J.M. and Sanger, J.W. 2000. Assembly of cytoskeletal proteins into cleavage furrows of tissue culture cells. *Microsc. Res. Tech.* 49:190–201.

Sanger, J.W., Sanger, J.M., and Jockusch, B.M. 1983. Differences in the stress fibers between fibroblasts and epithelial cells. *J. Cell Biol.* 96:961–969.

Sanger, J.W., Mittal, B., and Sanger, J.M. 1984a. Analysis of myofibrillar structure and assembly using fluorescently labeled contractile proteins. *J. Cell Biol.* 98:825–833.

Sanger, J.W., Mittal, B., and Sanger, J.M. 1984b. Formation of myofibrils in spreading chick cardiac myocytes. *Cell Motil. Cytoskel.* 4:405–416.

Sanger, J.M., Mittal, B., Pochapin, M.B., and Sanger, J.W. 1986a. Myofibrillogenesis in living cells microinjected with fluorescently labeled alpha-actinin. *J. Cell Biol.* 102:2053–2066.

Sanger, J.M., Mittal, B., Pochapin, M.B., and Sanger, J.W. 1986b. Observations of microfilament bundles in living cells microinjected with fluorescently labeled alpha-actinin. *J. Cell Sci. Suppl.* 5:17–44.

Sanger, J.M., Mittal, B., Meyer, T.W., and Sanger, J.W. 1989a. Use of fluorescent probes to study myofibrillogenesis. In: *Cellular and Molecular Biology of Muscle Development*, Eds. L. Kedes and F. Stockdale, Alan R. Liss Inc., New York, pp. 221–235.

Sanger, J.M., Mittal, B., Dome, J.S., and Sanger, J.W. 1989b. Analysis of cell division using fluorescently labeled actin and myosin in living PtK2 cells. *Cell Motil. Cytoskel.* 14:201–219.

Sanger, J.M., Dabiri, G., Mittal, B., Kowalski, M.A., Haddad, J.G., and Sanger, J.W. 1990. Disruption of microfilament organization in living nonmuscle cells by microinjection of plasma vitamin D-binding protein or DNase I. *Proc. Natl. Acad. Sci. USA* 87:5474–5478.

Sanger, J.M., Rhee, D., and Sanger, J.W. 1993. Cleavage furrows and premyofibrils in embryonic cardiomyocytes. *Mol. Biol. Cell* 4:53a.

Sanger, J.M., Dome, J.S., Hock, R.S., Mittal, B., and Sanger, J.W. 1994a. Occurrence of fibers and their association with talin in the cleavage furrows of PtK2 cells. *Cell Motil. Cytoskel* 27:26–40.

Sanger, J.M., Rhee, D., Leonard, M., Price, M., Zhukarev, V., Shuman, H., and Sanger, J.W. 1994b. Assembly of myofibrils and cleavage furrows in cardiomyocytes. *Mol. Biol. Cell* 5:165a.

Sanger, J.W., Zhukarev, V., and Sanger, J.M. 1997. Myofibrillogenesis and the surface attachments of Z-bands. *Mol. Biol. Cell* 8:375a.

Sanger, J.W., Ayoob, J.C., Yeh, C., Wei, Q., Adelstein, R.S., and Sanger, J.M. 1999. Study of cultured avian muscle cells transfected with a GFP plasmid encoding non-muscle myosin heavy chain IIB. *Mol. Biol. Cell* 10:168a.

Sanger, J.W., Ayoob, J.C., Chrowski, P., Zurawski, D., and Sanger, J.M. 2000. Assembly of myofibrils in cardiac muscle cells. In: *Elastic Filaments of the Cell*, Eds. G. Pollack and H. Granzier, Kluwer Academic/Plenum Press, New York, pp. 89–105.

Sorimachi, H., Freiburg, A., Kolmerer, B., Ishiura, S., Stier, G., Gregario, C.C., Labeit, D., Suzuki, K., and Labeit, S. 1997. Tissue-specific expression and alpha-actinin binding properties of the Z-disc titin: Implications for the nature of vertebrate Z-discs. *J. Mol. Biol.* 207:688–695.

Tapon, N. and Hall, A. 1997. Rho, Rac and Cdc GTPases regulate the organization of the actin cytoskeleton. *Curr. Opin. Cell Biol.* 9:86–92.

Tullio, A.N., Accili, D., Ferrans, V.J., Yu, Z.-X., Takeda, K.A., Grinberg, A., Westphal, H., Preston, Y.A., and Adelstein, R.S. 1997. Nonmuscle myosin II-B is required for normal development of the mouse heart. *Proc. Natl. Acad. Sci. USA* 94:12407–12412.

Turnacioglu, K.K., Mittal, B., Sanger, J.M., and Sanger, J.W. 1996. Partial characterization and DNA sequence of zeugmatin. *Cell Motil. Cytoskel.* 34:108–121.

Turnacioglu, K.K., Mittal, B., Dabiri, G.A., Sanger, J.M., and Sanger, J.W. 1997a. Zeugmatin is part of the Z-band region of titin. *Cell Struct. Funct.* 22:73–82.

Turnacioglu, K.K., Mittal, B., Dabiri, G.A., Sanger, J.M., and Sanger, J.W. 1997b. An N-terminal fragment of titin coupled to Green Fluorescent Protein localizes to Z-bands in living muscle cells: Overexpression leads to myofibril disassembly. *Mol. Biol. Cell* 8:705–717.

Yajima, H., Ohtsuka, H., Kawamura, Y., Kume, H., Murayama, T., Abe H., Kimuba, S., and Maruyama, K. 1996. A 11.5-kb 5′-terminal cDNA sequence of chicken connectin/titin reveals its Z line binding region. *Biochem. Biophys. Res. Commun.* 223:160–164.

Young, P.Y., Ferguson, C., Banuelos, S., and Gautel, M. 1998. Molecular structure of the sarcomeric Z-disk: two types of titin interactions lead to an asymmetrical sorting of alpha-actinin. *EMBO J.* 17:1614–1624.

CHAPTER **2**

Dynamics of Contractile Proteins Constituting Myofibrils in Living Muscle Cells

Yutaka Shimada, Tin Moe Nwe, Fukuko Hasebe-Kishi, and Homare Suzuki

The assembly of myofibrils is quite a complex process because this highly ordered structure is brought together by three sets of components: actin and myosin filaments, accessory proteins of actin and myosin filaments, and scaffolding proteins. Major advances in the understanding of this process have been made based on the antibody staining of cultured myogenic cells at different times of development. Various current descriptions of myofibrillogenesis are in agreement on the following process (Francini-Armstrong and Fischman, 1994).

At early phases, a structure containing many actin filaments resembling the stress fibers of fibroblasts is formed first. In these fibrils without cross-striations, α-actinin is initially distributed uniformly along their length (nonstriated fibrils or stress-fiber–like structures). It then forms dense punctate dots arranged linearly along them. These dots have a spacing shorter than the sarcomere periodicity ("minisarcomeres"—Sanger et al., 1986), and actin is not banded in these portions at this phase of myofibril assembly (premyofibrils). With increasing time, these dense dots (Z-bodies) evolve into more regular Z-lines with a spacing typical of the sarcomere, and periodic banding of actin becomes visible. During this process, myosin associates with these nascent structures, probably by the linking action of connectin (titin) (Handel et al., 1989, 1991; Komiyama et al., 1990, 1993; Van der Ven, 1999).

It has been demonstrated that when fluorescently labeled contractile proteins are injected into living cells, these analogs become localized precisely in the areas where the endogenous counterparts are known to be concentrated. Further, these fluorescent analogs can participate in reorganization of existing structures as well as the de novo assembly of new structures (Wang et al., 1982; Taylor et al., 1984; Wang, 1992). Thus, injected analogs can function as accurate tracers for the distribution and behavior of endogenous proteins.

Using this technique of fluorescent analog cytochemistry, Glacy (1983) elegantly observed, after microinjection of rhodamine-labeled actin into cultured cardiomyocytes, the rapid incorporation of this fluorescent analog into myofibrils of these cells, which displayed fluorescent bands that correspond to the sarcomeres. In subsequent years, in addition to actin, injected fluorescently labeled α-actinin, myosin, and tropomyosin have been shown to become localized precisely in the areas where these proteins are known to be concentrated (McKenna et al., 1985a, 1985b; Sanger et al., 1986; Dome et al., 1988; Johnson et al., 1988).

Various studies using radioactive precursors revealed incorporation and turnover of proteins in myofibrils of embryonic and adult myogenic tissues (Morkin et al., 1973; Anversa et al., 1975; Zak et al., 1977; Dadounce, 1980). Application of the combined techniques of microinjection, fluorescence microscopy, and image intensification with time-lapse recording (Wang, 1992) has enabled us to follow visually the fate of incorporated proteins in real time in the living cell. Further, a combination of these techniques with fluorescence recovery (FR) after photobleaching with a laser pulse allows examination of the relative exchangeability of proteins associated with various structures in cultured living cells. At appropriate intensities, the laser beam only bleaches the fluorophore, without disrupting the integrity of the structure (Jacobson et al., 1983; Saxton et al., 1984). Examination of such dynamic properties of proteins constituting myofibrils will assist in the understanding of the mechanisms of myofibrillogenesis.

In this chapter, we review the recent works of microinjection of fluorescently labeled myofibrillar proteins, which are incorporated into the forming myofibrils, in combination with FR after photobleaching with and without staining of antibodies against muscle scaffolding proteins.

DYNAMICS OF ACTIN

When actin labeled with a fluorescent dye (rhodamine [rh] or fluorescein isothiocyanate [FITC]) is injected into living cardiac or skeletal myocytes, the cells are filled with diffuse cytoplasmic fluorescence. As the diffuse fluorescence diminishes, the fluorescent conjugate associated with cellular structures becomes clearly visible within 2 hours. In younger areas near the advancing and lateral edges of cells where new myofibrils are known to assemble, the exogenously introduced actin is distributed in continuous thin lines (nonstriated fibrils), in punctate structures arranged in a linear pattern exhibiting minisarcomeres (premyofibrils), and in regular cross-striated bands with a single brighter zone in the center of each fluorescent band (mature myofibrils) (Figure 2.1a).

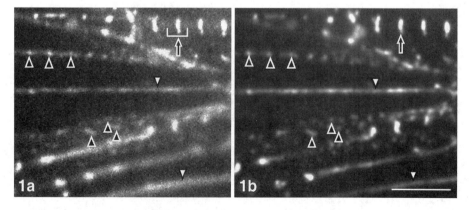

FIGURE 2.1. Images of a cardiomyocyte coinjected with FITC rh-actin (a) and rh-α-actinin (b). FITC-actin formed nonstriated (white arrowheads), punctate (black arrowheads), and striated patterns (black arrow, a). α-actinin was seen at the same sites as those of FITC-actin in the former two portions (black and white arrowheads) and in the center of striated bands in the latter portion (arrow, b). Bar = 5 μm.

Staining of rh-actin injected cells with anti-α-actinin antibody reveals that these antibody reactive sites (Z-lines) bisect the wide fluorescent bands in the rh-actin incorporated myofibrils at their striated portions and are colocalized with the central brighter zones (Figure 2.2). Staining with fluorescently labeled phalloidin, which specifically binds F-actin, shows that wide fluorescent bands in the micro-injected cells correspond to actin-containing I-bands, and the central brighter zone also appears in the shape of a bright line (Figure 2.3). An increased incorporation of actin monomers into nascent Z-bands (McKenna et al., 1985a; Dome et al., 1988; Suzuki et al., 1998; Nwe et al., 1999) appears to be extensive incorporation at barbed (fast-exchanging) ends of actin filaments, the portions known to possess a higher affinity for actin under physiological conditions (Wegner, 1976; Schafer and Cooper, 1995).

Exchangeability of Actin

Photobleaching of the fluorescent conjugate incorporated structures at appropriate intensities with an argon ion laser (200 to 500-ms laser pulse at 10 mW) does not seem to damage or destroy the structure. Staining of photobleached cells with anti-α-actinin and FITC-phalloidin indicated that the continuity of the structures is maintained across the bleached spots (Figures 2.2 and 2.3). Frequently, photobleached myofibrils continue to contract throughout the later experiments. The exchangeability of actin in nascent myofibrils can be analyzed both qualitatively by recording the images of the cell and quantitatively by measuring the average fluorescence intensity of the bleached segment.

FR after photobleaching of nascent myofibrils incorporated with fluorescent actin decreases significantly in the order of nonstriated, punctated, and striated portions

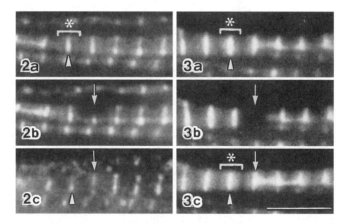

FIGURES 2.2 AND 2.3. Images of cardiomyocytes injected with rh-actin (a) before and (b) immediately after photobleaching (200 to 500 laser pulses at 10 mW), with FITC-α-actinin (Figure 2.2c) and FITC-phalloidin (Figure 2.3c) images of the same cells after bleaching. Labeled actin localized in myofibrils and formed striated patterns (asterisks, a), the centers of which were brighter (arrowheads, a). Anti-α-actinin stained the center of wide fluorescent bands (arrowhead, Figure 2.2c). FITC-phalloidin stained wide fluorescent bands (asterisk, Figure 2.3c) and especially their central brighter zones (arrowhead, Figure 2.3c). After photobleaching, continuity of structures was maintained across the bleached spot (arrows, c). Bar = 5 μm. (Reproduced by permission of Churchill Livingstone, Edinburgh, from Suzuki et al., 1998).

of myofibrils. Within 2 hours, nearly 95% recovery is seen in nonstriated portions, about 60% in punctate portions, and about 50% in striated portions (Shimada et al., 1997; Suzuki et al., 1998; Hasebe-Kishi and Shimada, 1999; Nwe et al., 1999) (Figures 2.4a, 2.5, 2.6, 2.7, and 2.8). Thus, as nascent myofibrils mature, actin becomes gradually stabilized within this structure; actin molecules cannot be readily exchanged at mature myofibrillar portions.

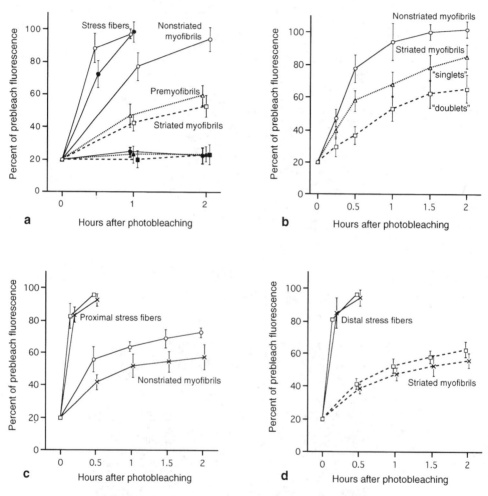

FIGURE 2.4. FR after photobleaching of myofibrils and stress fibers as a function of time. Muscle actin: open circles, triangles, and rectangles; nonmuscle actin: crosses; α-actinin: closed circles, triangles, and rectangles. FR of stress fibers with muscle actin, nonmuscle actin, and α-actinin is significantly faster than that of myofibrils at any phase (a, c, d). In myofibrils, FR of nonstriated (open circles, solid line), punctate (open triangles, dotted line), and striated portions (open rectangles, dashed line) decreases in this order (a). In nonstriated (proximal and terminal) portions of myofibrils, FR of muscle actin (open circles) is significantly faster than that of nonmuscle actin (crosses) (c); in striated portions it is insignificant (d). In stress fibers, FR between muscle (open rectangles) and nonmuscle (crosses) actins at both proximal and terminal portions is insignificant (c, d). Antinebulin immunostaining showed that FR becomes slower in the order of nonstriated myofibrils (open circles, solid line), myofibrils with nebulin singlets (open triangles, dotted line), and myofibrils with nebulin doublets (open rectangles, dashed line) (b). Values are mean ± SD.

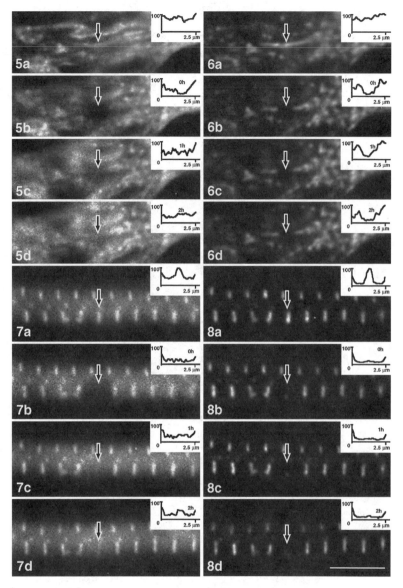

FIGURES 2.5–2.8. Fluorescence recovery (FR) after photobleaching of nascent myofibrils in cardiomyocytes coinjected with FITC-actin (Figures 2.5 and 2.7) and rh-α-actinin (Figures 2.6 and 2.8). Images of (a) before, (b) immediately after, (c) 1.0 hour after, and (d) 2 hours after nonstriated (Figures 2.5 and 2.6) and striated portions (Figures 2.7 and 2.8) of myofibrils. Arrows indicate the location of the bleached spot. Bar = 5 μm. Inset of each figure: intensity profiles of fluorescence (ordinate) across the bleached zone (abscissa, scale 2.5 μm) are shown.

Exchangeability of Nonmuscle Actin

Cardiac myocytes are incapable of discriminating between the incorporation of muscle and nonmuscle actin isoforms because both muscle (skeletal) and nonmuscle (brain and platelet) isoproteins can be incorporated identically into myofibrils (McKenna et al., 1985a; Shimada et al., 1997; Suzuki et al., 1998). Photobleaching experiments show that, in nonstriated portions of myofibrils, the FR rate of non-muscle actin is slow but that of muscle actin is significantly faster than that of non-muscle actin (Figure 2.4c). In striated portions of myofibrils, FR of both muscle and non-muscle actins is slow and insignificant between these two isoactins (Figure 2.4d) (Shimada et al., 1997; Suzuki et al., 1998). This shows that actin molecules in immature portions of cardiac myofibrils cannot be readily exchanged by heterotypic non-muscle actin.

In differentiating muscle cells, many of the contractile proteins, including actin, undergo a multitude of isoprotein transitions (Ordahl, 1986; Bandman, 1992). In the case of avian cardiac tissue, nonmuscle β-actin has been shown to be expressed in the embryo, and it is replaced by sarcomeric α-actin during subsequent development (Obinata et al., 1981; Ruzicka and Schwartz, 1988). It thus appears that different exchange rates of actin isoforms may account for the sorting of isoproteins; those with higher binding affinities (muscle actin) replace previous proteins, thereby increasing the stability of the resulting structures. Further, it has been indicated that the actin isoform (α-actin) predominantly found in the adult striated muscle is more closely related to the striated muscle form of α-actinin than γ- and β-actins found in embryonic muscles and nonmuscle tissues (Obinata et al., 1981; Endo and Masaki, 1984). This nature of α-actinin also facilitates the transition of actin isoforms to adult type.

Role of Nebulin for Actin Assembly in Skeletal Myofibrils

Nebulin is a giant structural protein of skeletal muscles and serves as an actin binding protein. It is considered to be associated with actin filaments and to serve as a molecular ruler either by restricting the lengths of the actin filaments or by stabilizing the actin filaments at uniform length (Wang and Wright, 1988; Kruger et al., 1991). Moncman and Wang (1996) have postulated that nebulin plays an important role in restricting the lengths of the actin filaments during myofibrillogenesis.

We have analyzed the temporal and spatial patterns of distribution of nebulin relative to those of rh-actin incorporated into cellular structures by antinebulin immunostaining of skeletal myotubes (Nwe et al., 1999). The sites where injected actin shows punctate structures arranged linearly are also stained with antinebulin antibodies as punctate dots similarly arranged in a linear fashion. Between these bundle-like lines, numerous smaller dots of nebulin are scattered. On these smaller dots, rh-actin fluorescence is not seen. In the myofibrillar areas where injected rh-actin forms striations, antinebulin antibodies form either single bands ("singlets") or double subbands ("doublets"). Myofibrils possessing nebulin singlets predominate initially (2 to 6 hours after injection), but those with doublets are more prevalent later on (6 hours or more after injection) (Figures 2.9–2.11).

Nebulin and/or its related protein(s) are known to possess actin monomer binding region(s) (Labeit et al., 1991; Luo et al., 1997). Thus, it appears that dots and lines of dots of nebulin and/or its related protein(s) construct a scaffold upon which actin monomers accumulate. Because smaller nebulin dots are devoid of rh-actin fluorescence, it is possible, as soon as rh-actin associates with these smaller nebulin dots, that they coalesce to arrange rh-actin into dots (actin oligomers or polymers) and then link these dots in chains to form nonstriated bundle-like structures.

A photobleaching experiment (Nwe et al., 1999) demonstrated that FR of myofibrils with nebulin singlets is faster than that of those with nebulin doublets (Figures 2.12 and 2.13). It is possible that early nebulin filaments exhibiting singlets are not

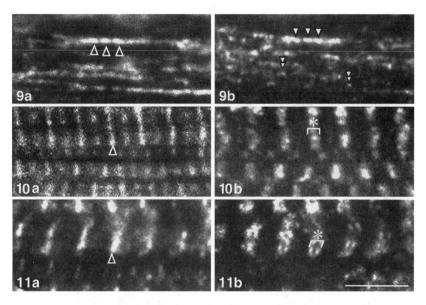

FIGURES 2.9–2.11. (a) Images of skeletal muscle cells injected with rh-actin. (b) The same cells stained for nebulin with NB2 (Sigma Chemical Co., St. Louis, MO). In premyofibrillar areas (Figure 2.9), rh-actin dots (black arrowheads) are arranged linearly to form nonstriated fibrils. Nebulin dots (white arrowheads) colocalize with rh-actin dots. Small nebulin dots (white double arrowheads) are seen between these fibrils. In developing myofibrils, nebulin forms singlets (asterisk, Figure 2.10b) or doublets (asterisk, Figure 2.11b) at the level of IZI regions (Z-line, arrowheads, Figures 2.10a and 2.11a). Bar = 5 μm.

tightly associated with actin filaments, and thus incorporation or exchange of actin occurs rapidly in the former myofibrillar areas. After the formation of a close linkage between nebulin and actin filaments during development, as is thought to be indicated by the presence of nebulin doublets, actin filaments become stabilized and the rate of actin dynamics becomes slower. Nonstriated fibrillar portions formed by a linear arrangement of rh-actin dots have much faster FR rates than striated myofibrillar portions do (Figure 2.4b). In these nonstriated portions, nebulin has a punctate appearance and has not yet formed striations (singlets or doublets). It thus appears that the association of such immature nebulin molecules/filaments with rh-actin allows the fast exchange of actin.

In the 1990s, it was reported that nebulin does not appear until the stage at which major myofibrillar proteins have emerged (Komiyama et al., 1992; Begum et al., 1998). It has been suggested that the absence of nebulin at early stages facilitates the transition of isoproteins to adult types during muscle maturation (Begum et al., 1998). After the emergence and assembly of nebulin into nascent myofibrils, the loose association of this cytoskeletal protein with actin filaments may further facilitate the exchange of nonadult-type isoform present in actin filaments into adult isoform, allowing uniform distribution of actin and other protein isoforms along actin filaments in myofibrils (Nwe and Shimada, 2000) (Figure 2.14).

Nebulin is absent from embryonic and adult cardiac muscles (Itoh et al., 1988; Wang and Wright, 1988). Instead, nebulin-like protein, nebulette, has been reported to exist in cardiac muscles. It has striking sequence conservation with the carboxy-terminal portion of skeletal muscle nebulin. It is also located at the IZI-complex of

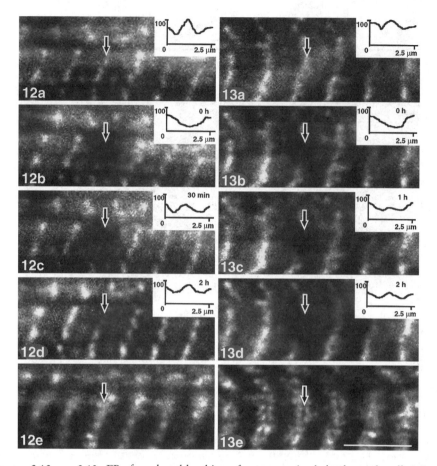

FIGURES 2.12 AND 2.13.   FR after photobleaching of structures in skeletal muscle cells injected with rh-actin. Images of (a) before, (b) immediately after, (c) 30 minutes (Figure 2.12), 1.0 hour (Figure 2.13), (d) 2.0 hours after bleaching. (e) Images of the same cells stained with antinebulin NB2 after FR. Striated myofibrils with nebulin singlets, Figure 2.12e; those with nebulin doublets, Figure 2.13e. Arrows indicate the location of the bleached spot. Bar = 5 μm. Insets of Figures 2.12a–d and 2.13a–d: intensity profiles of fluorescence (ordinate) across the bleached zone (abscissa) are shown.

the cardiac myofibrils but extends only about 25% of the actin filament length from the Z-line, in contrast to the entire length in skeletal muscle. This protein has been suggested to be involved in the early events of myofibril assembly (Moncman and Wang, 1995). It remains to be clarified if and how this small nebulin-like protein plays a role(s) in the dynamics of nascent cardiac myofibrils.

Inhibition of Nebulin

In order to prove whether nebulin is actually related to the process of actin filament assembly and dynamics during skeletal myofibrillogenesis, we examined this hypothesis by preventing the actin binding activity of nebulin in developing myotubes in culture by injection of a monoclonal antibody against nebulin. Several hours after the injection of the antibody, the same cells were injected with rh-actin. Observation of

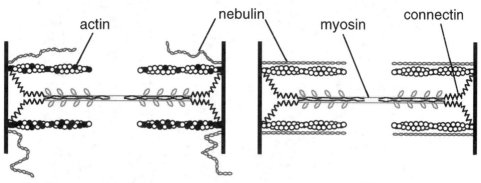

a. Early actin and nebulin filaments   b. Mature actin and nebulin filaments

FIGURE 2.14. A schematic diagram of (a) immature and (b) mature sarcomeres. Early actin filaments consist of actin molecules of adult (white circles) and nonadult isoforms (black circles). Early nebulin filaments are not tightly associated with actin filaments. (Reproduced by permission of Harcourt Health Sciences; Edinburgh, from Nwe and Shimada, 2000.)

these cells revealed that, although injected actin is seen as irregularly distributed amorphous patches or bright foci inside the cells, no incorporation of rh-actin into early and striated myofibrils is found until 24 hours after rh-actin injection. The injected antibody binds at linearly arranged dots and the I-band region forming singlets and doublets (Nwe and Shimada, 2000) (Figures 2.15–2.17). Thus, blockage of the actin binding sites of nebulin appears to inhibit the association of actin monomers to the preexisting nebulin scaffold.

Connectin (Titin) and Actin Assembly

Connectin is a long flexible filamentous protein in striated (cardiac and skeletal) muscles. Connectin filaments are believed to link myosin filaments to Z-bands (Maruyama, 1986; Horowits, 1999). During myofibrillogenesis, the possible role of connectin has been suggested for integrating myosin filaments with the early formed IZI-complexes of myofibrils (Handel et al., 1989; Komiyama et al., 1990, 1993; Van der Ven, 1999). Thus, although connectin does not appear to play a role in the accessibility of actin in the structures of cardiomyocytes, this was examined by microinjection of fluorescently labeled actin into cultured skeletal myocytes and subsequent staining with antibody against connectin.

Around the dots formed by injected actin, connectin dots were found (Figure 2.18). Thus, connectin dots do not colocalize with actin/nebulin dots. It has been reported that although connectin has the ability to bind F-actin, it does not bind actin monomers (Maruyama, 1994). Further, in mature muscles, connectin filaments link the Z-line to myosin filaments and do not directly associate with actin filaments (Maruyama, 1986). Therefore, injected actin appears to be associated with nebulin dots but not with connectin dots; connectin seems to have no significant relation to the accessibility of actin molecules into nascent myofibrils. On nonstriated fibrils formed by the linear arrangement of actin dots, connectin dots are already situated on both sides of the actin dots forming doublets (Figure 2.19). The earlier striation (doublets) formation (maturation) of connectin than actin and nebulin seems to show

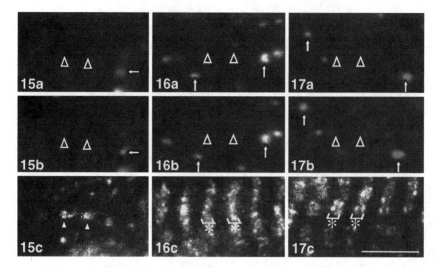

FIGURES 2.15–2.17. Images of skeletal muscle cells injected with monoclonal antinebulin (NB2), and after 6 hours with rh-actin. Although injected actin is seen as amorphous patches (arrows, a, b), no incorporation of rh-actin is seen after 6 (a) and 24 hours (b) into early (Figure 2.15) and striated myofibrils (Figures 2.16 and 2.17). Staining with FITC-labeled antimouse IgG showed that the injected antibody has bound at linearly arranged dots (Figure 2.15c) and the I-band region forming singlets (Figure 2.16c) or doublets (Figure 2,17c). Bar = 5 μm.

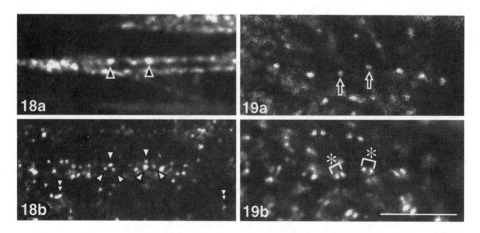

FIGURES 2.18 AND 2.19. (a) Images of skeletal muscle cells injected with rh-actin. (b) The same cells stained for connectin with 9D10 (Developmental Hybridoma Bank maintained by the University of Iowa, Department of Biological Sciences, Iowa City, IA, under Contract NO1-HD-7-3263 from NICHD). On and around rh-actin dots (black arrowheads), small connectin dots (white arrowheads) are seen (Figure 2.18). On nonstriated fibrils formed by linear arrangement of rh-actin dots (arrows), connectin doublets revealed with 9D10 (asterisks) are found on both sides of actin dots (Figure 2.19).

that connectin does not have any significant relation to the accessibility of actin molecules into nascent myofibrils.

Inhibition of Connectin

In order to rule out the possibility that connectin is involved in the process of actin filament assembly, we inhibited connectin activity by injection of anticonnectin antibody and the subsequent injection of rh-actin into the same myotubes (Nwe and Shimada, 2000). The result showed that, in spite of the anticonnectin treatment, rh-actin is incorporated into linearly arranged dot and I-band regions of myofibrils; in the former regions, the injected antibody has bound to the sites around these dots, and in the latter it has formed doublets at the level of IZI-complexes (Figures 2.20 and 2.21).

Immunoelectron Microscopy

The immunoelectron microscopic method will provide sufficient resolution and sensitivity to allow the detection of the sites of subunit addition to nascent myofibrils on a molecular level. We injected biotin-labeled actin into cardiomyocytes, and at selected time intervals reacted with 5-nm gold-coinjected antirabbit IgG (Komiyama et al., 1993; Kouchi et al., 1993; Shimada et al., 1996).

Gold labeling is first found around the A-band level of nascent myofibrils (Figure 2.22a). This observation suggests that incorporation/polymerization of actin and/or the addition of nearly formed actin filaments occurs preferentially in association with myosin filaments to increase the myofibrillar girth. At the distal portions of devel-

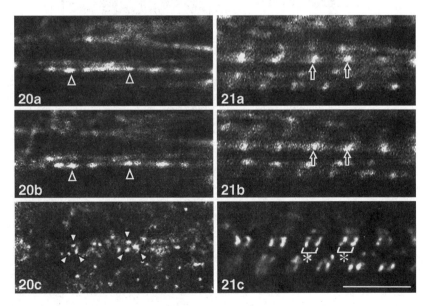

FIGURES 2.20 AND 2.21. Images of skeletal muscle cells injected with monoclonal anticonnectin (9D10) and after 6 hours with rh-actin. Six (a) and 24 hours (b) after the rh-actin injection, injected actin is seen to form dotted (arrowheads, Figure 2.20a, b) and striated patterns (arrows, Figure 2.21a, b). Staining with FITC-labeled antimouse IgM shows that dots of rh-actin are either surrounded by connectin dots (white arrowheads, Figure 2.20c) or sandwiched by connectin doublets (asterisks, Figure 2.21c).

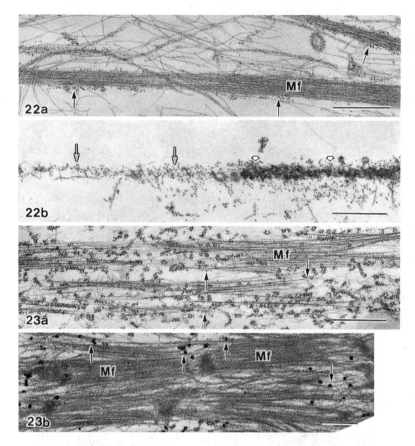

FIGURES 2.22 AND 2.23. Immunogold electron micrographs of cardiomyocytes injected with biotin-actin (Figure 2.22) and biotin-MLC (Figure 2.23) and reacted with 5-nm gold-labeled antibiotin. Silver enhancement was performed on the cell in Figure 2.23b. Injected actin is preferentially incorporated around myofibrils (Mf) at the A-band level (arrows, Figure 2.22a). Gold particles are also seen beyond the apices of the myofibrillar terminal (white arrows) in addition to terminal portions of myofibrils ending within the subsarcolemmal dense material (open arrows). Injected myosin is found primarily at the A-I junctional regions around myofibrils (arrows in Figure 2.23). Bar = 0.5 μm. (Figure 2.22 reproduced by permission of Kluwer Academic Publishers, Dordrecht, from Kouchi et al., 1993.)

oping myofibrils, their terminal ends are labeled, suggesting that continued reorganization and/or de novo formation of myofibrils occurs at these locations. Gold labeling is also seen along the termini of growing myofibrils (Figure 2.22b), indicating that actin subunits are added at the membrane-associated ends of preexisting actin filaments to increase the length of myofibrils.

DYNAMICS OF α-ACTININ

Injected fluorescently labeled α-actinin becomes incorporated into cellular structures in cardiomyocytes with a distribution similar to that of fluorescent actin. As in the case of actin, injected α-actinin is distributed in thin continuous lines, in punctate structures arranged in a linear pattern exhibiting minisarcomeres, and in regular cross-striated bands. In mature cross-striated myofibrils, injected α-actinin is localized in narrow well-defined Z-lines (Figure 2.1b) (McKenna et al., 1985b, 1986; Sanger et al., 1986).

Exchangeability of α-Actinin

The FR rates of α-actinin are extremely low and do not differ significantly among the three portions exhibiting different incorporation patterns of this fluorescent conjugate, namely nonstriated, punctate, and striated portions (Hasebe-Kishi and Shimada, 2000). Almost no FR is seen within the 6- to 7-hour observation period (Figure 2.4a). This finding using rabbit skeletal muscle α-actinin is not in agreement with that of McKenna et al. (1985b) using chicken smooth muscle α-actinin, in which FR in all portions of myofibrils occurred much faster. This difference appears to be due to the different isoform used (skeletal versus smooth muscle form). It seems that in nascent myofibrils of cardiac myocytes, α-actinin, apparently homologous to the skeletal muscle isoform, becomes tightly integrated into Z-lines and thus is stable, whereas the heterologous protein (the smooth muscle form) is unstable, probably in preparation for replacement by the skeletal muscle form during development. The difference in taxonomy (rabbit and chicken) does not seem to matter because α-actinin does not differ significantly in amino acid sequence among different animal species.

Mobility of α-Actinin Dots

Time-lapse fluorescent image observation shows frequent increases and decreases in the intervals of α-actinin dots along the longitudinal axis of nascent myofibrils. During this process, a few adjacent α-actinin dots come close and align with Z-bands of neighboring striated myofibrils. Thus, such a movement of α-actinin dots along the long axis of nascent myofibrils plays a role in making Z-bodies larger, apparently by coalescence of dots and alignment of their positions with those of more mature Z-bands of neighboring myofibrils (Hasebe-Kishi and Shimada, 2000). Once regular striations are formed, these dense bodies grow laterally by the addition of small beads of α-actinin to existing Z-bodies and by the merging of small Z-bands of neighboring myofibrils (McKenna et al., 1986; Sanger et al., 1986; Dabiri et al., 1997) (Figure 2.24).

Comparison of Exchangeabilities between Actin and α-Actinin

This can be examined by the coinjection of actin and α-actinin labeled with different fluorescent dyes into cultured cardiomyocytes. The possibility of fluorophore-related differences in fluorescence distribution has also been excluded previously in cardiomyocytes coinjected with muscle and brain actins labeled either with rh and FITC or with FITC and rh, respectively (McKenna et al., 1985b). Further, the dynamics of the single injected protein do not seem to inhibit those of the other simultaneously injected because the FR of actin does not differ significantly, irrespective of the presence or absence of α-actinin (Hasebe-Kishi and Shimada, 2000).

Both actin and α-actinin are similarly distributed in early myofibrils; they form bundle-like lines and linearly arranged punctate structures at the same location. In cross-striated portions of more mature myofibrils, α-actinin is incorporated into the center of each striation formed by actin (Figure 2.1).

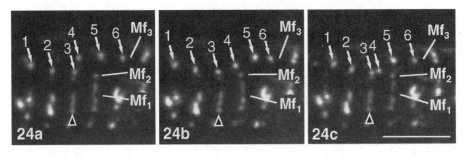

FIGURE 2.24. Time-lapse fluorescent images of a cardiomyocyte injected with rh-α-actinin at (a) 0 hours, (b) 2.5 hours, and (c) 4 hours. Myofibrils (Mf_1) with regular α-actinin striations are stable during these periods. In early myofibrils (Mf_2 and Mf_3), intervals of α-actinin dots change (see those between 1 and 2 and between 5 and 6). Adjoining dots (3 and 4) of neighboring myofibrils (Mf_2 and Mf_3, respectively) come closer and align with Z-bands (arrowhead) of the myofibril with regular cross-striations (Mf_1). Bar = 5 μm.

In these coinjected cardiomyocytes, although the exchangeability of actin decreases markedly, that of α-actinin is continuously low during myofibrillar development (Figures 2.4a, 2.6, and 2.8). During this phase of myofibrillogenesis, transition from a lumpy dense body of α-actinin to a structured Z-disk is seen. Thus, the difference in exchangeability between actin molecules in filaments and α-actinin molecules in merging dense bodies may be related to the mechanism of the IZI brush formation and sarcomere lengthening; fusing dots of low-exchangeable α-actinin may provide favorable situations for exchangeable actin in filaments to elongate (Rhee et al., 1994) and/or rearrange (Almenar-Queralt et al., 1999).

DYNAMICS OF MYOSIN

Association of microinjected myosin with nascent myofibrils in living muscle cells was examined by Johnson et al. (1988). After microinjection into skeletal myotubes, the fluorescent myosin analog can be incorporated into continuous and periodic fibrillar structures of immature muscle cells and into periodic bands of myofibrils in mature myotubes. This localization of injected myosin closely mimics that of endogenous myosin. Although light meromyosin can associate with A-bands, as does whole myosin, fluorescent heavy meromyosin is not incorporated into myofibrillar structures after injection. Thus, the light meromyosin portion of the molecule appears to be mainly responsible for the incorporation.

Exchangeability of Myosin

The same authors (Johnson et al., 1988) observed a slow recovery of fluorescence after photobleaching of fluorescent myosin that has become incorporated into striated portions of myofibrils. The bleached spot remains detectable as long as 18 hours after photobleaching. The recovery of myosin appears to be less than half that of actin if their data are normalized to the same level as ours. Thus, myosin molecules once incorporated into the thick filaments of mature myofibrils are relatively stably associated. This finding is consistent with the turnover rates determined for myosin in myofibrillar structures using isotopes (Zak et al., 1977). It remains to be clarified whether different exchangeabilities are found in younger myofibrillar portions and with embryonic and nonmuscle isoforms.

Immunoelectron Microscopy

We have examined myosin subunit incorporation in nascent myofibrils by monitoring a microinjected biotin-labeled myosin light chain (MLC) in cultured cardiomyocytes by immunogold electron microscopic methods with or without silver enhancement (Shimada et al., 1996). The results show that gold particles are found predominantly at A-I junctional regions around myofibrils (Figure 2.23), suggesting that myosin filaments are assembled at or near this region. Connectin filaments are believed to link myosin filaments to Z-bands (Maruyama et al., 1985; Fürst et al., 1988). Together with the previous fluorescence microscopic observation that the development and distribution of connectin is tightly linked with that of myosin (Handel et al., 1989; Komiyama et al., 1990, 1993; Van der Ven, 1999), it is plausible that newly synthesized myosin molecules and/or myosin filaments are associated with connectin filaments radiating from Z-bands to be added into preexisting nascent myofibrils.

DYNAMICS OF CONTRACTILE PROTEINS IN CARDIAC FIBROBLASTS

FR after photobleaching shows that stress fibers in cardiac fibroblasts exhibit a significantly higher exchange rate of actin and α-actinin than myofibrils at any phase of development. A "virtually" complete recovery of actin is seen by 30 to 40 minutes and that of α-actinin by 1 hour (McKenna, 1985b; Shimada et al., 1997; Suzuki et al., 1998; Hasebe-Kishi and Shimada, 1999) (Figure 2.4a, c, d). These results are consistent with the FR of actin associated with stress fibers of gizzard cells (Kreis et al., 1982). Higher exchangeabilities of actin monomers in stress fibers in fibroblasts may be due to the absence of actin filament-stabilizing proteins, such as capZ, tropomodulin, connectin, and nebulin/nebulette (Gregorio et al., 1995; Gregorio and Fowler, 1995; Moncman and Wang, 1995, 1996; Littlefield and Fowler, 1998); that of α-actinin molecules may be related directly or indirectly to the absence of some of the proteins listed earlier possessing actin binding activity. The high exchangeabilities of both proteins indicate that stress fibers are constantly exchanging their components for motile and other vital functions of these cells.

A similar exchange rate is found with muscle and nonmuscle actins in stress fibers of fibroblasts (Shimada et al., 1997; Suzuki et al., 1998) (Figure 2.4c, d). This appears to be related to the more primitive nature of stress fibers than that of myofibrils.

By comparison of the exchangeability in the terminal portions of stress fibers where they attach to the inner surface of the fibroblast cell membrane and the proximal portion of these fibers, it is noted that although FR of actin is similar between these two portions (Figure 2.4c, d), that of α-actinin in the former portion is faster than in the latter portion (McKenna et al., 1985b; Suzuki et al., 1998). The difference in the dynamics of actin and α-actinin at the areas of adhesion plaques may facilitate the reorganization of α-actinin to cope with the continuous movement of the cellular edges of fibroblasts.

Nonstriated fibrils in myocytes are similar in appearance and composition to stress fibers in many cell types and have frequently been called stress-fiber–like structures

(Dlugosz et al., 1984). However, these portions of myofibrils in myocytes are obviously different in dynamics from stress fibers in fibroblasts.

CONCLUSIONS

The fate of incorporated myofibrillar proteins in living muscle cells has been able to be followed by application of the combined techniques of microinjection, fluorescence microscopy, and image intensification with time-lapse recording. Further, the combination of these techniques with FR after photobleaching allows the examination of the relative exchangeability of proteins associated with various portions in cultured living cells.

1. Fluorescently labeled actin and α-actinin, when injected into cultured living muscle cells, become incorporated into cellular structures with a similar time course and distribution. Both of these fluorescently labeled analogs distribute in thin continuous lines (stress-fiber–like structures), in punctate structures arranged linearly (premyofibrils with minisarcomeres), and in cross-striated bands (mature myofibrils).

2. Exchangeability of actin decreases in the order of nonstriated fibrils, premyofibrils, and cross-striated myofibrils, whereas that of α-actinin is low and stable at all portions. During the transition phase from punctate to regular sarcomere structures of these proteins, short-spaced α-actinin dots evolve into more regular Z-lines. It appears that both the difference in exchangeability between actin and α-actinin molecules and the merger of α-actinin dots during this phase of myofibrillogenesis are related to sarcomere lengthening and IZI brush formation.

3. In skeletal myotubes, exogenously introduced actin is initially associated with nebulin dots. The dots of nebulin and/or its associated protein(s) may thus represent a preformed scaffold upon which actin monomers accumulate. In cross-striated myofibrils, exchangeability of actin is higher in the portions exhibiting nebulin singlets than in those with doublets. It seems that early nebulin filaments exhibiting singlets are not tightly associated with actin filaments and that this loose association allows myofibrils to exchange nonadult isoforms of actin and other proteins into adult types.

4. Connectin (titin) forms a striated pattern (doublets) before the formation of actin/nebulin striations. The earlier striation (doublets) formation (maturation) of connectin than actin and nebulin seems to show that connectin does not have any significant relation to or role in the accessibility of actin into nascent myofibrils.

5. The exchangeability of muscle actin is faster than that of nonmuscle actin in immature nonstriated myofibrillar portions. This indicates that actin molecules in cardiac myofibrils cannot readily be exchanged by heterotypic nonmuscle actin. It appears that isoforms with higher binding affinities (muscle actin) replace previous proteins, thus increasing the stability of the resulting structures (actin filaments).

6. In cardiac fibroblasts, the exchangeability of actin and α-actinin is much higher than that in myofibrils at any phase of development. Thus, these fibrils are clearly different from nonstriated stress-fiber–like structures of nascent myofibrils. A lack of muscle scaffolding proteins in fibroblasts may account for the high exchangeability of actin and α-actinin.

REFERENCES

Almenar-Queralt, A., Gregorio, C.C., and Fowler, V.M. 1999. Tropomodulin assembles early in myofibrillogenesis in chick skeletal muscle: evidence that thin filaments rearrange to form striated myofibrils. *J. Cell Sci.* 112:1111–1123.

Anversa, P., Vitali-Mazza, L., and Loud, A.V. 1975. Morphometric and autoradiographic study of developing ventricular and atrial myocardium in fetal rats. *Lab. Invest.* 33:696–705.

Bandman, E. 1992. Contractile protein isoforms during muscle development. *Dev. Biol.* 154:273–283.

Begum, S., Komiyama, M., Toyota, N., Obinata, T., Maruyama, K., and Shimada, Y. 1998. Differentiation of muscle-specific proteins in chicken somites as studied by immunofluorescence microscopy. *Cell Tissue Res.* 293:305–311.

Dabiri, G.A., Turnacioglu, K.K., Sanger, J.M., and Sanger, J.W. 1997. Myofibrillogenesis visualized in living embryonic cardiomyocytes. *Proc. Natl. Acad. Sci. USA* 94:9493–9498.

Dadounce, J.P.A. 1980. Protein turnover in muscle cells as visualized by autoradiography. *Int. Rev. Cytol.* 67:215–257.

Dlugosz, A.A., Antin, P.B., Nachmias, V.T., and Holtzer, H. 1984. The relationship between stress fiber-like structures and nascent myofibrils in cultured cardiac myocytes. *J. Cell Biol.* 99:2268–2278.

Dome, J.S., Mittal, B., Pochapin, M.B., Sanger, J.M., and Sanger, J.W. 1988. Incorporation of fluorescently labeled actin and tropomyosin into muscle cells. *Cell Differ.* 23:37–52.

Endo, T. and Masaki, T. 1984. Differential expression and distribution of chicken skeletal- and smooth-muscle-type α-actinins during myogenesis in culture. *J. Cell Biol.* 99:2322–2332.

Franzini-Armstrong, C. and Fischman, D.A. 1994. Morphogenesis of skeletal muscle fibers. In: *Myology: Basic and Clinical*, 2nd ed. Eds. A.G. Engel and C. Franzini-Armstrong, McGraw-Hill, Inc., New York, pp. 74–96.

Fürst, D.O., Osborn, M., Nave, R., and Weber, K. 1988. The organization of titin filaments in the half-sarcomere revealed by monoclonal antibodies in immunoelectron microscopy: a map of ten nonrepetitive epitopes starting at the Z line extends close to the M line. *J. Cell Biol.* 106:1563–1572.

Glacy, S.D. 1983. Pattern and time course of rhodamine-actin incorporation in cardiac myocytes. *J. Cell Biol.* 96:1164–1167.

Gregorio, C.C. and Fowler, V.M. 1995. Mechanism of thin filament assembly in embryonic chick cardiac myocytes: tropomodulin requires tropomyosin for assembly. *J. Cell Biol.* 129:683–695.

Gregorio, C.C., Weber, A., Bondad, M., Pennise, C.R., and Fowler, V.M. 1995. Requirement of pointed-end capping by tropomodulin to maintain actin filament length in embryonic chick cardiac myocytes. *Nature* 377:83–86.

Handel, S.E., Wang, S.-M., Greaser, M.L., Schultz, E., Bulinski, J.C., and Lessard, J.L. 1989. Skeletal muscle myofibrillogenesis as revealed with a monoclonal antibody to titin in combination with detection of the α- and γ-isoforms of actin. *Dev. Biol.* 132:35–44.

Handel, S.E., Greaser, M.X.L., Schultz, E., Wang, S.-M., Bulinski, J.C., and Lessard, J.L. 1991. Chicken cardiac myofibrillogenesis studied with antibodies specific for titin and the muscle and nonmuscle isoforms of actin and tropomyosin. *Cell Tissue Res.* 263:419–430.

Hasebe-Kishi, F. and Shimada, Y. 2000. Dynamics of actin and α-actinin in nascent myofibrils and stress fibers. *J. Muscle Res. Cell Motil.* 21:717–724.

Horowits, R. 1999. The physiological role of titin in striated muscle. *Rev. Physiol. Biochem. Pharmacol.* 138:57–96.

Itoh, Y., Matsuura, T., Kimura, S., and Maruyama, K. 1988. Absence of nebulin in cardiac muscles of the chicken embryo. *Biomed. Res.* 9:331–333.

Jacobson, K., Elson, E., Koppel, D., and Webb, W. 1983. International workshop on the application of fluorescence photobleaching techniques to problems in cell biology. *Fed. Proc.* 42:72–79.

Johnson, C.S., McKenna, N.M., and Wang, Y.-L. 1988. Association of microinjected myosin and its subfragments with myofibrils in living muscle cells. *J. Cell Biol.* 107:2213–2221.

Komiyama, M., Kouchi, K., Maruyama, K., and Shimada, Y. 1993. Dynamics of actin and assembly of connectin (titin) during myofibrillogenesis in embryonic chick cardiac muscle cells in vitro. *Dev. Dyn.* 196:291–299.

Komiyama, M., Maruyama, K., and Shimada, Y. 1990. Assembly of connectin (titin) in relation to myosin and α-actinin in cultured cardiac myocytes. *J. Muscle Res. Cell Motil.* 11:419–428.

Komiyama, M., Zhou, Z.-H., Maruyama, K., and Shimada, Y. 1992. Spatial relationship of nebulin relative to other myofibrillar proteins during myogenesis in embryonic chick skeletal muscle cells in vitro. *J. Muscle Res. Cell Motil.* 13:48–54.

Kouchi, K., Takahashi, H., and Shimada, Y. 1993. Incorporation of microinjected biotin-labelled actin into nascent myofibrils of cardiac myocytes: an immunoelectron microscopic study. *J. Muscle Res. Cell Motil.* 14:292–301.

Kreis, T.E., Geizer, B., and Schlessinger, J. 1982. Mobility of microinjected rhodamine actin within living chicken gizzard cells determined by fluorescence photobleaching recovery. *Cell* 29:835–845.

Kruger, M., Wright, J., and Wang, K. 1991. Nebulin as a length regulator of thin filaments of vetebrate skeletal muscles: correlation of thin filament length, nebulin size and epitope profiles. *J. Cell Biol.* 115:97–107.

Labeit, S., Gibson, T., Lakey, A., Leonard, K., Zeviani, M., Knight, P., Wardale, J., and Trinick, J. 1991. Evidence that nebulin is a protein-ruler in muscle thin filaments. *FEBS Lett.* 182:313–316.

Littlefield, R. and Fowler, V.M. 1998. Defining actin filament length in striated muscle: rulers and caps or dynamic stability? *Annu. Rev. Cell Dev. Biol.* 14:487–525.

Luo, G., Zhang, J.Q., Nguyen, T.-P., Herrera, A.H., Paterson, B., and Horowits, R. 1997. Complete cDNA sequence and tissue localization of N-RAP, a novel nebulin-related protein of striated muscle. *Cell Motil. Cytoskel.* 38:75–90.

Maruyama, K. 1986. Connectin, an elastic filamentous protein of striated muscle. *Int. Rev. Cytol.* 104:81–114.

Maruyama, K. 1994. Connectin, an elastic protein of striated muscle. *Biophys. Chem.* 50:73–85.

Maruyama, K., Yoshioka, T., Higuchi, H., Ohashi, K., Kimura, S., and Natori, R. 1985. Connectin filaments link thick filaments and Z lines in frog skeletal muscle as revealed by immunoelectron microscopy. *J. Cell Biol.* 101:2167–2172.

McKenna, N., Meigs, J.B., and Wang, Y.-L. 1985a. Identical distribution of fluorescently labeled brain and muscle actins in living cardiac fibroblasts and myocytes. *J. Cell Biol.* 100:292–296.

McKenna, N., Meigs, J.B., and Wang, Y.-L. 1985b. Exchangeability of alpha-actinin in living cardiac fibroblasts and muscle cells. *J. Cell Biol.* 101:2223–2232.

McKenna, N.M., Johnson, C.S., and Wang, Y.-L. 1986. Formation and alignment of Z lines in living chick myotubes microinjected with rhodamine-labeled alpha-actinin. *J. Cell Biol.* 103:2163–2171.

Moncman, C.L. and Wang, K. 1995. Nebulette: a 107 KD nebulin-like protein in cardiac muscle. *Cell Motil. Cytoskel.* 32:205–225.

Moncman, C.L. and Wang, K. 1996. Assembly of nebulin into the sarcomeres of avian skeletal muscle. *Cell Motil. Cytoskel.* 34:167–184.

Morkin, E., Yazaki, Y., Katagiri, T., and LaRaia, P.J. 1973. Comparison of the synthesis of the light and heavy chains of adult skeletal myosin. *Biochim. Biophys. Acta.* 324:420–429.

Nwe, T.M., Maruyama, K., and Shimada, Y. 1999. Relation of nebulin and connectin (titin) to dynamics of actin in nascent myofibrils of cultured skeletal muscle cells. *Exp. Cell Res.* 252:33–40.

Nwe, T.M. and Shimada, Y. 2000. Inhibition of nebulin and connectin (titin) for assembly of actin filaments during myofibrillogenesis. *Tissue Cell* 32:223–227.

Obinata, T., Maruyama, K., Sugita, H., Kohama, K., and Ebashi, S. 1981. Dynamic aspects of structural proteins in vertebrate skeletal muscle. *Muscle Nerve* 4:456–488.

Ordahl, C.P. 1986. The skeletal and cardiac α-actin genes are coexpressed in early embryonic striated muscle. *Dev. Biol.* 117:488–492.

Rhee, D., Sanger, J.M., and Sanger, J.W. 1994. The premyofibril: evidence for its role in myofibrillogenesis. *Cell Motil. Cytoskel.* 28:1–24.

Ruzicka, D.L. and Schwartz, R.J. 1988. Sequential activation of α-actin genes during avian cardiogenesis: vascular smooth muscle α-actin gene transcripts mark the onset of cardiomyocyte differentiation. *J. Cell Biol.* 107:2575–2586.

Sanger, J.M., Mittal, B., Pochapin, M.B., and Sanger, J.W. 1986. Myofibrillogenesis in living cells microinjected with fluorescently labeled alpha-actinin. *J. Cell Biol.* 102:2053–2066.

Saxton, W.M., Stemple, D.L., Leslie, R.L., Salmon, E.D., Zavortink, M., and McIntosh, J.R. 1984. Tubulin dynamics in cultured mammalian cells. *J. Cell Biol.* 99:2175–2186.

Schafer, D.A. and Cooper, J.A. 1995. Control of actin assembly at filament ends. *Annu. Rev. Cell Biol.* 11:497–518.

Shimada, Y., Komiyama, M., Begum, S., and Maruyama, K. 1996. Development of connectin/titin and nebulin in striated muscles of chicken. *Adv. Biophys.* 33:223–234.

Shimada, Y., Suzuki, H., and Konno, A. 1997. Dynamics of actin in cardiac myofibrils and fibroblast stress fibers. *Cell Struct. Funct.* 22:59–64.

Suzuki, H., Komiyama, M., Konno, A., and Shimada, Y. 1998. Exchangeability of actin in cardiac myocytes and fibroblasts as determined by fluorescence photobleaching recovery. *Tissue Cell* 30:274–280.

Taylor, D.L., Amato, P.A., Luby-Phelps, K., and McNeil, P. 1984. Fluorescent analog cytochemistry. *Trends Biochem. Sci.* 9:88–91.

Van der Ven, P.F.M., Ehler, E., and Perriard, J.-C. 1999. Thick filament assembly occurs after the formation of a cytoskeletal scaffold. *J. Muscle Res. Cell Motil.* 20:569–579.

Wang, K. and Wright, J. 1988. Architecture of the sarcomere matrix of skeletal muscle: immunoelectron microscopic evidence that suggests a set of parallel inextensible nebulin filaments anchored at the Z line. *J. Cell Biol.* 107:2199–2212.

Wang, Y.-L. 1992. Fluorescence microscopic analysis of cytoskeletal organization and dynamics. In: *The Cytoskeleton: A Practical Approach.* Eds. K.L. Carraway and A.C. Carraway, Oxford University Press, Oxford, UK, pp. 1–22.

Wang, Y.-L., Heiple, J.M., and Taylor, D.L. 1982. Fluorescent analog cytochemistry of contractile proteins. *Meth. Cell Biol.* 25:1–11.

Wegner, A. 1976. Head-tail polymerization of actin. *J. Mol. Biol.* 108:139–150.

Zak, R., Martin, A.F., Prior, G., and Rabinowitz, M. 1977. Comparison of turnover of several myofibrillar proteins and critical evaluation of double isotope method. *J. Biol. Chem.* 252:3430–3435.

CHAPTER 3

Emergence of the First Myofibrils and Targeting Mechanisms Directing Sarcomere Assembly in Developing Cardiomyocytes

Elisabeth Ehler and Jean-Claude Perriard

INTRODUCTION

The heart of higher vertebrates becomes functional during early development and is essential to support embryonic life. When other tissues are still in the process of determination, cardiac cells are already functional and contract. Any impairment of their performance is fatal for the embryo, as shown by the many cases of embryonic lethality in knockout mice due to malfunction of the heart. Cardiomyocytes are highly specialized cells, able to convert chemical energy into mechanical contraction, and their work is essential to move body fluids already in the very primitive embryo. During the cardiac developmental program, cells accumulate sarcomeric proteins in a characteristic mixture in the cytoplasm. Their cytoarchitecture develops and myofibrils, with their basic functional units—the sarcomeres—are assembled with astonishing precision. A sarcomere spans between two Z-disks, where the thin (actin) filaments are anchored. In the middle of the sarcomere is a defined structure called the M-band, which serves for the integration of the thick (myosin) filaments. The third filament system is composed of individual titin molecules that stretch from the Z-disk all the way to the M-band. Although the general organization appears to be identical in heart and skeletal muscle, there are slight adaptations to ensure optimal function.

The cytoarchitecture of the differentiated cardiomyocytes is unique. The individual cardiomyocyte is rod-shaped and most of the cytoplasmic space is filled with the myofibrils, which are in register and aligned exactly in the longitudinal axis of the cells. At the bipolar ends, a specific type of cell–cell contact is found, the intercalated disk. There the myofibrils are inserted, ensuring mechanical coupling within the working myocardium.

Little is known about where and how these highly organized structures are assembled and where and how the instruction for assembly is stored. Investigations on cultured cardiomyocytes have led to proposals that scaffold structures serve as precursors, which are subsequently replaced by definitive sarcomeres. Other theories suggest that an integral component of the sarcomere may serve as both blueprint and building block at the same time. The aim of this review is to discuss the hypotheses on myofibrillogenesis derived from investigations on cultured cardiomyocytes and to compare them with observations made during sarcomere assembly in the developing heart in situ. In addition, molecular mechanisms of the integration of sarcomeric proteins will be discussed.

MYOFIBRILLOGENESIS IN CULTURED CARDIOMYOCYTES

Cardiomyocytes have been isolated from different vertebrate species, and their adaptation to culture conditions has been studied extensively (Claycomb and Palazzo, 1980; Jacobson and Piper, 1986; Eppenberger et al., 1988; Wang et al., 1988; Schultheiss et al., 1990; Messerli et al., 1993; Rhee et al., 1994; Clark et al., 1998; Harder et al., 1998; Rothen-Rutishauser et al., 1998). There are marked differences between species in the ability to adapt to life in the culture dish (e.g., whereas cardiomyocytes from adult rats can be cultured, those isolated from adult mice do not adhere readily to generally used substrates) (Eppenberger et al., 1988; Ehler and Perriard, unpublished observations). Additionally, there are differences between developmental stages in the way the cardiomyocytes adapt to culture conditions. For example, cardiomyocytes that were isolated from embryonic or neonatal heart spread readily on the substrate and maintain their myofibrils (Atherton et al., 1986; Wang et al., 1988; Lin et al., 1989; Rhee et al., 1994), whereas in cardiomyocytes isolated from adult rat heart, for example, a disassembly–reassembly process of the myofibrillar apparatus takes place in the presence of serum (Claycomb and Palazzo, 1980; Guo et al., 1986; Jacobson and Piper, 1986; Eppenberger et al., 1988; Messerli et al., 1993; Rothen-Rutishauser et al., 1998). The transition of the rod-shaped cardiomyocyte to a typical cultured cardiomyocyte was studied in rat and hamster adult ventricular cardiomyocytes (Messerli et al., 1993; Horackova and Byczko, 1997). This regeneration process is accompanied by the reexpression of fetal genes, such as ANF (atrial natriuretic factor), α-vascular actin, and β-MyHC (myosin heavy chain) (Eppenberger et al., 1994). Some of these fetal genes are regulated by factors such as IGF-1 and bFGF (Donath et al., 1994; Harder et al., 1996; Eppenberger-Eberhardt et al., 1997) or hormones such as thyroxine (Gosteli-Peter et al., 1996; Schaub et al., 1998). IGF-1 treatment downregulated the expression of fetal proteins and promoted myofibrillar assembly, whereas bFGF and T3 induced the expression of fetal markers such as α-vascular actin and ANF and caused a restriction of the myofibrils to the perinuclear region.

A striking feature of myofibrillogenesis in cultured cardiomyocytes is that defined subregions of myofibril assembly can be identified. Whereas the most mature myofibrils are always found in the perinuclear region in the center of the cell, nonsarcomeric cytoskeletal elements predominate in the cellular periphery. Regions of myofibrillogenesis are located in-between (Messerli et al., 1993; Rhee et al., 1994). These differences are also reflected by a distinct organization of sarcomeric proteins. The actin filaments are striated in the cell's center, with nonstriated extensions toward the periphery of the cells. These extensions resemble stress fibers, a typical form of actin organization in cultured cells, and were therefore termed SFLS (stress-fiber–like structures; Dlugosz et al., 1984; Antin et al., 1986). α-actinin, a component of the Z-disk, is also found in a cross-striated organization in the center, which changes to a more punctuated pattern with a shorter periodicity in the transition zone. The M-band protein myomesin, on the other hand, can only be detected in striations in the perinuclear region and is not associated with the filamentous structures of the cellular periphery.

Several hypotheses have been developed to explain how myofibril assembly might occur in cultured cardiomyocytes. The first hypothesis involves IZI-complexes, composed of α-actinin, α-actin, and titin, that assemble to so-called NSMFs (nonstriated myofibrils). Thick filaments, consisting mainly of myosin, are only integrated into these NSMFs at a later stage (Dlugosz et al., 1984; Antin et al., 1986; Lin et al., 1989; Schultheiss et al., 1990; Handel et al., 1991). The second hypothesis proposes that so-called premyofibrils that are composed of minisarcomeres, namely narrower spaced Z-disks that do not contain titin but nonmuscle isoforms of contractile proteins such as nonmuscle myosin IIB, might play an important role. After the integration of titin, the distance between individual α-actinin dots increases and the nonmuscle myosin IIB gets replaced by the muscle myosin isoform (Rhee et al., 1994; LoRusso et al., 1997).

Do Other Cytoskeletal Proteins Play a Role During Myofibrillogenesis?

The two other cytoskeletal systems in addition to actin filaments consist of microtubuli that are essential during cell division and for intracellular transport or intermediate filaments that seem to be important for the maintenance of cell shape. Information on the contribution of these two systems to myofibrillogenesis is still scarce. Experiments where the microtubuli were destroyed by nocodazole have highlighted important differences between cultured neonatal and adult cardiomyocytes in their requirement for this system. Whereas myofibrillogenesis is more or less unaffected in neonatal rat cardiomyocytes, adult rat cardiomyocytes are absolutely unable to adapt to the culture conditions and to assemble myofibrils if their microtubular network was destroyed (Rothen-Rutishauser et al., 1998).

The major intermediate filament protein in the heart is desmin (Lazarides, 1982), although vimentin is expressed in precardiac cells as well. In neonatal cardiomyocytes cultured for a short period, desmin is organized in long filamentous structures that are similar to intermediate filament organization observed in other cultured cell types. Only after several days in vitro does a reorganization take place and the desmin filaments become concentrated around the Z-disks, resembling their organization in the mature heart. A similar delay in association with the myofibrils can be observed when myofibrillogenesis is investigated in the heart of chicken embryos, where desmin, although expressed rather early, is also one of the last components to attain its mature position at the Z-disk (Ehler et al., 1999). Mice that are homozygous null for desmin show no defects in myofibrillogenesis per se (Li et al., 1996; Milner et al., 1996); however, desmin seems to play an important role for stabilizing the myofibrillar apparatus during the contractions because these mice show ultrastructural defects that finally result in a cardiomyopathy with increasing age (Thornell et al., 1997; Milner et al., 1999). This defect appears to be specific for desmin because mice lacking vimentin develop without an obvious phenotype and heart development proceeds normally (Colucci-Guyon et al., 1994).

Are Cell–Matrix or Cell–Cell Contacts Important?

Several experiments have pointed out that contacts between the cell and its surrounding extracellular matrix might play an important role during myofibril assembly. First, cell spreading on a substrate is a prerequisite for myofibrillogenesis to happen in cultured cardiomyocytes (Marino et al., 1987). If the spreading is disturbed by addition of cytochalasin D to the cell culture medium, for example, no assembly of new myofibrils occurs, although preexisting myofibrils in neonatal cardiomyocytes are unaffected (Rothen-Rutishauser et al., 1998). Second, incubation of the cardiomyocytes with, for example, antisense oligonucleotides to cell–matrix contact proteins such as vinculin inhibits myofibril assembly (Shiraishi et al., 1997a). Third, if cardiomyocytes are differentiated from embryonic stem cells homozygous null for β-1 integrin, a severe retardation of myofibrillogenesis can be observed as well (Fässler et al., 1996). However, these proteins seem not to be essential because sarcomeres can be assembled in vinculin knockout mice and also in β-1 integrin-deficient cardiomyocytes, although to a reduced extent and with a severe delay (Xu et al., 1998; Guan et al., 1999). If antibodies to N-cadherin, a molecule confined to cell–cell contacts and therefore also in intercalated disks, are included in the cell culture medium, they also have an inhibitory effect on the development and alignment of myofibrils in cultured cardiomyocytes (Goncharova et al., 1992; Imanaka-Yoshida et al., 1998). Deletion of plakoglobin, another cell–cell contact protein, results in embryonic lethality due to heart defects (Ruiz et al., 1996); however, the function of the protein appears to be complemented by another component at the anchorage site of the intercalated disks because myofibrils can be assembled (Isac et al., 1999).

MYOFIBRILLOGENESIS IN THE EMBRYONIC HEART IN VIVO

Although cultured cardiomyocytes offer many advantages for the study of myofibrillogenesis because of their easy accessibility to microscopy techniques as well as to functional experiments such as transfection or investigating the influence of growth factors or drugs on myofibrillogenesis, one must always bear in mind that they might not represent a faithful picture of myofibrillogenesis as it happens during cardiac development. First, by the time cardiomyocytes are isolated for in vitro culture, they already possess mature myofibrils and were beating for some time. Therefore, it cannot be excluded that one does not only study de novo myofibrillogenesis but also recycling of preexisting sarcomeric structures. Second, the environment in the culture dish differs substantially from the situation in situ. The cardiomyocytes in the heart are three-dimensional rod-shaped bipolar cells surrounded by an extracellular matrix except for the direct contacts with other cardiomyocytes at the intercalated disks. After isolation and seeding in culture dishes, they flatten and spread on the culture substrate, assuming a rather two-dimensional morphology. Additionally, the composition of the extracellular matrix components is chosen by the investigator, and most of the cardiomyocyte surface is exposed to the culture medium instead of to other cardiomyocytes. Thus, some of the observations made, including changes in the organization of cytoskeletal proteins, possibly represent adaptations to the different environment and might only be typical for the situation in culture.

With the advance of confocal microscopy, it had become possible to study thicker specimens by immunofluorescence techniques and among the first objects studied was also myofibrillogenesis. In their pioneering studies, Tokuyasu and Maher established a staining protocol for chicken heart whole mount preparations and investigated the localization of key components of the sarcomere such as α-actinin and titin in hearts taken around the time when first contractions can be observed, which is around the 9 somite stage in the chicken (Tokuyasu and Maher, 1987a, 1987b). They could show that the first organized complexes were composed of α-actinin and titin while myosin was still distributed diffusely throughout the cytoplasm (Tokuyasu and Maher, 1987a, 1987b). Further investigations on chick heart whole mount preparations analyzed the distribution of other components such as vinculin, N-cadherin, fibronectin, and phospho-tyrosine (Tokuyasu, 1989; Shiraishi et al., 1993, 1995, 1997b). These techniques were not only employed in the chick but also in the axolotl, for example, where the three-dimensional distribution of α-actinin was compared between wild-type and cardiac mutant animals. Despite the fact that normal levels of α-actinin are expressed in the mutants, myofibrils fail to assemble (Lemanski et al., 1997). Ectopic expression of tropomyosin can partially rescue this defect in myofibrillogenesis in the cardiac mutant (Zajdel et al., 1999).

Recently, we set out to define the series of events that take place during myofibrillogenesis in situ by combining confocal microscopy on whole mount preparations with triple-immunofluorescence analysis in order to be able to compare the localization of different antigens in defined cardiomyocytes in situ at a given time. Using these techniques, we were able to come up with a hypothesis on the process of myofibrillogenesis in the heart in situ and to highlight important differences as well as similarities to myofibrillogenesis as it is known from cultured cardiomyocytes (Ehler et al., 1999). In the chicken heart, first contractions can be observed at the 9 somite stage, which is about 36 hours after the start of incubation. Just a few hours later, the myofibrillar system is already so far developed that the contracting cardiomyocytes can pump blood throughout the embryo. By immunoblotting, it could be shown that all the major components of the sarcomere are already expressed in the 6 somite stage embryo and that the assembly of myofibrils is an extremely rapid event (Ehler et al., 1999). Two major stages can be distinguished in this process. In the 8 somite heart, namely just before beating starts, first organized complexes can be observed, which consist of α-actinin, the N-terminus of titin and of actin (actin shown in red in Figure 3.1A; see Color Plate 1). These complexes show a close association with the cardiomyocyte plasma membrane. At that time, all the thick filament components, including the M-band protein myomesin as well as the C-terminus of titin, are not yet organized and only show a diffuse distribution in the cytoplasm (MyBP-C, myosin binding protein C in green in Figure 3.1A). Only about 2 hours later, at the 9 somite stage, most of the sarcomeric proteins have already been incorporated at their proper positions into the myofibril and the first contractions can be observed. The myofibrils have detached from the membrane and now run throughout the cytoplasm. By the 12-somite stage, more myofibrils have been assembled and although the individual myofibrils are not yet as bundled as they are in the mature cardiomyocyte, they get aligned in the same direction by this time. Their structural units are indistinguishable from adult sarcomeres, and MyBP-C is incorporated into the A-band (Figure 3.1B; (see color insert) actin in red, MyBP-C in green). Another striking observation at the 12 somite stage is that the myofibrils seem to transgress from one cell to its neighbor without clear interruptions at the sites of cell–cell contacts (Figure 3.1B; Manasek, 1970; Ehler et al., 1999).

The initial complexes as observed in the 8 somite heart are rather reminiscent of the so-called IZI-complexes in cultured cardiomyocytes. Investigations have shown that α-actinin seems to act as an integration molecule between actin filaments and titin filaments, and therefore these complexes might act as nucleation centers and serve to organize the N-termini of titin in opposite directions (Gregorio et al., 1998; Young et al., 1998). However, there are important differences between myofibrillogenesis in cultured cardiomyocytes and in the heart in situ. In cultured cells, α-actinin shows precise myofibrillar striations in the perinuclear region of the cell, and as the myofibril runs toward the periphery the thin filaments change to SFLS that are dotted very densely with α-actinin (Figure 3.2A, arrows). A similar transition cannot be observed when an individual myofibril is followed in a 3-D representation of a cardiomyocyte in a whole mount (Figure 3.1B) or when the α-actinin localization is compared between different stages of myofibrillogenesis in situ. At the 8 somite stage, α-actinin is localized in a punctuated pattern, which runs along the plasma membrane (Figure 3.2B); however, the spacing is never as narrow in the whole mount preparations of chick heart embryos as it is in the peripheral regions of cultured cardiomyocytes. By the 9 somite stage, the first cross-striations appear (Figure 3.2C), and by the 12 somite stage α-actinin is found exclusively in a striated pattern (Figure 3.2D).

Another difference between myofibrillogenesis in cultured cardiomyocytes and in the developing heart is that nonmuscle myosin IIB, which was proposed to act as a space-holder for sarcomeric myosin in the premyofibril hypothesis (Rhee et al., 1994), could never be detected between initial IZI-complexes in whole mount preparations (Ehler et al., 1999). Therefore, nm myosin IIB does not seem to play an important role during sarcomere assembly in situ, as already suggested by the results from

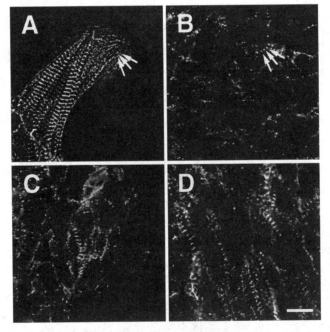

FIGURE 3.2. Single confocal sections of cultured embryonic chicken cardiomyocytes (A) and whole mount preparations of embryonic chicken heart from the 8 somite (B), the 9 somite (C), and the 12 somite (D) stages stained with monoclonal antibodies against sarcomeric α-actinin. Whereas in cultured cardiomyocytes a transition in the α-actinin pattern can be observed, with cross-striations in the cellular center and α-actinin localized in a more punctuate pattern toward the periphery (A, arrows), in whole mount preparations similar transitions cannot be observed. Before the start of beating, α-actinin is also localized in a punctuate pattern, although the distances between the individual dots are wider than in the peripheral areas of cultured cells (B, arrows); as soon as the first sarcomeres have been assembled, α-actinin is mainly found in striations as well (C, D).

knockout mice for nonmuscle myosin IIB that are able to assemble initial sarcomeres and only show heart defects later during development (Tullio et al., 1997). Nm myosin IIB might rather be required for adaptation to the in vitro environment to ensure cellular spreading mediated by the nonsarcomeric cytoskeleton, providing the tension so that myofibrillogenesis can proceed in cultured cardiomyocytes.

Titin as an Organizer?

Because of its localization in the sarcomere, with its N-terminus anchored in the Z-disk and the individual molecule stretching all the way to the middle of the sarcomere, where its C-termini overlap at the M-band, titin has always been regarded as an organizer of the sarcomere (Fürst and Gautel, 1995; Trinick, 1996). Evidence from many different experiments indicates that titin may indeed play a crucial role during sarcomere assembly. As described earlier, its N-terminus is organized from the earliest stages on. Therefore, it was important to find out whether the C-terminus of titin becomes organized that early as well. In cultured cardiomyocytes, there is a delay in the organization of N- and C-terminal epitopes. Whereas N-terminal epitopes can already be detected in concentrated spots in the peripheral regions of the cardiomyocytes, C-terminal epitopes could only be visualized once the myofibrils show already clear striations (van der Ven et al., 1999). A similar delay in the degree of organization can also be observed in heart whole mount preparations. Whereas N-terminal titin shows clear spots at the 8 somite stage (Figure 3.3A), most of the staining that is seen with an antibody to a C-terminal epitope is completely diffuse at this stage and only a few foci of epitopes can be detected (Figure 3.3C, arrows). However, the C-terminus of titin also gets organized rather quickly, as can be seen in the 9 somite stage when antibodies directed against both epitopes show a clear cross-striation (Figure 3.3B and D; see also Ehler et al., 1999).

These observations suggest that titin might indeed provide a kind of scaffold, which is necessary for the integration of other sarcomeric components. Additional evidence that titin acts as a blueprint for sarcomere architecture comes from the observation that slight differences of Z-disk width or sarcomere length, for example, can be accounted for by different splice variants of titin (Gautel et al., 1996; Kolmerer et al., 1996; Linke et al., 1999). Studies have suggested that titin might even play another role in addition to acting as an internal scaffold for myofibrillogenesis. Close to its C-terminus, titin possesses a kinase domain, which has been shown to be an auto-regulated serine kinase (Mayans et al., 1998). So far, the function of titin kinase is not quite clear because the only substrate identified so far is telethonin/Tcap, a protein that appears to be associated with the Z-disk rather than with the M-band region (Gregorio et al., 1998; Mues et al., 1998). Due to the enormous size of the titin molecule, it is rather difficult to envisage the C-terminus of titin changing its position so dramatically, and the significance of titin kinase for the regulation of myofibril assembly remains to be established.

The Role of the M-Band

In addition to the overlapping C-termini of titin, another component of the M-band is probably crucial for the integration of thick filaments, namely myomesin (Eppenberger et al., 1981; Grove et al., 1984). This protein, which is expressed in all kinds of striated muscle examined so far from early on, possesses binding sites for titin as well as for myosin (Obermann et al., 1995; Auerbach et al., 1999). Analysis

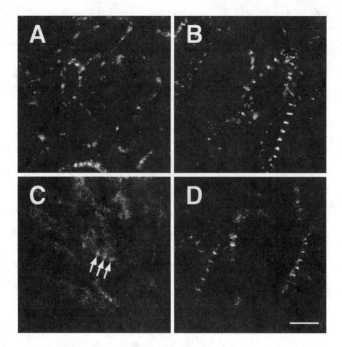

FIGURE 3.3. Single confocal sections of whole mount preparations from embryonic chicken hearts at the 8 somite (A, C) and the 9 somite (B, D) stages stained with monoclonal antibodies against an N-terminal epitope of titin (clone T12, panels A, B) or against a C-terminal epitope of titin (clone T41, panels C, D). At the 8 somite stage, N-terminal epitopes are already organized into discrete spots (A); with C-terminal epitopes, this is less prominent (arrows) and most of the signal is still diffuse at that time (C). By the 9 somite stage, both epitopes already show clear striations in the myofibrils that have just been assembled (B, D).

of myomesin localization in cultured cardiomyocytes revealed that although clear striations can only be detected in the perinuclear region of the cells where mature myofibrils are situated, a concentration of the signal for myomesin is always apparent once sarcomeric myosin has become organized to bipolar filaments (van der Ven et al., 1999). In chicken heart whole mount preparations of the 8 somite stage, myomesin is still distributed in a diffuse fashion. However, when the organization of myomesin is compared with that of MyBP-C, an A-band component, in the first myofibrils, striations of myomesin always seem to precede the appearance of double bands of the MyBP-C staining. Association of M-CK (muscle creatine kinase), another component of the mammalian M-band, appears even later during development (Carlsson et al., 1982). Therefore, myomesin might play a similar role in the M-band as α-actinin does in the Z-disk—namely to integrate titin filaments with the contractile filament system, in this case the thick myosin filaments. We propose the existence of a basic cytoskeletal framework of the sarcomere, composed of α-actinin at the Z-disk and myomesin at the M-band with titin filaments running in-between, that serve as integration points for the thin filaments and the thick filaments, respectively (Ehler et al., 1999). If the function of any of these components was ablated, myofibrillogenesis should be severely impaired. So far, this has only been shown partially in the case of titin: if a Z-disk segment of titin is overexpressed in cultured cardiomyocytes, myofibrillogenesis is affected; additionally, antisense oligonucleotides to Z-disk titin have a similar effect (Turnacioglu et al., 1997; Jutta Schaper, personal communication).

In chicken heart, several isoforms of myomesin are coexpressed during embryonic development. By alternative splicing, a specific domain can be included in the middle of the molecule, resulting in the predominant myomesin isoform in the embryonic heart of birds as well as of mammals. Around the time of birth, this isoform, which we have termed EH (embryonic heart)-myomesin, had become downregulated almost completely, and we were also unable to detect its expression in skeletal muscle of any developmental stage (Agarkova et al., 2000). In addition to EH-myomesin, which is the first myomesin isoform to be expressed during cardiac development, H (heart)-myomesin and to a much lesser amount also S (skeletal muscle)-myomesin, both characterized by a specific C-terminal extension, become expressed at later stages of embryonic development, and their expression is maintained in the adult chicken heart. Mammals have lost the expression of a heart-specific isoform of myomesin and only express EH-myomesin before birth and S-myomesin afterward (Agarkova et al., 2000). So far, the function of the additional EH-domain still remains obscure; according to sequence analysis, it could provide an additional more flexible element in the myomesin molecule that might be required at the earliest stages of myofibrillogenesis.

Thick and Thin Filament Assembly Occur Independently

When the myofibrils of cultured cardiomyocytes are analyzed, clear striations of Z-disk, M-band, and A-band components can already be observed when the actin filaments still do not yet seem to be segmented to proper I-bands. Similar observations can be made when the actin organization is studied in whole mount preparations with phalloidin to visualize F-actin (Shiraishi et al., 1992). Whereas at the 8 somite stage actin filaments can be found in close proximity to the membrane, they detach from the membrane and run throughout the cytoplasm by the 9 somite stage. However, clear segmentation into I-bands is not readily visible at this stage and only becomes apparent by the 12 somite stage (Shiraishi et al., 1992). Because phalloidin does not discriminate among different actin isoforms, we wanted to determine eventual isoform diversities. At the early stages of cardiac development, several actin isoforms, α-smooth muscle actin, α-cardiac actin, and α-skeletal actin are coexpressed, as shown on the mRNA as well as the protein level (Ordahl, 1986; Ruzicka and Schwartz, 1988; Ehler et al., manuscript in preparation). At the 8 somite stage, they seem to localize in an indistinguishable fashion because stainings with antibodies specific to either α-cardiac or α-smooth muscle actin always revealed a pattern similar to phalloidin, namely in filaments proximal to the membrane (Ehler et al., manuscript in preparation). However, by the 9 somite stage, an important difference can be observed. Whereas phalloidin staining still shows actin filaments that are not yet properly segregated, with antibodies specific to α-cardiac actin a clear staining pattern restricted to the I-band can be visualized in the same myofibril. This suggests that actin filaments composed of α-cardiac actin seem to have already attained a fixed length at the 9 somite stage and that actin filaments composed of other isoforms coexist that are still extending further toward the M-band (Ehler et al., manuscript in preparation). By the 12 somite stage, these thin filaments have been reduced to their proper length or have disappeared. Because nebulin is not expressed in cardiomyocytes and thus cannot function as a thin filament length regulator, other mechanisms must exist to determine the exact length of the α-cardiac actin filaments already at these early stages. In adult cardiomyocytes, the pointed-end filament capping protein tropomodulin seems to be responsible (for a review, see Littlefield and Fowler, 1998); however, clear tropomodulin striations in the sarcomere are only apparent at the 12 somite stage; therefore other factors must play a role as well (Ehler et al., manuscript in preparation). One hypothesis is the length regulation by optimal working output. The length of thin filaments composed of α-cardiac actin might be stabilized from early stages on because this isoform is optimal for interaction with the myosin heads of cardiac thick filaments; thin filaments composed of other actin isoforms might be subject to dynamic instability and therefore have a high variability in length. Recent evidence has shown also that an effect of titin on thin filament length cannot be excluded. Although interaction of I-band titin with the actin filaments has been shown only for the first part of the I-band (Trombitas et al., 1997), overexpression of an I-band fragment of cardiac titin, N2B, had drastic consequences on thin filament stability, at least in cultured cardiomyocytes (Linke et al., 1999). Additionally, there might be other factors yet to be identified that specify thin filament length at these early stages.

MOLECULAR MECHANISMS DIRECTING SARCOMERE ASSEMBLY

A striking specificity of protein targeting to unique sites within the myofibril was observed for many sarcomeric proteins (e.g., myomesin, M protein, α-actinin, MyBP-C). Additionally, a preferential incorporation of sarcomeric isoforms of contractile proteins was apparent. The knowledge on the sequence elements that carry the information for correct assembly is still limited, but functional studies are necessary in order to identify mechanisms of myofibrillogenesis at the molecular level. For these experiments, the development of epitope-tagging techniques has been crucial, and in a pioneering study it was possible to show that the isoforms of the essential myosin light chains MLC 3nm (nonmuscle) and MLC 3f (fast muscle) have different affinities for incorporation into sarcomeres in cultured cardiomyocytes (Soldati and Perriard, 1991). The cDNA encoding a sarcomeric protein or a fragment thereof is fused with sequences encoding for a viral epitope that has no equivalent in mammalian cells and is cloned into an expression plasmid under the control of a ubiquitously active eukaryotic promoter (e.g., the CMV promoter). After transient transfection, the expressed fusion proteins can be visualized by antibodies that specifically recognize the viral epitope, and their localization can be compared with that of endogenous proteins by double immunofluorescence. Because most viral epitope tags are rather short (5 to 20 amino acids), they are not known to interfere with protein folding or function. A disadvantage of this technique is that the transfected cells must be fixed in order to visualize the epitopes by immunofluorescence.

Recently, the advance of GFP (green fluorescent protein) and its color variants has brought the exciting possibility of watching the epitope-tagged proteins also in living cells (Auerbach et al., 1997, 1999; Dabiri et al., 1997; Turnacioglu et al., 1997; Helfman et al., 1999; Linke et al., 1999). Surprisingly, even the rather large tag GFP (27 kd) does not appear to interfere with the function of most proteins, as shown by the regular contractions of myofibrils in cardiomyocytes that were transfected with MLC 3f-GFP, myomesin-domain2-GFP, and tropomyosin isoproteins (Figure 3.4A, B, and C, respectively; Auerbach et al., 1999; Helfman et al., 1999; Leu et al., 1999). Using these techniques, it was possible both to investigate isoform diversities in incorpora-

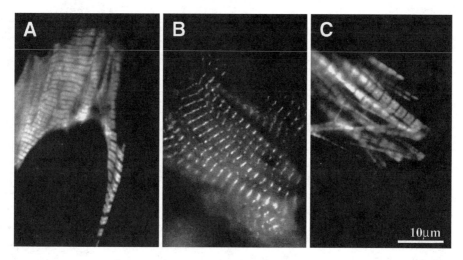

FIGURE 3.4. Video stills of cultured neonatal rat cardiomyocytes that were transiently trans-fected with MLC3f-eGFP (A), myomesin-domain2-eGFP (B), or α-skeletal tropomyosin-eGFP (C). The fusion proteins are incorporated in their correct pattern into the sarcomere, and the cardiomyocytes continue to beat. Note that these cells are live and have not been fixed for photography. The pictures demonstrate the high fidelity of incorporation with little back-ground of unincorporated material (picture courtesy of Martin Leu, Department of Biology, Swiss Federal Institute of Technology).

tion into myofibrils as well as to map parts of sarcomeric proteins that are indis-pensable for proper assembly.

Isoform Diversity

Many sarcomeric proteins are expressed in different isoforms (for a review, see Schiaffino and Reggiani, 1996). In order to find out whether the information for targeting to the myofibrils is specific for muscle isoforms, double epitope-tagging experiments were performed with members of the isoform family of the essential myosin light chains (e.g., MLC 3nm and MLC 3f). By cotransfection of expression constructs encoding different viral epitopes, it could be shown that the muscle isoform MLC 3f localizes perfectly to the sarcomeres, showing double bands as it associates with the myosin heads of the bipolar thick filaments. The nonmuscle MLC isoform, on the other hand, seems to contain a weaker targeting information because it is mostly distributed in a diffuse fashion throughout the cytoplasm. Exchange of the different viral epitopes gave identical results (Komiyama et al., 1996). Experiments with chimeric constructs could then nail down the targeting information to the N-terminal lobe in the MLC 3f or 1f molecule (Komiyama et al., 1996). Various com-binations of cDNAs of two different MLC isoforms allowed determination of the relative affinities for assembly into myofibrils, ranging from MLC 3nm as the weakest and MLC 3f as the strongest interaction partner for assembly (Komiyama et al., 1996). This approach was also used to investigate the ability of mutant myosin heavy-chain isoforms that are known to cause familial hypertrophic cardiomyopathy to integrate into myofibrils in the presence of a competing wild-type protein (Becker et al., 1997). Depending on the position of the point mutation, striking differences were observed. Whereas mutations in the consensus ATP binding sequence of a human β-myosin heavy chain (K184R) caused disruption of the myofibrillar apparatus, other mutations (e.g., R249Q and R403Q) were competent to target to the myofibrils and did not affect myofibril structure (Becker et al., 1997).

For myosin light chains, no obvious effects on cytoarchitecture or cardiomyocyte function were observed after expression of nonmuscle isoforms; this was not the case when different actin isoforms were compared. All muscle actin isoforms (striated as well as smooth muscle) are able to incorporate correctly into the sarcomeres. However, if nonmuscle actins are transfected into cardiomyocytes, the myofibrillar apparatus is severely affected because overexpression of β- or γ-cytoplasmic actin results in a dominant-negative effect on myofibrillar structure. The cardiomyocytes stop beating while the overall myofibrillar pattern persists except for the displaced thin filaments. Actin accumulates at membranous sites, and the cells have a completely changed morphology with long filopodial extensions (von Arx et al., 1995). Because similar expression levels of muscle actin isoforms do not affect myofibril integrity, these experiments highlight important intrinsic differences among distinct actin isoforms (von Arx et al., 1995).

Mapping Targeting Sites in Proteins

The technique of epitope tagging has also facilitated studies to determine at the molecular level which sequences of a protein are required for proper targeting to the sarcomere. For example, the M-band components myomesin and M protein have a nearly identical domain structure (Bantle et al., 1996; Steiner et al., 1998). By generating expression constructs of fusions between different domains of M protein or of myomesin with an epitope tag, it could be shown that the M-band targeting site in M protein resides in the domains Mp2–Mp3, but that in myomesin the domain My2 is on its own sufficient to target to the M-band (Auerbach et al., 1997, 1999; Obermann et al., 1998; Auerbach, Perriard, and Perriard, unpublished observations). In the case of myomesin, M-band targeting could thus be separated from myosin binding because the principal myosin binding site seems to reside in the head domain of myomesin (Obermann et al., 1997; Auerbach et al., 1999).

Similar studies were performed to identify the targeting sites in MyBP-C, α-actinin, and the N-terminal part of titin (Gilbert et al., 1996, 1999; Turnacioglu et al., 1997; Lin et al., 1998; Ayoob et al., 2000). Interestingly, overexpression of titin fragments can have a deleterious effect on myofibril structure. Whereas overexpression of an N-terminal part of titin leads to the arrest of myofibrillogenesis (Turnacioglu et al., 1997), overexpression of an I-band segment of titin, encoding for the cardiac-specific N2B insertion, results in disruption of the thin filaments without affecting the thick filaments (Linke et al., 1999). Future studies along these lines should provide important insights on the mechanisms that are important for myofibrillogenesis on the basis of the assembly of individual sarcomeric components.

SUMMARY

Analysis of myofibrillogenesis in whole mount preparations of embryonic chicken heart, taken at the time when beating sets in, reveals important differences but also similarities when compared to myofibril assembly in cultured cardiomyocytes. No stress-fiber–like structures or premyofibrils can be detected in the developing heart in situ, structures that are typical for myofibrillogenesis in cultured cardiomyocytes.

On the other hand, the first organized complexes in situ consist also of sarcomeric α-actinin, the N-terminus of titin together with filamentous actin, making up dense body-like structures. The next step during myofibrillogenesis seems to involve the setup of a basic cytoskeletal framework of the sarcomere, consisting of α-actinin at the Z-disk, myomesin at the M-band, and titin filaments stretching in-between. Myomesin seems to be required for the integration of the thick filaments with the M-band end of titin. Assembly of thick and thin filaments occurs independently. At the time when mature A-bands can already be observed, thin filaments still must attain their definitive length. The association of the pointed-end capping protein tropomodulin and the concentration of desmin filaments around the Z-disks mark the final steps of myofibrillogenesis.

Epitope-tagging experiments allow pinpointing of the sequences that are necessary for targeting to the sarcomere at the molecular level. With the advance of tags such as GFP and its color variants, which allow one to follow the fate of tagged proteins in living cardiomyocytes, an exciting future can be envisaged. Additionally, information from mice that are homozygous null for proposed key players during sarcomere assembly or transgenic for deletions should shed more light on what is really important during myofibrillogenesis.

ACKNOWLEDGMENTS

We are very grateful to the team "Biogenesis of Cytoarchitecture" at the Institute of Cell Biology ETH and acknowledge the enthusiastic help of its members Irina Agarkova, Daniel Auerbach, Pierre Giro, Alain Hirschy, Martin Leu, Sara Meier, Mohamed Nemir, and Evelyne Perriard. We also want to thank the research team of Dr. Hans M. Eppenberger for supporting interest and collaborative efforts. The support of the Forschungskommission ETH for several predoctoral training grants, of the Swiss National Science Foundation (# 31.37537/93 and # 31.52417/97), the Swiss Heart Foundation, and the Swiss Foundation for Research on Muscle Diseases was fundamental for the realization of the research summarized in this review.

REFERENCES

Agarkova, I., Auerbach, D., Ehler, E., and Perriard, J.-C. 2000. A novel marker for vertebrate embryonic heart: the EH-myomesin isoform. *J. Biol. Chem.* 275:10256–10264.

Antin, P.B., Tokunaka, S., Nachmias, V.T., and Holtzer, H. 1986. Role of stress fiber-like structures in assembling nascent myofibrils in myosheets recovering from exposure to ethyl methanesulfonate. *J. Cell Biol.* 102:1464–1479.

Atherton, B.T., Meyer, D.M., and Simpson, D.G. 1986. Assembly and remodelling of myofibrils and intercalated disks in cultured neonatal rat heart cells. *J. Cell Sci.* 86:233–248.

Auerbach, D., Rothen-Rutishauser, B., Bantle, S., Leu, M., Ehler, E., Helfman, D., and Perriard, J.-C. 1997. Molecular mechanisms of myofibril assembly in heart. *Cell Struct. Funct.* 22:139–146.

Auerbach, D., Bantle, S., Keller, S., Hinderling, V., Leu, M., Ehler, E., and Perriard, J.-C. 1999. Different domains of the M-band protein myomesin are involved in myosin binding and M-band targeting. *Mol. Biol. Cell* 10:1297–1308.

Ayoob, J.C., Turnacioglu, K.K., Mittal, B., Sanger, J.M., and Sanger, J.W. 2000. Targeting of cardiac muscle titin fragments to the Z-bands and dense bodies of living muscle and non-muscle cells. *Cell Motil. Cytoskel.* 45:67–82.

Bantle, S., Keller, S., Haussmann, I., Auerbach, D., Perriard, E., Mühlebach, S., and Perriard, J.-C. 1996. Tissue-specific isoforms of chicken myomesin are generated by alternative splicing. *J. Biol. Chem.* 271:19042–19052.

Becker, K.D., Gottshall, K.R., Hickey, R., Perriard, J.-C., and Chien, K.R. 1997. Point mutations in human beta cardiac myosin heavy chain have differential effects on sarcomeric structure and assembly: an ATP binding site change disrupts both thick and thin filaments,
whereas hypertrophic cardiomyopathy mutations display normal assembly. *J. Cell Biol.*
137:131–140.

Carlsson, E., Kjorell, U., Thornell, L.E., Lambertsson, A., and Strehler, E. 1982. Differentiation of the myofibrils and the intermediate filament system during postnatal development
of the rat heart. *Eur. J. Cell Biol.* 27:62–73.

Clark, W.A., Decker, M.L., Behnke Barclay, M., Janes, D.M., and Decker, R.S. 1998. Cell
contact as an independent factor modulating cardiac myocyte hypertrophy and survival in
long-term primary culture. *J. Mol. Cell Cardiol.* 30:139–155.

Claycomb, W. and Palazzo, M. 1980. Culture of the terminally differentiated adult cardiac
muscle cell: a light and scanning electron microscope study. *Dev. Biol.* 80:466–482.

Colucci-Guyon, E., Portier, M.M., Dunia, I., Paulin, D., Pournin, S., and Babinet, C. 1994.
Mice lacking vimentin develop and reproduce without an obvious phenotype. *Cell*
79:679–694.

Dabiri, G.A., Turnacioglu, K.K., Sanger, J.M., and Sanger, J.W. 1997. Myofibrillogenesis
visualized in living embryonic cardiomyocytes. *Proc. Natl. Acad. Sci. USA* 94:9493–
9498.

Dlugosz, A.A., Antin, P.B., Nachmias, V.T., and Holtzer, H. 1984. The relationship between
stress fiber-like structures and nascent myofibrils in cultured cardiac myocytes. *J. Cell Biol.*
99:2268–2278.

Donath, M.Y., Zapf, J., Eppenberger-Eberhardt, M., Froesch, E.R., and Eppenberger,
H.M. 1994. Insulin-like growth factor I stimulates myofibril development and decreases
smooth muscle alpha-actin of adult cardiomyocytes. *Proc. Natl. Acad. Sci. USA* 91:1686–
1690.

Ehler, E., Rothen, B.M., Hämmerle, S.P., Komiyama, M., and Perriard, J.-C. 1999. Myofibrillogenesis in the developing chicken heart: assembly of Z-disk, M-line and the thick filaments.
J. Cell Sci. 112:1529–1539.

Eppenberger, H.M., Perriard, J.-C., Rosenberg, U., and Strehler, E.E. 1981. The Mr 165,000
M-protein myomesin: a specific protein of cross-striated muscle cells. *J. Cell Biol.* 89:
185–193.

Eppenberger, H.M., Hertig, C., and Eppenberger-Eberhardt, M. 1994. Adult rat cardiomyocytes in culture: A model system to study the plasticity of the differentiated cardiac
phenotype at the molecular and cellular levels. *Trends Cardiovasc. Med.* 4:187–192.

Eppenberger, M.E., Hauser, I., Baechi, T., Schaub, M.C., Brunner, U.T., Dechesne, C.A., and
Eppenberger, H.M. 1988. Immunocytochemical analysis of the regeneration of myofibrils
in long-term cultures of adult cardiomyocytes of the rat. *Dev. Biol.* 130:1–15.

Eppenberger-Eberhardt, M., Aigner, S., Donath, M., Kurer, V., Walther, P., Zuppinger, C.,
Schaub, M.C., and Eppenberger, H.M. 1997. IGF-I and bFGF differentially influence atrial
natriuretic factor and alpha-smooth muscle actin expression in cultured atrial compared to
ventricular adult rat cardiomyocytes. *J. Mol. Cell Cardiol.* 29:2027–2039.

Fässler, R., Rohwedel, J., Maltsev, V., Bloch, W., Lentini, S., Guan, K., Gullberg, D., Hescheler,
J., Addicks, K., and Wobus, A.M. 1996. Differentiation and integrity of cardiac muscle cells
are impaired in the absence of beta1 integrin. *J. Cell Sci.* 109:2989–2999.

Fürst, D.O. and Gautel, M. 1995. The anatomy of a molecular giant: how the sarcomere
cytoskeleton is assembled from immunoglobulin superfamily molecules. *J. Mol. Cell
Cardiol.* 27:951–959.

Gautel, M., Goulding, D., Bullard, B., Weber, K., and Fürst, D.O. 1996. The central Z-disk
region of titin is assembled from a novel repeat in variable copy numbers. *J. Cell Sci.*
109:2747–2754.

Gilbert, R., Kelly, M.G., Mikawa, T., and Fischman, D.A. 1996. The carboxyl terminus of
myosin binding protein C (MyBP-C, C-protein) specifies incorporation into the A-band of
striated muscle. *J. Cell Sci.* 109:101–111.

Gilbert, R., Cohen, J.A., Pardo, S., Basu, A., and Fischman, D.A. 1999. Identification of the A-band localization domain of myosin binding proteins C and H (MyBP-C, MyBP-H) in skeletal muscle. *J. Cell Sci.* 112:69–79.

Goncharova, E.J., Kam, Z., and Geiger, B. 1992. The involvement of adherens junction components in myofibrillogenesis in cultured cardiac myocytes. *Development* 114:173–183.

Gosteli-Peter, M., Harder, B., Eppenberger, H.M., Zapf, J., and Schaub, M.C. 1996. Triiodothyronine induces over-expression of alpha-smooth muscle actin, restricts myofibrillar expansion and is permissive for the action of basic fibroblast growth factor and insulin-like growth factor I in adult rat cardiomyocytes. *J. Clin. Invest.* 98:1737–1744.

Gregorio, C.C., Trombitas, K., Centner, T., Kolmerer, B., Stier, G., Kunke, K., Suzuki, K., Obermayr, F., Herrmann, B., Granzier, H., Sorimachi, H., and Labeit, S. 1998. The NH2 terminus of titin spans the Z-disc: its interaction with a novel 19-kD ligand (T-cap) is required for sarcomeric integrity. *J. Cell Biol.* 143:1013–1027.

Grove, B.K., Kurer, V., Lehner, C., Doetschman, T.C., Perriard, J.-C., and Eppenberger, H.M. 1984. Monoclonal antibodies detect new 185,000 dalton muscle M-line protein. *J. Cell Biol.* 98:518–524.

Guan, K., Fürst, D.O., and Wobus, A.M. 1999. Modulation of sarcomere organization during embryonic stem cell-derived cardiomyocyte differentiation. *Eur. J. Cell Biol.* 78:813–823.

Guo, J.X., Jacobson, S.L., and Brown, D.L. 1986. Rearrangement of tubulin, actin, and myosin in cultured ventricular cardiomyocytes of the adult rat. *Cell Motil. Cytoskel.* 6:291–304.

Handel, S.E., Greaser, M.L., Schultz, E., Wang, S.M., Bulinski, J.C., Lin, J.J., and Lessard, J.L. 1991. Chicken cardiac myofibrillogenesis studied with antibodies specific for titin and the muscle and nonmuscle isoforms of actin and tropomyosin. *Cell Tissue Res.* 263:419–430.

Harder, B.A., Schaub, M.C., Eppenberger, H.M., and Eppenberger Eberhardt, M. 1996. Influence of fibroblast growth factor (bFGF) and insulin-like growth factor (IGF-I) on cytoskeletal and contractile structures and on atrial natriuretic factor (ANF) expression in adult rat ventricular cardiomyocytes in culture. *J. Mol. Cell Cardiol.* 28:19–31.

Harder, B.A., Hefti, M.A., Eppenberger, H.M., and Schaub, M.C. 1998. Differential protein localization in sarcomeric and nonsarcomeric contractile structures of cultured cardiomyocytes. *J. Struct. Biol.* 122:162–175.

Helfman, D.M., Berthier, C., Grossman, J., Leu, M., Ehler, E., Perriard, E., and Perriard, J.-C. 1999. Nonmuscle tropomyosin-4 requires coexpression with other low molecular weight isoforms for binding to thin filaments in cardiomyocytes. *J. Cell Sci.* 112:371–380.

Horackova, M. and Byczko, Z. 1997. Differences in the structural characteristics of adult guinea pig and rat cardiomyocytes during their adaptation and maintenance in long-term cultures: confocal microscopy study. *Exp. Cell Res.* 237:158–175.

Imanaka-Yoshida, K., Knudsen, K.A., and Linask, K.K. 1998. N-cadherin is required for the differentiation and initial myofibrillogenesis of chick cardiomyocytes. *Cell Motil. Cytoskel.* 39:52–62.

Isac, C.M., Ruiz, P., Pfitzmaier, B., Haase, H., Birchmeier, W., and Morano, I. 1999. Plakoglobin is essential for myocardial compliance but dispensable for myofibril insertion into adherens junctions. *J. Cell Biochem.* 72:8–15.

Jacobson, S. and Piper, H. 1986. Cell cultures of adult cardiomyocytes as models of the myocardium. *J. Mol. Cell Cardiol.* 18:661–678.

Kolmerer, B., Olivieri, N., Witt, C.C., Herrmann, B.G., and Labeit, S. 1996. Genomic organisation of M line titin and its tissue-specific expression in two distinct isoforms. *J. Mol. Biol.* 256:556–563.

Komiyama, M., Soldati, T., von Arx, P., and Perriard, J.-C. 1996. The intracompartmental sorting of myosin alkali light chain isoproteins reflects the sequence of developmental expression as determined by double epitope-tagging competition. *J. Cell Sci.* 109:2089–2099.

Lazarides, E. 1982. Intermediate filaments: a chemically heterogeneous, developmentally regulated class of proteins. *Annu. Rev. Biochem.* 51:219–250.

Lemanski, S.F., Kovacs, C.P., and Lemanski, L.F. 1997. Analysis of the three-dimensional distributions of alpha-actinin, ankyrin, and filamin in developing hearts of normal and cardiac mutant axolotls (*Ambystoma mexicanum*). *Anat. Embryol.* 195:155–163.

Leu, M., Auerbach, D., Helfman, D., and Perriard, J.-C. 1999. Green fluorescent protein in living cardiomyocytes. *Trends Cell Biol.* CD GFP in Motion.

Li, Z., Colucci-Guyon, E., Pincon, R.M., Mericskay, M., Pournin, S., Paulin, D., and Babinet, C. 1996. Cardiovascular lesions and skeletal myopathy in mice lacking desmin. *Dev. Biol.* 175:362–366.

Lin, Z.X., Holtzer, S., Schultheiss, T., Murray, J., Masaki, T., Fischman, D.A., and Holtzer, H. 1989. Polygons and adhesion plaques and the disassembly and assembly of myofibrils in cardiac myocytes. *J. Cell Biol.* 108:2355–2367.

Lin, Z., Hijikata, T., Zhang, Z., Choi, J., Holtzer, S., Sweeney, H.L., and Holtzer, H. 1998. Dispensability of the actin-binding site and spectrin repeats for targeting sarcomeric alpha-actinin into maturing Z bands in vivo: implications for in vitro binding studies. *Dev. Biol.* 199:291–308.

Linke, W.A., Rudy, D.E., Centner, T., Gautel, M., Witt, C., Labeit, S., and Gregorio, C.C. 1999. I-band titin in cardiac muscle is a three-element molecular spring and is critical for maintaining thin filament structure. *J. Cell Biol.* 146:631–644.

Littlefield, R. and Fowler, V.M. 1998. Defining actin filament length in striated muscle: rulers and caps or dynamic stability? *Annu. Rev. Cell Dev. Biol.* 14:487–525.

LoRusso, S.M., Rhee, D., Sanger, J.M., and Sanger, J.W. 1997. Premyofibrils in spreading adult cardiomyocytes in tissue culture: evidence for reexpression of the embryonic program for myofibrillogenesis in adult cells. *Cell Motil. Cytoskel.* 37:183–198.

Manasek, F.J. 1970. Histogenesis of the embryonic myocardium. *Am. J. Cardiol.* 25:149–168.

Marino, T.A., Kurseryk, L., and Lauva, I.K. 1987. Role of contraction in the structure and growth of neonate rat cardiocytes. *Am. J. Physiol.* 253:H1391–1399.

Mayans, O., van der Ven, P.F., Wilm, M., Mues, A., Young, P., Fürst, D.O., Wilmanns, M., and Gautel, M. 1998. Structural basis for activation of the titin kinase domain during myofibrillogenesis. *Nature* 395:863–869.

Messerli, J.M., Eppenberger-Eberhardt, M.E., Rutishauser, B.M., Schwarb, P., von Arx, P., Koch-Schneidemann, S., Eppenberger, H.M., and Perriard, J.-C. 1993. Remodelling of cardiomyocyte cytoarchitecture visualized by three-dimensional (3D) confocal microscopy. *Histochemistry* 100:193–202.

Milner, D.J., Weitzer, G., Tran, D., Bradley, A., and Capetanaki, Y. 1996. Disruption of muscle architecture and myocardial degeneration in mice lacking desmin. *J. Cell Biol.* 134: 1255–1270.

Milner, D.J., Taffet, G.E., Wang, X., Pham, T., Tamura, T., Hartley, C., Gerdes, A.M., and Capetanaki, Y. 1999. The absence of desmin leads to cardiomyocyte hypertrophy and cardiac dilation with compromised systolic function. *J. Mol. Cell Cardiol.* 31:1063–1076.

Mues, A., van der Ven, P.F., Young, P., Fürst, D.O., and Gautel, M. 1998. Two immunoglobulin-like domains of the Z-disc portion of titin interact in a conformation-dependent way with telethonin. *FEBS Lett.* 428:111–114.

Obermann, W.M., Plessmann, U., Weber, K., and Fürst, D.O. 1995. Purification and biochemical characterization of myomesin, a myosin-binding and titin-binding protein, from bovine skeletal muscle. *Eur. J. Biochem.* 233:110–115.

Obermann, W.M., Gautel, M., Weber, K., and Fürst, D.O. 1997. Molecular structure of the sarcomeric M band: mapping of titin and myosin binding domains in myomesin and the identification of a potential regulatory phosphorylation site in myomesin. *EMBO J.* 16: 211–220.

Obermann, W.M., van der Ven, P.F., Steiner, F., Weber, K., and Fürst, D.O. 1998. Mapping of a myosin-binding domain and a regulatory phosphorylation site in M-protein, a structural protein of the sarcomeric M band. *Mol. Biol. Cell* 9:829–840.

Ordahl, C.P. 1986. The skeletal and cardiac alpha-actin genes are coexpressed in early embryonic striated muscle. *Dev. Biol.* 117:488–492.

Rhee, D., Sanger, J.M., and Sanger, J.W. 1994. The premyofibril: evidence for its role in myofibrillogenesis. *Cell Motil. Cytoskel.* 28:1–24.

Rothen-Rutishauser, B.M., Ehler, E., Perriard, E., Messerli, J.M., and Perriard, J.-C. 1998. Different behaviour of the non-sarcomeric cytoskeleton in neonatal and adult rat cardiomyocytes. *J. Mol. Cell Cardiol.* 30:19–31.

Ruiz, P., Brinkmann, V., Ledermann, B., Behrend, M., Grund, C., Thalhammer, C., Vogel, F., Birchmeier, C., Gunthert, U., Franke, W.W., and Birchmeier, W. 1996. Targeted mutation of plakoglobin in mice reveals essential functions of desmosomes in the embryonic heart. *J. Cell Biol.* 135:215–225.

Ruzicka, D.L. and Schwartz, R.J. 1988. Sequential activation of alpha-actin genes during avian cardiogenesis: vascular smooth muscle alpha-actin gene transcripts mark the onset of cardiomyocyte differentiation. *J. Cell Biol.* 107:2575–2586.

Schaub, M.C., Hefti, M.A., Harder, B.A., and Eppenberger, M.E. 1998. Triiodothyronine restricts myofibrillar growth and enhances beating frequency in cultured adult rat cardiomyocytes. *Basic Res. Cardiol.* 93:391–395.

Schiaffino, S. and Reggiani, C. 1996. Molecular diversity of myofibrillar proteins: gene regulation and functional significance. *Physiol. Rev.* 76:371–423.

Schultheiss, T., Lin, Z.X., Lu, M.H., Murray, J., Fischman, D.A., Weber, K., Masaki, T., Imamura, M., and Holtzer, H. 1990. Differential distribution of subsets of myofibrillar proteins in cardiac nonstriated and striated myofibrils. *J. Cell Biol.* 110:1159–1172.

Shiraishi, I., Takamatsu, T., Minamikawa, T., and Fujita, S. 1992. 3-D observation of actin filaments during cardiac myofibrinogenesis in chick embryo using a confocal laser scanning microscope. *Anat. Embryol.* 185:401–408.

Shiraishi, I., Takamatsu, T., and Fujita, S. 1993. 3-D observation of N-cadherin expression during cardiac myofibrillogenesis of the chick embryo using a confocal laser scanning microscope. *Anat. Embryol.* 187:115–120.

Shiraishi, I., Takamatsu, T., and Fujita, S. 1995. Three-dimensional observation with a confocal scanning laser microscope of fibronectin immunolabeling during cardiac looping in the chick embryo. *Anat. Embryol.* 191:183–189.

Shiraishi, I., Simpson, D.G., Carver, W., Price, R., Hirozane, T., Terracio, L., and Borg, T.K. 1997a. Vinculin is an essential component for normal myofibrillar arrangement in fetal mouse cardiac myocytes. *J. Mol. Cell Cardiol.* 29:2041–2052.

Shiraishi, I., Takamatsu, T., Price, R.L., and Fujita, S. 1997b. Temporal and spatial patterns of phosphotyrosine immunolocalization during cardiac myofibrillogenesis of the chicken embryo. *Anat. Embryol.* 196:81–89.

Soldati, T. and Perriard, J.C. 1991. Intracompartmental sorting of essential myosin light chains: molecular dissection and in vivo monitoring by epitope tagging. *Cell* 66:277–289.

Steiner, F., Weber, K., and Fürst, D.O. 1998. Structure and expression of the gene encoding murine M-protein, a sarcomere-specific member of the immunoglobulin superfamily. *Genomics* 49:83–95.

Thornell, L., Carlsson, L., Li, Z., Mericskay, M., and Paulin, D. 1997. Null mutation in the desmin gene gives rise to a cardiomyopathy. *J. Mol. Cell Cardiol.* 29:2107–2124.

Tokuyasu, K.T. 1989. Immunocytochemical studies of cardiac myofibrillogenesis in early chick embryos. III. Generation of fasciae adherentes and costameres. *J. Cell Biol.* 108:43–53.

Tokuyasu, K.T. and Maher, P.A. 1987a. Immunocytochemical studies of cardiac myofibrillogenesis in early chick embryos. I. Presence of immunofluorescent titin spots in premyofibril stages. *J. Cell Biol.* 105:2781–2793.

Tokuyasu, K.T. and Maher, P.A. 1987b. Immunocytochemical studies of cardiac myofibrillogenesis in early chick embryos. II. Generation of α-actinin dots within titin spots at the time of the first myofibril formation. *J. Cell Biol.* 105:2795–2801.

Trinick, J. 1996. Titin as a scaffold and spring. *Curr. Biol.* 6:258–260.

Trombitas, K., Greaser, M.L., and Pollack, G.H. 1997. Interaction between titin and thin filaments in intact cardiac muscle. *J. Muscle Res. Cell Motil.* 18:345–351.

Tullio, A.N., Accili, D., Ferrans, V.J., Yu, Z.X., Takeda, K., Grinberg, A., Westphal, H., Preston, Y.A., and Adelstein, R.S. 1997. Nonmuscle myosin II-B is required for normal development of the mouse heart. *Proc. Natl. Acad. Sci. USA* 94:12407–12412.

Turnacioglu, K.K., Mittal, B., Dabiri, G.A., Sanger, J.M., and Sanger, J.W. 1997. An N-terminal fragment of titin coupled to green fluorescent protein localizes to the Z-bands in living muscle cells: overexpression leads to myofibril disassembly. *Mol. Biol. Cell* 8:705–717.

van der Ven, P.F.M., Ehler, E., Perriard, J.-C., and Fürst, D.O. 1999. Thick filament assembly occurs after the formation of a cytoskeletal scaffold. *J. Muscle Res. Cell Motil.* 20:569–579.

von Arx, P., Bantle, S., Soldati, T., and Perriard, J.-C. 1995. Dominant negative effect of cytoplasmic actin isoproteins on cardiomyocyte cytoarchitecture and function. *J. Cell Biol.* 131:1759–1773.

Wang, S.M., Greaser, M.L., Schultz, E., Bulinski, J.C., Lin, J.J., and Lessard, J.L. 1988. Studies on cardiac myofibrillogenesis with antibodies to titin, actin, tropomyosin, and myosin. *J. Cell Biol.* 107:1075–1083.

Xu, W., Baribault, H., and Adamson, E.D. 1998. Vinculin knockout results in heart and brain defects during embryonic development. *Development* 125:327–337.

Young, P., Ferguson, C., Banuelos, S., and Gautel, M. 1998. Molecular structure of the sarcomeric Z-disk: two types of titin interactions lead to an asymmetrical sorting of α-actinin. *EMBO J.* 17:1614–1624.

Zajdel, R., Dube, D., and Lemanski, L. 1999. The cardiac mutant Mexican axolotl is a unique animal model for evaluation of cardiac myofibrillogenesis. *Exp. Cell Res.* 248:557–566.

CHAPTER 4

Tropomodulin: An Important Player in Cardiac Myofibrillogenesis

Catherine McLellan and Carol C. Gregorio

Tropomodulin has been studied intensively since it was first reported in 1987 as a tropomyosin binding protein associated with the erythrocyte membrane skeleton (Fowler, 1987). The study of tropomodulin's role in heart myofibrillogenesis was initiated after the discovery that in mature skeletal muscle tropomodulin was localized to the pointed (slow-growing) ends of the thin filaments (Fowler et al., 1993). Subsequent in vitro studies demonstrated that tropomodulin is a potent actin filament pointed-end capping protein, functioning to prevent actin filaments from elongating or shrinking (Weber et al., 1994). The results of the studies that will be discussed here demonstrate that tropomodulin plays a critical role in maintaining the length of thin filaments within sarcomeres, in the physiologic function (contractile activity) of cardiac myocytes, and in the organogenesis of the heart. Investigations on the properties of this molecule continue to provide new insights into cytoskeletal protein dynamics.

TROPMODULIN ISOFORMS

Tropomodulin is an integral cytoskeletal component found in muscle and nonmuscle tissues in vertebrates as well as in *C. elegans* and in *Drosophila melanogaster* (Sanpodo). Its expression is developmentally regulated and tissue specific. Tropomodulin has been identified in several terminally differentiated cell types in vertebrates, including erythrocytes (Fowler, 1987; Fowler, 1990), lens fiber cells (Woo and Fowler, 1994; Sussman et al., 1996; Fischer et al., 2000; Lee et al., 2000), neurons (Sussman et al., 1994b; Watakabe et al., 1996), and skeletal (Fowler et al., 1993; Almenar-Queralt et al., 1999a) and cardiac myocytes (Sussman et al., 1994a; Gregorio and Fowler, 1995; Gregorio et al., 1995). Several isoforms of tropomodulin have been found in vertebrates to date: erythrocyte tropomodulin (E-Tmod) (Fowler, 1987; Sung et al., 1992; Babcock and Fowler, 1994; Ito et al., 1995), neural tropomodulin (N-Tmod/TMOD2) (Watakabe et al., 1996; Cox and Zoghbi, 2000), skeletal tropomodulin (Sk-Tmod/TMOD4) (Almenar-Queralt et al., 1999b; Cox and Zoghbi, 2000), and a new tropomodulin isoform that appears to be ubiquitously expressed in all tissues (TMOD3) (Cox and Zoghbi, 2000). The vertebrate isoforms are approximately 60% identical (70% similar) to one another at the amino acid level and are encoded by separate genes (Watakabe et al., 1996; Almenar-Queralt et al., 1999b; Cox and Zoghbi, 2000). Larger, more distantly related proteins with limited regions of similarity to portions of the ~40 kd tropomodulins have also been identified, including an ~64 kd protein expressed in smooth and extraocular muscle (Dong et al., 1991; Conley and Fowler, 1999) and an ~60 kd protein from *C. elegans* (for a review, see Fowler and Conley, 1999). E-Tmod is the only isotype characterized so far in the vertebrate heart; the expression pattern of the ubiquitous isoform (TMOD3) in the heart awaits further study. The genomic organization of human and mouse E-Tmods has been reported. Interestingly, the gene spans approximately 60 kilobases, with 9 exons and 10 introns (Chu et al., 1999b).

THIN FILAMENT STRUCTURE AND TROPOMODULIN ACTIVITY

In mature cardiac myocytes, the pointed (slow-growing) ends of the actin filaments terminate in the A-band, where they are capped by E-Tmod and overlap with the bipolar myosin-containing thick filaments. The barbed (fast-growing) ends are inserted into the Z-disk, where they are capped by capZ and crosslinked by α-actinin to each other and to titin filaments (see Figure 4.1; see Color Plate 1). The nomenclature "barbed" and "pointed" end refers to the arrowhead appearance visualized by electron microscopy of myosin 1 subfragments when they bind to actin filaments. The actin filaments in cardiac myocytes are coated with rodlike tropomyosin molecules that are associated with each other head-to-tail, forming two polymers per thin filament (see Figure 4.1). Each tropomyosin molecule binds one troponin complex (composed of troponins T, I, and C); together they mediate the calcium regulation of myosin ATPase (Weber and Murray, 1973). The binding of tropomyosin stabilizes the actin filaments and blocks the binding of other actin binding proteins (Goll et al., 1972; Wegner, 1979; Broschat, 1990). Whether the pointed ends of the actin filaments are capped by one or two E-Tmods per actin filament awaits further clarification. The stoichiometry as determined biochemically in skeletal myofibrils is equivocal at 1.2 to 1.6 tropomodulins per filament (Fowler et al., 1993). The stoichiometry determined by measuring capping activity predicts one tropomodulin per filament end (Weber et al., 1999).

Our understanding of the role of E-Tmod in cardiac muscle has been greatly assisted by the use of complementary in vitro and in vivo systems. As a result, the study of E-Tmod function using in vitro models of actin filament dynamics has formed the basis for many of the ideas that are now being tested in cardiac model systems. For example, extensive in vitro actin filament elongation and depolymerization assays have revealed that E-Tmod can cap actin filaments alone, but it does so with a much lower affinity and, consequently, less effectively than when it caps tropomyosin-coated actin filaments (Weber et al., 1994). For these assays, the barbed (fast-growing) ends of the short actin seeds are capped first with gelsolin, so that only events at the pointed ends are measured. E-Tmod is then added (with actin monomers) to actin seeds, and the effect of E-Tmod on the polymerization/depolymerization properties of the filaments and on the critical concentration of the monomers is studied. E-Tmod binds to the pointed ends of pure actin filaments with a low affinity (Kd ~0.3 μm). Transient binding of E-Tmod to all the pointed filament ends results in slower monomer addition, giving ADP.P$_i$-actin time to convert to ADP-actin at the filament ends. The result is that the affinity of the actin monomers for the ends of the filaments is lower, resulting in an increase in the critical concentration (Weber et al., 1999). Interestingly, if tropomyosin is added to the filaments (analogous to filaments found in striated muscle in vivo), then the affinity of E-Tmod for the pointed end increases 1000-fold (Kd < 0.05 nM). Results from these in vitro capping assays suggest that E-Tmod and tropomyosin work effectively together in the context of a cell to regulate thin filament formation and length. Indeed, evidence from a variety of experimental systems suggests that tropomodulin performs a similar function in vivo, as will be discussed in more detail later.

TROPOMODULIN DOMAIN STRUCTURE

Tropomodulin is an actin and tropomyosin binding protein. The domains of E-Tmod that interact with tropomyosin have been mapped by analyzing the binding of tropomyosin to bacterially expressed fragments of E-Tmod. These studies demonstrate that the amino terminal region of E-Tmod contains two regions that are sufficient for binding to tropomyosin (see Figure 4.2; Babcock and Fowler, 1994). Skeletal muscle and erythrocyte tropomyosin bind to E-Tmod with similar affinities (Kd ~0.2 μm), but in different ways. Residues 6 to 94 of E-Tmod are sufficient for binding skeletal muscle tropomyosin; residues 94 to 184 are sufficient for binding erythrocyte tropomyosin. Although the two tropomyosin preparations show preferential binding to distinct regions, actual E-Tmod–tropomyosin interactions probably make use of this whole region (0 to 184) because each tropomyosin preparation can compete for both tropomyosin binding sites (although they compete more effectively for their own binding site). In other words, skeletal tropomyosin can displace the binding of erythrocyte tropomyosin but only at high concentrations, and vice versa. The purpose of these two binding regions and their relevance in cardiac myocytes is not clear. It has been suggested that they may allow some of E-Tmod's functions to be tissue-specific (Babcock and Fowler, 1994). It is not yet known whether tropomyosin isoforms use a common motif for binding to tropomodulin isoforms. The E-Tmod binding site on nonmuscle tropomyosin 5 (TM5) has been narrowed down to residues 7 to 14 of the N-terminal heptad repeat (Vera et al., 2000), but the binding site of other tropomyosin isoforms, such as muscle tropomyosin, has not been determined. Two regions of E-Tmod have been identified that are important for its actin filament capping activity; one is located in the carboxy-terminal end between amino acids 322 and 359 and the other is between amino acids

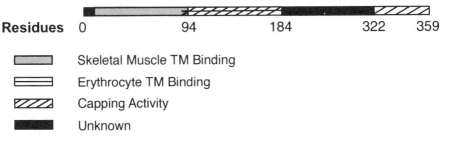

FIGURE 4.2. Schematic of E-Tmod domains. Note that some of the known functional domains appear to overlap.

90 and 184 (Figure 4.2; Fowler and Conley, 1999). The tropomodulin binding site on actin has not been mapped. Current work is focused on the identification of additional domains within tropomodulin that are important for other unique, tissue-specific activities.

TROPOMODULIN AND CARDIAC MYOFIBRILLOGENESIS

To begin to understand the role of E-Tmod during heart development, it was necessary to determine when E-Tmod is incorporated into myofibrils during assembly. The initial studies of this type were done using the well-utilized model of cardiac myofibril assembly, primary cultures of embryonic chick cardiac myocytes. At the early stages of assembly in this system, stress-fiber–like structures are observed. These are referred to as "nonstriated myofibrils" because filamentous actin, when stained with fluorescent-conjugated phalloidin, appears continuous by light microscopy, most likely as a result of actin filament overlap. At later stages, these nonstriated myofibrils are replaced by mature striated myofibrils (for reviews, see Epstein and Fischman, 1991; Gregorio, 1997; Gregorio and Antin, 2000). CapZ, the barbed-end capping protein, assembles early in nonstriated myofibrils (Schafer et al., 1993). In contrast, E-Tmod assembles at the pointed ends of the thin filaments only after the thin filaments are organized into their mature pattern in fully striated myofibrils. This indicates that tropomodulin itself is unlikely to specify the lengths of the thin filaments and that the capping of thin filament pointed ends is a late event in myofibrillogenesis (Gregorio and Fowler, 1995). Another interesting observation is that in this system E-Tmod is only seen on subsets of striated myofibrils. This does not seem to be due to a lack of available tropomodulin because at steady state ~40% of the tropomodulin in a cell is in the soluble pool throughout myofibril assembly. The model that emerged from these studies is that thin filaments in cardiac myocytes achieve their mature length and sarcomeric organization without E-Tmod, but that E-Tmod is required for maintaining thin filament length (see Figure 4.3 Color Plate 2; Gregorio and Fowler, 1995; Gregorio and Fowler, 1996).

An elegant study of chicken cardiac thin filament assembly in situ supports the assembly model developed in primary cultures of cardiac myocytes (E. Ehler, S.P. Hämmerle, B.M. Rothen, and J.-C. Perriard, personal communication). In this study, whole mount specimens of chicken heart rudiments were immunofluorescently labeled and studied by confocal microscopy. Using this method, sarcomeric proteins were observed diffusely (unorganized) in the embryo at the 8 somite stage. By the 9 somite stage, myofibrils can be seen and the first contractions are observed. Tropomodulin immunoreactivity was observed early but was not found to be assembled at the thin filament pointed ends until late, at the 12 to 13 somite stage, well after the onset of beating. As with the previously described culture system, the late incorporation of E-Tmod is not the result of the absence of E-Tmod protein, although the amount of E-Tmod does increase with respect to actin throughout development (E. Ehler, S.P. Hämmerle, B.M. Rothen, and J.-C. Perriard, personal communication).

The late and selective assembly of tropomodulin onto specific populations of myofibrils, as shown in these studies, is an intriguing phenomenon that poses many questions. Could this be due to the inhibition of its assembly by some other, as yet unidentified, pointed-end capping protein or the newly identified, ubiquitously expressed tropomodulin isoform (TMOD3) (Cox and Zoghbi, 2000)? In cardiac myocytes that were permeablized and extracted with 0.5 M KCl, which removes tropomodulin, exogenous E-Tmod could only be reconstituted onto a select population of myofibrils (Gregorio and Fowler, 1995). Therefore, if a new unidentified putative capping protein existed that prevented E-Tmod from binding to some of the myofibrils, it must be unextractable by salt, have different biochemical properties, and have perhaps little protein sequence homology to known tropomodulin isoforms because current monoclonal and polyclonal anti-E-Tmod antibodies have failed to detect more than one isoform of tropomodulin in the heart by Western blots of 2-D gels (Almenar-Queralt et al., 1999b). Alternatively, E-Tmod assembly may require another factor (or be prevented from binding by another factor), which has not yet been appreciated. For example, actin isoform expression has been shown to change throughout cardiac myofibrillogenesis during heart development; the assembly of the α-cardiac actin isoform might be a possible factor for promoting tropomodulin assembly (Ruzicka and Schwartz, 1988; E. Ehler, S.P. Hämmerle, B.M. Rothen, and J.-C. Perriard, unpublished results). Other possible mechanisms that might drive the assembly of Tmod are localization of E-Tmod by targeting of mRNA translation or by posttranslational modification. The former has previously been reported for other sarcomeric proteins in muscle cells (Fulton and Alftine, 1997). This hypothesis has been tested in rat cardiac myocytes using in situ hybridization, but the results suggest that this is not the case (Sussman et al., 1994a). With respect to posttranslational modifications, there is no evidence for modifications of E-Tmod in chick cardiac myocytes (Almenar-Queralt et al., 1999a). Furthermore, the observation that recombinant E-Tmod caps filaments in in vitro assays with purified components argues against this as a mechanism (Weber et al., 1994).

Although the incorporation of E-Tmod into the sarcomere after the thin and thick filaments are assembled suggests that E-Tmod is not required to specify the length of actin filaments in cardiac myocytes; disruption of the E-Tmod capping function shows that it is required for *maintenance* of actin filament length. This was demonstrated by microinjection of a monoclonal anti-tropomodulin antibody (mAb9) into embryonic chick cardiac myocytes. MAb9, which functionally blocks the actin filament capping activity of E-Tmod (but does not block Tmod/tropomyosin interactions) in vitro, recognizes an epitope that is on the carboxy-terminal end of tropomodulin (Gregorio et al., 1995), within amino acids 322 to 359 (V. Fowler, unpublished data). Immunolocalization of the antibody in injected cells showed that E-Tmod remained bound to the thin filaments (via tropomyosin?), whereas the actin filaments, but surprisingly not the tropomyosin polymers, elongated from the thin filament pointed ends across the H-zone. This latter observation suggests that tropomodulin may also cap the tropomyosin polymer, preventing its cooperative assembly along the new portion of the filaments (Wegner, 1979; Fowler, 1990). Microinjection of mAb9 also resulted in a greater than 80% reduction in the beating of the cardiac myocytes. Together, these studies suggest that E-Tmod is important for maintaining the length of actin filaments in cardiac myocytes. Additionally, maintenance of filament length and/or presence of the regulatory molecule, tropomyosin, along the entire length of the on actin filaments appears to be essential for proper contractile function (Gregorio et al., 1995).

Similar conclusions can be drawn from studies looking at the underexpression and overexpression of E-Tmod levels in rat cardiac myocytes. In these studies, either sense or antisense adenovirus constructs were used to alter E-Tmod expression levels

(Sussman et al., 1997). Overexpression of E-Tmod resulted in myofibril degeneration. Decreased E-Tmod levels in the antisense-treated myocytes resulted in abnormally long actin filaments, as deduced from the loss of actin filament striations in the peripheral areas of the cell. Both interventions affected contractile function, as evidenced by loss or aberrant beating of the cardiac myocytes (Sussman et al., 1997). Therefore, the levels of tropomodulin also appear to be important for the stabilization and proper organization of thin filaments in myofibrils. See Littlefield and Fowler (1998) and the following sections for further discussion.

To understand how E-Tmod functions within the context of the whole heart, transgenic mice were engineered that overexpress E-Tmod under the α-myosin heavy-chain gene promoter, which restricts expression to the myocardium (Sussman et al., 1998b). These postnatal transgenic overexpression of tropomodulin (TOT) mice experience cardiomyopathic changes after birth that are consistent with a phenotype of dilated cardiomyopathy in humans, a disease characterized by ventricular chamber enlargement, decreased contractile function, and loss of myofibril organization. Contractile function was impaired in the TOT mice, and analysis of muscle samples showed a loss of myofibrils and myofibrillar disarray (like that observed in the overexpression studies in primary cultures of rat cardiac myocytes, noted earlier) (Sussman et al., 1998b). Interestingly, dilation in TOT hearts results from the activation of the calcium-regulated phosphatase, calcineurin, secondary to the myofibril degeneration caused by tropomodulin overexpression (Sussman et al., 1998a). It is important to note that although genetic defects related to tropomodulin expression have not been directly correlated with cases of dilated cardiac myopathy, they have been implicated in some. A familial autosomal dominant form of dilated cardiac myopathy has been mapped to 9q13–9q22 (Krajinovic et al., 1995), which is in the vicinity of E-Tmod at 9q22 (Sung et al., 1996). The study of the TOT mice will continue to provide insight into E-Tmod's role in the heart as an organ and provides a useful model for the understanding of this deadly disorder (Sussman et al., 1998a, 1998b, 1999, 2000).

It is puzzling that overexpression of E-Tmod in the TOT mice, which appears normally to be present in excess throughout development (as inferred from studies in embryonic chick cardiac myocytes in vitro), can disrupt myofibrillogenesis. This suggests that sites for E-Tmod incorporation into myofibrils are either not saturated endogenously or that there is an aberrant E-Tmod interaction when it is present at high levels. If the former is true, then one possibility is that incorporation of E-Tmod is not dependent on the availability of E-Tmod but on a dynamic equilibrium with a cytosolic pool. This would also suggest that capping all the thin filaments is detrimental to cardiac function. Data from both in vitro and in vivo experiments are consistent with the idea that the presence of some actin filaments that are not capped by E-Tmod may be necessary for normal sarcomeric contraction.

The recent development of engineered mice that are homozygous for a null mutation in the E-Tmod gene is providing new insights into the requirement for E-Tmod during heart development. E-Tmod is essential for viability; the knockout mice did not survive past 10.5 days post coitus. Specifically, they first demonstrate abnormalities at 9.5 d.p.c. with very few or no intersomite vasculatures and blood cells. Their hearts did not loop and were missing the right ventricle (Chu et al., 1999a). Future investigation is necessary to decipher whether the arrest of vasculature is directly due to the Tmod null mutation or secondary due to the requirement of a functional heart and the mechanical force to produce its pumping activity.

TROPOMODULIN AND SKELETAL MYOFIBRILLOGENESIS

Much of what we know about myofibrillogenesis in general has come from the study of this process in skeletal muscle. Comparison of myofibrillogenesis in cardiac and skeletal muscle has been shown to provide valuable insights into the role of individual proteins in this process. Studies of the temporal incorporation of E-Tmod into myofibrils in skeletal muscle and the identification of a new isoform of tropomodulin, Sk-Tmod, have provided such an opportunity. In contrast to the late assembly of Tmod into cardiac myofibrils, E-Tmod seems to be one of the earliest components of the thin filaments and is found in nonstriated myofibrils in primary cultures of embryonic chick myogenic cells. It is also seen in actin filament bundles at the growing tips of young myotubes. In nonstriated myofibrils, E-Tmod is observed in an irregular pattern of dots, often found between dots of α-actinin, suggesting that, in skeletal muscle, thin filaments reach their mature (or close to their mature) length early and later align to form mature striated myofibrils (Almenar-Queralt et al., 1999a).

Interestingly, as chick development proceeds, Sk-Tmod replaces E-Tmod at the pointed ends of actin filaments in fast skeletal muscle (Almenar-Queralt et al., 1999b). Although in chickens, at the amino acid level (see earlier discussion) and biochemically, E-Tmod and Sk-Tmod appear to be quite similar (e.g., actin filament capping assays show that, like E-Tmod, Sk-Tmod requires tropomyosin for optimal capping activity), these isoforms are recruited to structures containing distinct actin filaments in adult fast skeletal muscle expressing both Sk-Tmod and E-Tmod. Although E-Tmod is eventually excluded from the sarcomere by Sk-Tmod, it is then found as a component of the myofibril to sarcolemma attachment sites, called costameres (Craig and Pardo, 1983; Almenar-Queralt et al., 1999b). This observation strongly suggests functional specificity for each tropomodulin isoform. Whether this is the result of differential binding to different tropomyosin isoforms or the presence of another factor (such as additional binding protein or different mechanical demands) awaits further study.

An observation from the studies of tropomodulin expression in skeletal muscle suggests that tropomodulin may, in fact, participate in thin filament length specification; that is, the expression of different tropomodulin isoforms in striated muscle correlates with changes in thin filament lengths. In chicken embryonic breast muscle and heart muscle, where E-Tmod is present at the thin filament pointed ends only, the lengths of thin filaments vary. In contrast, in adult fast muscle, where Sk-Tmod is present on the pointed ends of the thin filaments, the filament lengths are extremely uniform. Specifically, in rat cardiac sarcomeres, for example, the length distribution is from 0.6 to 1.1 μm, and in embryonic breast muscle this distribution is ~0.95 to 1.1 μm (Ohtsuki, 1979; Robinson and Winegrad, 1979; Almenar-Queralt et al., 1999b). In adult breast muscle, where Sk-Tmod caps the pointed ends, however, the distribution narrows remarkably so that all the thin filaments are precisely maintained at ~0.9 μm long (Ohtsuki, 1979; Almenar-Queralt et al., 1999b). A possible mechanism that has been proposed to account for the influence of particular tropomodulin isoforms on actin filament length specification is that the two isoforms may have different effects on actin dynamics, with Sk-Tmod allowing less exchange of monomers at the filament end. In this case, it was proposed that Sk-Tmod would be able to regulate filament lengths more precisely (Almenar-Queralt et al., 1999b). These results are quite exciting because observations on the properties of E-Tmod in skeletal muscle myofibrillogenesis, where E-Tmod behaves differently from that observed in cardiac myofibrillogenesis, is likely to contribute important clues regarding E-Tmod's functional roles.

SUMMARY

Much is known about the roles of tropomodulin in cardiac myofibrillogenesis. At the molecular level, we know that tropomodulin interacts with tropomyosin-coated actin filaments to prevent elongation and depolymerization from the pointed end. Specific domains of tropomodulin have been identified that are important for these interactions. At the cellular level, E-Tmod assembles onto the pointed ends of actin thin filaments after the heart has commenced beating. Disruption studies indicate that tropomodulin is critical for maintaining filament length and contractile activity in cardiac myocytes. Studies in skeletal muscle have revealed that E-tropomodulin assembles onto thin filaments very early, highlighting differences in myofibrillogenesis in the two striated muscle systems and providing the first clues about the functional and developmental specificity of individual tropomodulin isoforms in striated muscle. At the level of the organ, overexpression of tropomodulin in the hearts of transgenic mice causes malformation of the heart, which results in early death. Mice homozygous for a null mutation in the tropomodulin gene also indicate that tropomodulin plays an important role in heart development during embryogenesis. There are still many unanswered questions about tropomodulin's role in cardiac myofibrillogenesis. Continuing research on this multifunctional protein is expected to increase our understanding of actin filament dynamics, the process of cytoskeletal assembly, the etiology of heart disease, and cardiac myofibrillogenesis.

ACKNOWLEDGMENTS

The authors wish to thank Drs. Elisabeth Ehler, Jean-Claude Perriard (Institute of Cell Biology, Zurich, Switzerland), Velia M. Fowler (Scripps Research Institute, La Jolla, California), Mark A. Sussman (University of Cincinnati, Ohio), and L. Amy Sung (University of California at San Diego, California) for sharing unpublished data, Drs. Velia M. Fowler and Abby McElhinny (University of Arizona) for critical reading of the chapter, and David Carroll and Adam Geach for help with the figures. This work was supported by grants NIH HL57461 and HLO3985 (to C.C.G.) and a fellowship from the American Heart Association Dessert/Mountain Affiliate, Inc., #9804186P (to C.M.).

REFERENCES

Almenar-Queralt, A., Gregorio, C.C., and Fowler, V.M. 1999a. Tropomodulin assembles early in myofibrillogenesis in chick skeletal muscle: evidence that thin filaments rearrange to form striated myofibrils. *J. Cell Sci.* 112:1111–1123.

Almenar-Queralt, A., Lee, A., Conley, C.A., Ribas de-Pouplana, L., and Fowler, V.M. 1999b. Identification of a novel tropomodulin isoform, Sk-Tmod, that caps actin filament pointed ends in fast skeletal muscle. *J. Biol. Chem.* 274:28466–28475.

Babcock, G.G. and Fowler, V.M. 1994. Isoform-specific interactions of tropomodulin with skeletal muscle and erythrocyte tropomyosins. *J. Biol. Chem.* 269:27510–27418.

Broschat, K.O. 1990. Tropomyosin prevents depolymerization of actin filaments from the pointed end. *J. Biol. Chem.* 265:21323–21329.

Chu, X., Chen, J., Chien, K.R., Vera, C., and Sung, L.A. 1999a. Tropomodulin-null mutation arrests heart development, vasculogenesis, and hematopoiesis during embryogenesis. *Mol. Biol. Cell* 10:153a.

Chu, X., Thomson, D., Yee, L.J., and Sung, L.A. 1999b. Exon-intron organization of the mouse and human tropomodulin genes. *Mol. Biol. Cell* 10:92a.

Conley, C.A. and Fowler, V.M. 1999. Localization of the human 64 kD autoantigen D1 to myofibrils in a subset of extraocular muscle fibers. *Curr. Eye Res.* 19:313–322.

Cox, P.R. and Zoghbi, H.Y. 2000. Sequencing, expression analysis, and mapping of three unique human tropomodulin genes and their mouse orthologs. *Genomics* 63:97–107.

Craig, S.W. and Pardo, J.V. 1983. Gamma actin, spectrin, and intermediate filament proteins colocalize with vinculin at costameres, myofibril-to-sarcolemma attachment sites. *Cell Motil.* 3:449–462.

Dong, Q., Ludgate, M., and Vassart, G. 1991. Cloning and sequencing of a novel 64-kDa autoantigen recognized by patients with autoimmune thyroid disease. *J. Clin. Endocrinol. Metab.* 72:1375–1381.

Epstein, H.F. and Fischman, D.A. 1991. Molecular analysis of protein assembly in muscle development. *Science* 251:1039–1044.

Fischer, R.S., Lee, A., and Fowler, V.M. 2000. Tropomodulin and tropomyosin mediate lens cell actin cytoskeleton reorganization in vitro. *Invest. Ophthalmol. Vis. Sci.* 41:166–174.

Fowler, V.M. 1987. Identification and purification of a novel Mr 43,000 tropomyosin-binding protein from human erythrocyte membranes. *J. Biol. Chem.* 262:12792–12800.

Fowler, V.M. 1990. Tropomodulin: A cytoskeletal protein that binds to the end of erythrocyte tropomyosin and inhibits tropomyosin binding to actin. *J. Cell Biol.* 111:471–482.

Fowler, V.M. and Conley, C.A. 1999. Tropomodulin. In: *Guidebook to the Cytoskeletal and Motor Proteins*, 2nd edition. Eds. T.E. Kreis and R.D. Vale. Oxford University Press, Oxford, U.K. pp. 154–159.

Fowler, V.M., Sussman, M.A., Miller, P.G., Flucher, B.E., and Daniels, M.P. 1993. Tropomodulin is associated with the free (pointed) ends of the thin filaments in rat skeletal muscle. *J. Cell Biol.* 120:411–420.

Fulton, A.B. and Alftine, C. 1997. Organization of protein and mRNA for titin and other myofibril components during myofibrillogenesis in cultured chicken skeletal muscle. *Cell Struct. Funct.* 22:51–58.

Goll, D.E., Suzuki, A., Temple, J., and Holmes, G.R. 1972. Studies on purified α-actinin. I. Effect of temperature and tropomyosin on the α-actinin/F-actin interaction. *J. Mol. Biol.* 67:469–488.

Gregorio, C.C. 1997. Models of thin filament assembly in cardiac and skeletal muscle. *Cell Struct. Funct.* 22:191–195.

Gregorio, C.C. and Antin, P.B. 2000. At the heart of myofibril assembly. 10:355–362.

Gregorio, C.C. and Fowler, V.M. 1995. Mechanisms of thin filament assembly in embryonic chick cardiac myocytes: tropomodulin requires tropomyosin for assembly. *J. Cell Biol.* 129:683–695.

Gregorio, C.C. and Fowler, V.M. 1996. Tropomodulin function and thin filament assembly in cardiac myocytes. *Trends Cardiovasc. Med.* 6:136–141.

Gregorio, C.C., Weber, A., Bondad, M., Pennise, C.R., and Fowler, V.M. 1995. Requirement of pointed-end capping by tropomodulin to maintain actin filament length in embryonic chick cardiac myocytes. *Nature* 377:83–86.

Ito, M., Swanson, B., Sussman, M.A., Kedes, L., and Lyons, G. 1995. Cloning of tropomodulin cDNA and localization of gene transcripts during mouse embryogenesis. *Dev. Biol.* 167:317–328.

Krajinovic, M., Pinamonti, B., Sinagra, G., Vatta, M., Severini, G.M., Milasin, J., Falaschi, A., Camerini, F., and Giacci, M. 1995. Linkage of familial dilated cardiomyopathy to chromosome 9. *Am. J. Hum. Genetics* 57:846–852.

Lee, A., Fischer, R.S., and Fowler, V.M. 2000. Stabilization and remodeling of the membrane skeleton during lens fiber cell differentiation and maturation. *Dev. Dyn.* 217:257–270.

Littlefield, R. and Fowler, V.M. 1998. Defining actin filament length in striated muscle: rulers and caps or dynamic stability. *Annu. Rev. Cell Dev. Biol.* 14:487–525.

Ohtsuki, I. 1979. Molecular arrangement of troponin-T in the thin filament. *J. Biochem.* 85:1377–1378.

Robinson, T.F. and Winegrad, S. 1979. The measurement and dynamic implications of thin filament lengths in heart muscle. *J. Physiol.* 286:607–619.

Ruzicka, D. and Schwartz, R.J. 1988. Sequential activation of α-actin genes during avian cardiogenesis: vascular smooth muscle α-actin gene transcripts mark the onset of cardiomyocyte differentiation. *J. Cell Biol.* 107:2575–2586.

Schafer, D.A., Waddle, J.A., and Cooper, J.A. 1993. Localization of CapZ during myofibrillogenesis in cultured chicken muscle. *Cell Motil. Cytoskel.* 25:317–335.

Sung, L.A., Fan, Y.S., and Lin, C.C. 1996. Gene assignment, expression, and homology of human tropomodulin. *Genomics* 34:92–96.

Sung, L.A., Fowler, V.M., Lambert, K., Sussman, M.A., Karr, D., and Chien, S. 1992. Molecular cloning and characterization of human fetal liver tropomodulin. A tropomyosin-binding protein. *J. Biol. Chem.* 267:2616–2621.

Sussman, M.A., Baque, S., Uhm, C.S., Daniels, M.P., Price, R.L., Simpson, D., Terracio, L., and Kedes, L. 1997. Altered expression of tropomodulin in cardiomyocytes disrupts the sarcomeric structure of myofibrils. *Circ. Res.* 82:94–105.

Sussman, M.A., Lim, H.W., Gude, N., Taigen, T., Olson, E.N., Robbins, J., Colbert, M.C., Gualberto, A., Wieczorek, D.F., and Molkentin, J.D. 1998a. Prevention of cardiac hypertrophy in mice by calcineurin inhibition. *Science* 281:1690–1693.

Sussman, M.A., McAvoy, J.W., Rudisill, M., Swanson, B., Lyons, G.E., Kedes, L., and Blanks, J. 1996. Lens tropomodulin: developmental expression during differentiation. *Exp. Eye Res.* 63:223–232.

Sussman, M.A., Sakhi, S., Barrientos, P., Ito, M., and Kedes, L. 1994a. Tropomodulin in rat cardiac muscle: localization of protein is independent of messenger RNA distribution during myofibrillar development. *Circ. Res.* 75:221–232.

Sussman, M.A., Sakhi, S., Tocco, G., Najm, I., Baudry, M., Kedes, L., and Schreiber, S.S. 1994b. Neural tropomodulin: developmental expression and effect of seizure activity. *Dev. Brain Res.* 80:45–53.

Sussman, M.A., Welch, S., Cambon, N., Klevitsky, R., Hewett, T.E., Price, R., Witt, S.A., and Kimball, T.R. 1998b. Myofibril degeneration caused by tropomodulin overexpression leads to dilated cardiomyopathy in juvenile mice. *J. Clin. Invest.* 101:51–61.

Sussman, M.A., Welch, S., Gude, N., Khoury, P.R., Daniels, S.R., Kirkpatrick, D., Walsh, R.A., Price, R.L., Lim, H.W., and Molkentin, J.D. 1999. Pathogenesis of dilated cardiomyopathy: molecular, structural, and population analyses in tropomodulin-overexpressing transgenic mice. *Am. J. Pathol.* 155:2101–2113.

Sussman, M.A., Welch, S., Walker, A., Klevitsky, R., Hewett, T.E., Witt, S.A., Kimball, T.R., Price, R., Lim, H.W., and Molkentin, J.D. 2000. Hypertrophic defect unmasked by calcineurin expression in asymptomatic tropomodulin overexpressing transgenic mice. *Cardiovasc. Res.* 46:90–101.

Vera, C., Sood, A., Gao, K.-M., Yee, L.J., Lin, J.J.-C., and Sung, L.A. 2000. Tropomodulin-binding site mapped to residues 7 to 14 at the N-terminal heptad repeats of tropomyosin isoform 5. *Arch. Biochem. Biophys.* 378:16–24.

Watakabe, A., Kobayashi, R., and Helfman, D.M. 1996. N-tropomodulin: a novel isoform of tropomodulin identified as the major binding protein to brain tropomyosin. *J. Cell Sci.* 109:2299–2310.

Weber, A. 1999. Actin binding proteins that change extent and rate of actin monomer-polymer distribution by different mechanisms. *Mol. Cell. Biochem.* 190:67–74.

Weber, A. and Murray, J.M. 1973. Molecular control mechanisms in muscle contraction. *Physiol. Rev.* 53:612–673.

Weber, A., Pennise, C.C., Babcock, G.G., and Fowler, V.M. 1994. Tropomodulin caps the pointed ends of actin filaments. *J. Cell Biol.* 127:1627–1635.

Weber, A., Pennise, C.R., and Fowler, V.M. 1999. Tropomodulin increases the critical concentration of barbed end-capped actin filaments by converting ADP.P(i)-actin to ADP-actin at all pointed filament ends. *J. Biol. Chem.* 274:34637–34645.

Wegner, A. 1979. Equilibrium of the actin-tropomyosin interaction. *J. Mol. Biol.* 131:839–853.

Woo, M.K. and Fowler, V.M. 1994. Identification and characterization of tropomodulin and tropomyosin in the adult rat lens. *J. Cell Sci.* 107:1359–1367.

PART II

Maintenance and Manipulation of Myofibrillar Organization

CHAPTER 5

Maintaining the Fully Differentiated Cardiac Sarcomere

Daniel E. Michele and Joseph M. Metzger

The contractile proteins of the fully differentiated muscle cell are arranged in a highly ordered three-dimensional lattice. The functional interactions among these proteins are optimized to produce force and motion in response to rapid changes in intracellular calcium. In the muscle sarcomere, the structural interactions among different contractile proteins are complex, in some cases forming multimeric complexes (troponin, myosin), micron-long filaments (thick and thin filaments), or multicomponent anchoring networks (Z-line). The de novo formation of the muscle sarcomere (myofibrillogenesis) in a differentiating muscle cell is a very dynamic process, with coordinate regulation of muscle protein isoform gene expression, localization of the newly synthesized protein to the myofibrillar structures, and the organization of these structures into sarcomeres. The process of myofibrillogenesis is well documented in the chapters of this book and has been markedly advanced by recent genetic and immunocytochemistry approaches applied to embryonic, neonatal, and other types of differentiating muscle cells.

However, a less well-understood, and perhaps underappreciated, process is the maintenance of the fully differentiated muscle sarcomere. Because the sarcomere is a highly ordered protein complex, it is often viewed as stable and static. Yet, because adult cardiac muscle cells are postmitotic, there is some requirement that this complex sarcomeric structure remain dynamic, with contractile proteins being synthesized and incorporated into the sarcomere to replace older and potentially damaged contractile proteins (termed "sarcomere maintenance"). Indeed, isotopic labeling studies have been used to measure the turnover rate of the major contractile proteins, and, in the adult rat heart, the half-lives of these proteins range from ~3 days for troponin I (TnI), ~5 days for myosin, troponin C, and tropomyosin (Tm), and up to ~10 days for sarcomeric actin.[1] In other words, the sarcomere for the most part is completely remade every 2 to 3 weeks. It is somewhat puzzling, even under normal conditions, how the process of sarcomere maintenance occurs at all because the continuous beating at 50 to 800 times per minute, depending on species and activity, allows for no "rest" time for the adult cardiac muscle cell to repair sarcomeres. Clearly, with the alterations of cardiac sarcomeres seen under conditions of growth and pathology, understanding how normal and pathological sarcomere maintenance occurs may aid the understanding of clinically relevant cardiac muscle disorders.

The study of sarcomere maintenance in fully differentiated contractile cells has been limited by several factors. First, fully differentiated muscle cells, including adult cardiac myocytes, are notoriously difficult to culture. For instance, adult cardiac myocytes undergo a complex dedifferentiation process of myofilament degeneration, cell spreading, and formation of new sarcomeres (myofibrillogenesis) when cultured in the presence of fetal bovine serum.[2–5] Biochemical and immunofluorescent labeling of serum-treated cardiac myocytes has shown that the protein isoform expression and localization and the sarcomeric organization in these serum-treated cultures more closely resemble a differentiating embryonic or neonatal cardiac myocyte.[4,6] Second, genetic manipulation of adult muscle cells, so elegantly used to study myofibrillogenesis in embryonic and neonatal cardiac myocytes (see Chapter 3 by Ehler and Perriard), has proved difficult. Standard DNA transfection techniques have been shown to be either extremely toxic or incapable of transferring DNA into adult cardiac myocytes.[7]

Recently, it has been shown that replication-deficient recombinant adenovirus vectors provide an extremely efficient method (approaching 100% with reporter constructs) to transfer DNA into adult cardiac myocytes in vitro.[7,8] Importantly, methods have been developed to isolate, treat with viral vectors, and culture adult cardiac myocytes for up to seven days in vitro while maintaining the fully differentiated state of the cell.[7,9] This has created a window of opportunity for studying sarcomere maintenance, as well as structure/function for several of the more rapidly replaced cardiac contractile proteins in the context of a fully differentiated striated muscle cell. Using methods to uniquely identify the contractile protein of interest (i.e., an alternate contractile protein isoform or epitope tagging), it has been possible to follow the expression of the newly synthesized contractile protein using biochemical techniques and the incorporation of the specific protein into the cardiac sarcomere with high-resolution immunocytochemistry.[9–14] Thus, the process of sarcomere maintenance can be visualized under physiological conditions of protein transcription/translation, myofilament protein turnover, and sarcomeric protein incorporation in a fully differentiated cardiac muscle cell.

SARCOMERE MAINTENANCE: CONTRACTILE PROTEIN EXPRESSION

The process of sarcomere maintenance in fully differentiated contractile cells is complicated by the cell's rigorous requirement for preserving contractile protein stoichiometry. Because many of the proteins of the sarcomere form multimeric complexes, if one subunit is missing, the complex may become nonfunctional. Indeed, extraction of small amounts of contractile proteins from permeabilized cardiac muscle, such as the Ca^{2+} binding subunit troponin C, can have disastrous effects on Ca^{2+} regulation of force production.[15] Therefore, mechanisms must exist that regulate the amount of protein in the sarcomere such that myofilament protein stoichiometry is maintained (i.e., $1:1:7$ for the ratio of troponin complex : tropomyosin dimer : actin monomer in the thin filament of the sarcomere).

It has become clear from adenoviral mediated gene transfer to adult cardiac myocytes in vitro and from in vivo studies using transgenic mice (see the following) that mechanisms exist that are capable of maintaining contractile protein stoichiometry even in the presence of increased gene transcription. Although strong constitutive promoters have been used in expression cassettes for several contractile proteins, the proteins are not "overexpressed" at the level of the sarcomere. In fact, quantitative Western blotting of the expressed protein has shown that when TnI, troponin T (TnT), or Tm is expressed from these promoters, the total contractile protein stoichiometry remains unchanged (as shown schematically in Figure 5.1).[10–14]

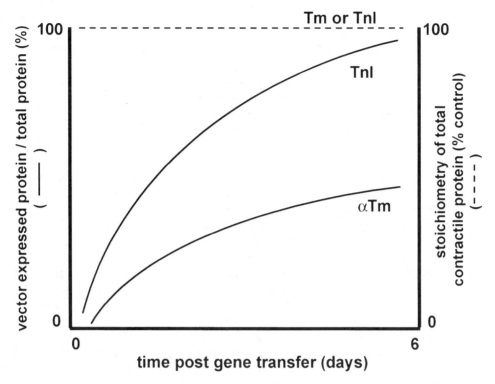

FIGURE 5.1. Generalized time course of expression of Tm and TnI following gene transfer into adult cardiac myocytes. Protein expression was determined by Western blotting and densitometry and describes qualitatively the results in references 10, 11, 13, and 14. As an example for the Tm results, the percent vector-expressed protein (left axis) was calculated as $(Tm_{VEP}/(Tm_{VEP} + Tm_{END}) \times 100$, where Tm_{VEP} is the vector-expressed (or gene transferred) Tm and Tm_{END} is the endogenous Tm isoform. Total contractile protein stoichiometry (right axis) was determined by normalizing the total contractile protein to actin or another myofibrillar protein (e.g., $(Tm_{VEP} + Tm_{END}/actin) \times 100$).

The amount of ectopic contractile protein expressed following gene transfer to adult cardiac myocytes also differs with different protein isoforms. Because of the high efficiency of adenoviral gene transfer,[7] it is possible to estimate the average amount of protein expression in each cardiac myocyte by examining the ratio of the vector-expressed contractile protein to the endogenous contractile protein in a sample of a large number of cardiac myocytes. Shown in Figure 5.1 is a schematic summary of several separate studies, using adenoviral gene transfer of different TnI and Tm isoforms and mutants into adult cardiac myocytes.[10,11,13,14] As shown, the endogenous TnI can be nearly completely replaced with the expressed TnI after 6 days post-TnI gene transfer, whereas the expressed Tm can only replace approximately 40% to 50% of total Tm over 6 days in culture following Tm gene transfer (Figure 5.1). Thus, in adult cardiac myocytes, it appears that the expression of contractile proteins, at least for TnI and Tm, is limited by the turnover rate or half-life (TnI ~3.2 days; Tm ~5.5 days) of the endogenous contractile protein. Indeed, attempts to further increase the rate of ectopic contractile protein expression and incorporation using higher doses of adenoviral vectors do not increase the rate of replacement of the endogenous contractile protein (D.M. and J.M., unpublished results and Reference 14).

How this process of maintenance of contractile protein stoichiometry occurs is still a matter of debate. Multiple studies on transgenic mice using the α-myosin heavy-chain promoter to drive expression of a variety of contractile proteins in the heart have shown that, in general, contractile protein stoichiometry is maintained when gene transcription of a specific contractile protein is increased.[16] This suggests that, for the most part, a posttranscriptional process is responsible for maintenance of contractile protein stoichiometry. The accumulating evidence, using biochemical and immunocytochemistry assays and gene transfer or transgenic technologies, shows that the expressed or endogenous sarcomeric protein does not accumulate significantly in the myocyte cytoplasm.[10–14,16] This leaves two possibilities: (1) stoichiometry is regulated at the translation level (translation limited) or (2) the proteins are translated but are rapidly degraded if they cannot find an appropriate "home" or vacant binding site in the sarcomere (turnover limited). Reduction of Tm gene transcription by gene targeting causes no change in total Tm protein and no changes in the amount of transcript bound to polysomes, which suggests that translational regulation is important.[17,18] In addition, the mRNA for cytoskeletal proteins and the sarcomeric proteins titin and nebulin in differentiating muscle cells is localized to the sarcomere. This suggests that these proteins may incorporate into the sarcomere while being translated.[19,20] However, mRNA for myosin in adult cardiac muscle is not localized in the sarcomere, indicating that there is not always a direct physical link between protein translation and incorporation into myofilaments.[21] Therefore, it is unclear exactly if and how the regulation of protein translation is coupled to the rate of turnover of the sarcomeric protein. Thus, the mechanism for maintaining sarcomere stoichiometry is still unclear, although translational regulation and myofilament protein turnover and degradation may play a role. In the future, adenoviral gene transfer into adult cardiac myocytes may be a useful tool, allowing greater experimental manipulation in vitro to address the problem of maintenance of contractile protein stoichiometry in fully differentiated contractile cells.

SARCOMERE MAINTENANCE: CONTRACTILE PROTEIN INCORPORATION INTO THE SARCOMERE

Combining contractile protein gene transfer techniques and new methods developed to uniquely identify expressed proteins in adult cardiac myocytes allows for the direct visualization of sarcomere maintenance in fully differentiated striated muscle cells. Epitope tagging (addition of small sequences of amino acids recognized by antibodies) of contractile proteins allows unique detection of the expressed contractile protein using immunocytochemistry and has been used recently to localize and follow the expression of contractile proteins in several types of differentiating muscle cells (see Chapter 3). By combining gene transfer, epitope tagging, immunocytochemistry, and high-resolution confocal microscopy, fundamental questions about sarcomere maintenance in the adult myocyte can be addressed under normal conditions of contractile protein gene transcription, translation, and sarcomere incorporation. These questions include: (1) When new contractile proteins are synthesized, where do they incorporate into the sarcomere? (2) Is the incorporation process ordered or stochastic? (3) Are sarcomere structures, such as the thin filaments, disassembled and reassembled as complete units or does maintenance of individual contractile proteins occur by separate mechanisms? (4) How is sarcomere maintenance altered under pathophysiological conditions?

Tropomyosin and Troponin I Incorporation

As one approach to examine sarcomere maintenance, gene transfer of Tm and TnI to adult cardiac myocytes was used, and the myofilament incorporation of these proteins was followed over time in culture.[10] The incorporation patterns of these two proteins were compared because the rates of expression and incorporation of TnI and Tm assessed by Western blot are different. As mentioned earlier, the expression of TnI replaces nearly all the endogenous TnI in 6 days; Tm only has 50% replacement in 6 days,[10,13,14] suggesting that TnI and Tm may be replaced by different mechanisms. Also, TnI and Tm differ structurally. Tm resides in a long polymerized chain of 24 coiled-coil dimers along the actin thin filament, whereas TnI forms a heterotrimeric globular complex with TnC and TnT and is anchored to actin and Tm (Figure 5.2A). Therefore, by comparing the replacement patterns of these two proteins, it should be possible to distinguish whether thin filaments are broken down and replaced as a whole or proteins are replaced individually by separate mechanisms.

High-resolution confocal microscope images showing the immunofluorescent localization of epitope-tagged Tm and TnI at day 2 post gene transfer, when the newly incorporated protein first becomes detectable in adult cardiac myocytes, are shown in Figure 5.3. In these images, the Z-line is identified by colocalization studies using α-actinin antibodies (arrows).[10] The resting sarcomere length is 1.8 to 1.9 µm so the thin filaments are in partial overlap. It is evident from these images that the replacement of Tm and TnI occurs uniformly throughout the entire length, width, and depth of the cardiac myocyte. Interestingly, newly synthesized Tm is preferentially incorporated at the pointed or free ends of the thin filaments (Figure 5.3A). By examining myocytes at later timepoints following treatment with viral vectors, Tm replacement has been shown to continue further toward the Z-line over time.[10] In marked contrast, the TnI immunofluorescence pattern is uniform throughout the length of the thin filament (Figure 5.3B) even though at these early timepoints the exogenous TnI has replaced only a portion of the total TnI (Figure 5.1). The pattern of TnI immunofluorescence does not change over more extended time intervals, as the expressed TnI replaces more of the endogenous TnI. These results suggest that the TnI replacement is random along the entire length of the thin filament.

The TnI and Tm results together suggest that distinct mechanisms are used for sarcomere maintenance of the different contractile proteins: an ordered mechanism for Tm and a stochastic mechanism for TnI. Interestingly, the pointed-end incorporation of Tm is not seen when epitope-tagged Tm is expressed in neonatal cardiac (differentiating) myocytes.[22] In differentiating skeletal muscle, it has been shown that newly synthesized sarcomeric protein incorporates around the periphery of the myofibril (the myofibrils get fatter as more thick and thin filaments are added). This feature of differentiating muscle may help explain why expressed Tm in neonatal myocytes incorporates along the entire length of the thin filament. In neonatal myocytes, new filaments added to the periphery polymerizing from the Z-line will also incorporate the newly expressed Tm and give a pattern of incorporation along the entire length of the thin filament.[23] The different results with Tm expression in neonatal and fully differentiated adult cardiac myocytes reinforce the idea that sarcomere maintenance and myofibrillogenesis probably occur by separate mechanisms.

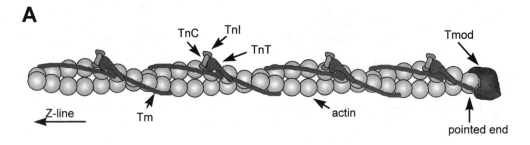

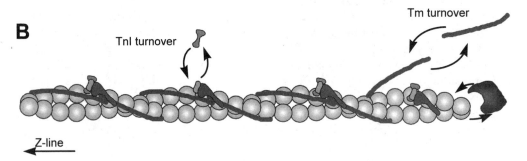

FIGURE 5.2. Model of thin filament sarcomere maintenance in fully differentiated adult cardiac myocytes. The contractile proteins are shown schematically and labeled (A). In the model shown (B), the thin filament remains intact during sarcomere maintenance with ordered (Tm) or stochastic (TnI) exchange of newly synthesized contractile protein with older proteins bound to the thin filament. As shown, individual Tm proteins are depicted as dimers and are replaced more readily at the pointed end due to structural properties of the Tm polymer and anchoring of the polymer at the Z-line. Tmod interactions with Tm and actin are dynamic, allowing pointed-end replacement of Tm. Over time, more Tm replacement occurs toward the Z-line. TnI is replaced stochastically along the entire length of the thin filament. (Adapted from Michele, D.E., et al. 1999. Thin filament protein dynamics in fully differentiated adult cardiacmyocytes: toward a model of sarcomere maintenance. *J. Cell Biol.* 145:1483–1495.)

Model of Thin Filament Sarcomere Maintenance

A model of sarcomere maintenance for the thin filament, based on the results of newly synthesized Tm and TnI expression and incorporation, is shown in Figure 5.2. The different patterns of replacement of Tm and TnI over time in adult cardiac myocytes suggest that the thin filaments of adult cardiac myocytes remain intact during sarcomere maintenance. The Tm incorporation results could be explained if the thin filament was broken down as a whole, most rapidly from the free end, and reformed by repolymerization. However, in that case one would expect the patterns of incorporation of TnI to be similar to Tm with similar rates. Similarly, if the thin filaments were replaced by the formation of new thin filaments nucleated from the Z-line, it would be expected that Tm and TnI incorporation would be similar and would likely begin from the Z-line, not the pointed end.

In the model shown in Figure 5.2, the thin filament is assumed to remain intact, and the replacement of a specific contractile protein within the sarcomere is very

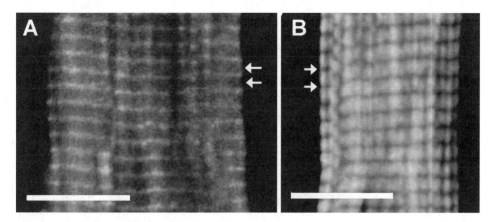

FIGURE 5.3. Comparison of the initial localization of newly synthesized epitope-tagged α-Tm (A) and epitope-tagged cardiac TnI (B) using confocal 3-D reconstructions of fully differentiated adult cardiac myocytes at day 2 post gene transfer. A and B show the immunofluorescence for epitope antibody labeled with a fluorescent secondary antibody. Arrows indicate the position of the Z-line that was revealed by coimmunofluorescence staining with an antibody for α-actinin. Bar = 10 μm. The resting sarcomere length is 1.8 to 1.9 μm. Thus, thin filaments are in overlap and segregated I-bands are not expected to be seen. Reproduced from *The Journal of Cell Biology*, 1999, 145:1483–1495 by copyright permission of The Rockefeller University Press.

rapid. In other words, as the sarcomere proteins turn over, the site occupied by the protein must not remain vacant for any significant period of time. As mentioned before, removal of specific contractile proteins by chemical extraction dramatically alters force production.[15] In addition, in the absence of Tm, actin filaments are unstable or do not readily form,[24,25] suggesting that the exchange of old for new Tm at the Tm binding site must be rapid. Precise control of contractile protein expression and stoichiometry would be a necessity in this environment. Therefore, it makes intuitive sense that the expression of contractile proteins seems to be limited by the turnover of the endogenous proteins (Figure 5.1).

The different patterns of replacement of Tm and TnI may be related to the structural characteristics of these proteins. It may not be that surprising that TnI replacement is stochastic along the length of the thin filament because TnI would only have to dissociate from the relatively well-spaced individual troponin complexes on actin. However, coiled-coil polymers of Tm (with each Tm protein being a Tm dimer) are connected head-to-tail along the entire length of the filament, and these head-to-tail interactions are stabilized by interactions with actin and TnT. Therefore, random Tm exchange would be difficult, and it may be more favorable to exchange Tm more readily on the ends of the polymer. One end of the Tm polymer may be the preferred site for exchange if one end is anchored more tightly than the other. For example, interactions with capping proteins such as α-actinin or capZ may tightly anchor Tm in the Z-line, leaving the pointed end as the preferred site of Tm replacement. Interestingly, tropomodulin (Tmod) is a key pointed-end actin capping protein that binds both actin and Tm and is thought to play an important role in the regulation of thin filament length.[26] If Tm replacement is most rapid at the pointed ends, then interactions of Tmod with actin and Tm must be, in turn, relatively dynamic. Indeed, modulation of the amount of Tmod in differentiating cardiac muscle cells seems to be able

to rapidly modify thin filament length.[27,28] Thus, the preferred pointed-end incorporation of Tm may be explained by a lower stability of the Tm and Tmod interactions
relative to the stability of the Tm–Z-line interaction. Binding of Tm to Tmod may
also recruit new Tm to the thin filament pointed end. However, it is still not clear
from the gene transfer studies and this model (Figure 5.2) how Tm gets replaced
further toward the Z-line over time. More studies are needed to determine whether
Tm is replaced as a monomer or preformed dimer or polymer. It is not known, for
instance, whether Tm monomers fold into a two Tm coiled-coil protein before incorporating or Tm is replaced in the thin filament as monomers, with one monomer
remaining attached to the thin filament. It is also possible that replacement of the
second Tm dimer requires removal of the first and second Tm dimers and rapid
replacement with a new two Tm long polymer or two individual Tm dimers. This
process might become less favorable as longer polymers are removed, in part explain-
ing the preferred pointed-end incorporation of Tm. Another hypothesis is that the
Tm filament treadmills toward the Z-line and old Tm proteins are removed near the
barbed end. Further experiments using gene transfer of Tm with vital epitope tags
and time-resolved confocal microscopy of living cells may be able to distinguish
among these mechanisms. In addition, the binding or "parking" of Tm on reconstituted thin filaments, as a function of the length of Tm, Tm head-to-tail interactions,
and interactions of Tm with troponin, has been carefully examined biochemically.[29]
It would be of great interest to determine which of these aspects of Tm structure are
important in determining the precise mechanism of sarcomere maintenance of Tm
within adult cardiac myocytes.

The preferred pointed-end incorporation of Tm in adult muscle cells may have
functional implications. For instance, it is known that the sensitivity of force production in cardiac muscle to intracellular Ca^{2+} is markedly sarcomere-length–
dependent.[30] Previous immunocytochemistry studies have suggested that the Tm
closest to the Z-line is more highly phosphorylated than the rest of the Tm along the
thin filament.[31] This difference in the posttranslational modification of Tm may play
a structural role in making Tm replacement at the Z-line less favorable than at the
pointed end. Conversely, it is tantalizing to hypothesize that "age"-dependent (age
of the protein) posttranslational modification of Tm along the thin filament modifies
its regulation of contractile function and contributes to the sarcomere-length dependence of myofilament Ca^{2+} sensitivity.

Other Contractile Proteins

The number of previous studies examining sarcomere maintenance is limited. This is
in part due to the lack of effective gene delivery tools and limitations for studying
fully differentiated adult muscle cells, as described earlier. However, one earlier study
utilized thyroid hormone manipulation to "turn off" and then "turn back on" the α-
myosin heavy-chain promoter.[32] Following propylthiouracil treatment to render the
animals hypothyroid, the sarcomere was composed entirely of β-myosin heavy chain.
Four days following thyroid hormone replacement treatment, the nascent α-myosin
heavy chain was found to preferentially incorporate into the free ends of the thick
filament (at the ends of the A-band).[32] Because thyroid treatment results in a growth
response via a generalized increase in protein synthesis,[32] it would be interesting to
determine whether similar results would be obtained using gene transfer into fully
differentiated adult cardiac myocytes in vitro.

How sarcomere maintenance of the other contractile proteins occurs in fully differentiated cardiac myocytes is currently unknown. Therefore, future experiments applying gene transfer to adult cardiac myocytes to examine the other components of the cardiac sarcomere will be important to our understanding of sarcomere maintenance. For example, troponin T binds the troponin complex to the thin filament by binding tightly to Tm (Figure 5.2). However, TnT half-life measured in adult rat heart by radioisotope experiments is more similar to TnI, suggesting that it may turn over more rapidly than Tm.[1] Attempts to replace adult cardiac TnT with embryonic cardiac TnT using gene transfer into adult cardiac myocytes results in replacement of about 35% to 40% of the endogenous TnT at 6 days post gene transfer.[12] Therefore, it is possible that TnT may turn over with rates similar to Tm and incorporate with newly synthesized Tm into the myofilament. Alternatively, the embryonic TnT may not compete as well as adult cardiac TnT for TnT binding sites. Careful examination of sarcomere maintenance of the troponin complex using epitope-tagged adult cardiac TnT and cardiac TnC and following localization of newly synthesized protein into the adult sarcomere would provide important additional information about how the thin filaments are maintained in adult cardiac myocytes.

The sarcomere maintenance of actin in adult cardiac myocytes is also an interesting question that is still not fully understood. Microinjection of fluorescently labeled actin into adult muscle cells results in the rapid incorporation (on the order of minutes) of the protein along the entire length of the thin filament[33] (also see Chapter 1). However, the half-life of actin measured in adult rat heart in vivo with radioisotopes is ≈10 days, suggesting that the turnover of actin is slow. Thus, it is not clear whether the microinjection experiments are measuring the same process as sarcomere maintenance. Future experiments using gene transfer of actin into adult cardiac myocytes would be informative because the amount of protein expressed and incorporated into the sarcomere can be both quantified and visualized under conditions of gene transcription, translation, and sarcomere incorporation.

SARCOMERE MAINTENANCE AND HYPERTROPHIC CARDIOMYOPATHY

Familial hypertrophic cardiomyopathy (FHC) is an inherited disorder characterized by ventricular hypertrophy in the absence of stimuli that normally trigger cardiac hypertrophy (i.e., hypertension).[34] Another hallmark of FHC is disarray of the cardiac myocytes and disarray of the sarcomeres within the myocyte itself. Thus, at some point in the development of the FHC phenotype, normal sarcomere maintenance is altered, leading to aberrant sarcomeric structure. Thus far, all the mutations associated with FHC reside in contractile proteins of the cardiac sarcomere, including the β-myosin heavy chain and associated light chains, myosin binding C protein, actin, TnT, TnI, and Tm.[35,36] Because each of these proteins plays a structural as well as functional role in the cardiac sarcomere, it is unclear whether the sarcomere disarray is a direct result of the mutation in the specific contractile proteins. Conversely, it is uncertain whether functional changes in the heart muscle are due to alterations in the structure of the sarcomere or whether the altered function of contractile protein itself is directly causing changes in contractile function. In addition, in most cases FHC is an autosomal dominant disorder and only one allele of the contractile protein is mutated. Therefore, it is also unknown whether sarcomere maintenance of the mutated protein is normal and the protein is incorporated normally into the sarcomere. Conversely, the alterations in structure and function may be due to the lack of the mutant protein incorporating into the sarcomere, resulting in haploinsufficiency and alterations in contractile protein stoichiometry.

Gene transfer into adult cardiac myocytes in vitro combined with single-cell measurements of contractile function offers a unique approach to help understand the molecular basis of FHC pathogenesis. As shown earlier, this approach can be used to assess directly the expression and incorporation of these mutant proteins into the adult cardiac sarcomere. The gene transfer approach has the advantage of examining the direct, initial effects of mutant protein on the structure and function of the adult cardiac sarcomere without potential compensatory and secondary effects that may occur in animal models in vivo.

Gene transfer of FHC mutant Tm and TnT to adult rat cardiac myocytes in vitro has been used to assess the effect of these mutant proteins on myocyte structure and function.[11,12] There have been four mutations identified in α-Tm associated with HCM: A63V, K70T, D175N, and E180G.[35] The FHC mutant α-Tm protein was expressed and incorporated into sarcomeres at rates similar to normal Tm protein incorporation rates such that $\approx40\%$ of total Tm was the mutated Tm after 5 to 6 days post gene transfer.[11] In addition, epitope tagging and confocal microscopy directly showed that all four mutant α-Tm proteins are capable of incorporating normally into the adult cardiac sarcomere without producing any sarcomere disarray. Interestingly, most of the FHC mutant α-Tm proteins produced a marked increase in the Ca^{2+} sensitivity of force production. In contrast, the two FHC mutant TnT proteins, I79N and R92Q TnT, could not be expressed and incorporated into the myofilaments at rates similar to wild-type TnT.[12] The mutant TnT proteins only replaced 4% to 8% of total TnT in sarcomeres of adult cardiac myocytes at 5 to 6 days post gene transfer. However, the mutant TnT proteins produced a significant decrease in the Ca^{2+} sensitivity of force production without producing any sarcomere disarray. Other laboratories have shown that gene transfer of R92Q TnT to feline adult cardiac myocytes also produces changes in contractile function without significant sarcomere disarray.[37] These results from gene transfer suggest that the primary effect of mutations in TnT and Tm proteins associated with FHC is alteration of contractile performance, and disruption of sarcomere maintenance and formation of sarcomere disarray are secondary effects of FHC mutations in contractile proteins. However, whereas some FHC mutant proteins (Tm) are capable of undergoing normal sarcomere maintenance, some FHC mutant proteins (TnT) may not be able to compete as well with normal contractile proteins for available sites in the contractile apparatus. Therefore, it remains unclear for some FHC mutations, such as FHC mutant TnT, what extent of mutant protein incorporation can be expected in patients that are heterozygous for the FHC alleles (one normal and one mutant gene, presumably with similar levels of transcription).

This does not preclude that some mutations in contractile proteins associated with FHC do directly alter sarcomere structure. For example, myosin binding protein C (My BP-C) is thought to play a key role in the structure of the thick filament, but its functional significance is not well understood. Several mutations in My BP-C result in severe truncations, suggesting that these mutations may behave as nulls.[35] In fact, transgenic mice expressing an FHC-truncated My BP-C showed that the protein accumulated in the cytoplasm, reduced sarcomeric My BP-C stoichiometry, and caused sarcomere disarray.[38] Thus, FHC mutations in My BP-C may have direct effects on thick filament structure. In addition, gene transfer of R403Q β-MyHC in feline adult cardiac myocytes was shown to cause sarcomere disarray in approximately half of the cardiac myocytes treated with vector, although incorporation of the mutant protein was not directly established.[39] However, expression of R403Q and other FHC mutant β-MyHC in neonatal rat cardiac myocytes results in their normal incorporation into the sarcomere without causing sarcomere disarray.[40] For the most part, the expression and incorporation of FHC mutant proteins in myocytes in vitro do not appear to directly alter the structure of the cardiac sarcomere, suggesting that the sarcomere alterations seen in FHC patients may be an in vivo secondary consequence resulting from functional alterations caused by FHC mutant contractile proteins.

Nevertheless, sarcomere disarray does occur in response to FHC mutant protein expression in several in vivo gene-targeted and transgenic mouse models of hypertrophic cardiomyopathy.[38,41,42] The identity of the precise signals for misregulation of sarcomere maintenance in FHC are currently unknown. It is also unclear how large an impact sarcomere disarray (once it does occur) has on myocardial function and FHC pathogenesis. It is possible that alterations in the function of the myocardium directly lead to load-dependent sarcomere damage. It is also possible that the sarcomere disarray is a generalized response to changes in intracellular signaling that trigger inappropriately high amounts of cellular hypertrophy. Future approaches, using inducible transgenes and careful monitoring of the time course of changes in cardiac function, signaling, and ultrastructural changes in the sarcomere, may help elucidate the direct causes of alterations of sarcomere structure in FHC.

CONCLUDING REMARKS

Adenovirus mediated gene transfer of genes encoding contractile proteins to the adult cardiac myocyte offers a unique opportunity to examine the process of maintaining the fully differentiated muscle sarcomere. With future gene transfer experiments on multiple contractile proteins and spatiotemporal examination of rates of gene transcription, translation and protein replacement, protein trafficking, and localization of newly synthesized contractile protein incorporation into the sarcomere, the secrets of sarcomere maintenance in the fully differentiated contractile cell will be revealed.

ACKNOWLEDGMENTS

We thank Drs. M.V. Westfall and P.A. Wahr for helpful discussion of this work. We thank Dr. V.M. Fowler for bringing to our attention the possible role of Tmod in the mechanism of Tm replacement. This work was funded by grants from the American Heart Association and the National Institutes of Health.

REFERENCES

1. Martin, A.F. 1981. Turnover of cardiac troponin subunits. *J. Biol. Chem.* 256:964–968.
2. Jacobson, S.L. 1977. Culture of spontaneously contracting myocardial cells from adult rats. *Cell Struct. Funct.* 2:1–9.
3. Moses, R.L. and Claycomb, W.C. 1982. Disorganization and reestablishment of cardiac cell ultrastructure in adult rat ventricular muscle cells. *J. Ultrastruct. Res.* 81:358–374.
4. Eppenberger, M.E., Hauser, I., Bächi, T., et al. 1988. Immunocytochemical analysis of the regeneration of myofibrils in long term cultures of adult cardiac myocytes of the rat. *Dev. Biol.* 130:1–15.

5. Messerli, J.M., Eppenberger-Eberhardt, M.E., Rutishauser, B.M., et al. 1993. Remodeling of the cardiomyocyte architecture visualized with three-dimensional confocal microscopy. *Histochemistry* 100:193–202.

6. LoRusso, S.M., Rhee, D., Sanger, J.M., and Sanger, J.W. 1997. Premyofibrils in spreading adult cardiac myocytes in tissue culture: Evidence for reexpression of the embryonic program for myofibrillogenesis in adult cells. *Cell Motil. Cytoskel.* 37:183–198.

7. Rust, E.M., Westfall, M.V., and Metzger, J.M. 1998. Stability of the contractile assembly and Ca^{2+} activated tension in adenovirus infected adult cardiac myocytes. *Mol. Cell Biochem.* 181:143–155.

8. Kirshenbaum, L.A., MacLellen, W.R., Mazur, W., French, B.A., and Schneider, M.D. 1993. Highly efficient gene transfer into adult ventricular myocytes by recombinant adenovirus. *J. Clin. Invest.* 92:381–387.

9. Westfall, M.V., Rust, E.M., Albayya, F., and Metzger, J.M. 1997. Adenovirus mediated myofilament gene transfer into adult cardiac myocytes. *Methods Cell Biol.* 52:307–322.

10. Michele, D.E., Albayya, F.P., and Metzger, J.M. 1999. Thin filament protein dynamics in fully differentiated adult cardiac myocytes: toward a model of sarcomere maintenance. *J. Cell Biol.* 145:1483–1495.

11. Michele, D.E., Albayya, F.P., and Metzger, J.M. 1999. Expression of FHC mutant tropomyosins increases the sensitivity of Ca^{2+} activated force production in adult cardiac myocytes. *Biophys. J.* 76:A274.

12. Rust, E.M., Albayya, F.P., and Metzger, JM. 1999. Identification of a contractile deficit in adult cardiac myocytes expressing hypertrophic cardiomyopathy-associated mutant troponin T proteins. *J. Clin. Invest.* 103:1459–1467.

13. Westfall, M.V., Rust, E.M., and Metzger, J.M. 1997. Slow skeletal troponin I gene transfer, expression, and myofilament incorporation enhances adult cardiac myocyte contractile function. *Proc. Natl. Acad. Sci. USA* 94:5444–5449.

14. Westfall, M.V., Albayya, F.P., and Metzger, J.M. 1999. Functional analysis of troponin I regulatory domains in the intact myofilament of adult cardiac myocytes. *J. Biol. Chem.* 274:22508–22516.

15. Moss, R.L. 1992. Ca^{2+} regulation of mechanical properties of striated muscle: mechanistic studies using extraction and replacement of regulatory proteins. *Circ. Res.* 70:865–884.

16. James, J. and Robbins, J. 1997. Molecular remodeling of cardiac contractile function. *Am. J. Physiol.* 273:H2105–H2118.

17. Rethinasamy, P., Muthuchamy, M., Hewett, H., et al. 1998. Molecular and physiological effects of alpha-tropomyosin ablation in the mouse. *Circ. Res.* 82:116–123.

18. Blanchard, E.M., Iizuka, K., Christe, M., et al. 1997. Targeted ablation of the alpha tropomyosin gene. *Circ. Res.* 81:1005–1010.

19. Morris, E.J. and Fulton, A.B. 1994. Rearrangement of mRNAs for costamere proteins during costamere development in cultured skeletal muscle from chicken. *J. Cell Sci.* 107:377–386.

20. Fulton, A.B. and Alftine, C. 1997. Organization of protein and mRNA for titin and other myofibril components during myofibrillogenesis in cultured skeletal muscle. *Cell Struct. Funct.* 22:51–58.

21. Russell, B., Wenderoth, M.P., and Goldspink, P.H. 1992. Remodeling of myofibrils: subcellular distribution of myosin heavy chain mRNA and protein. *Am. J. Physiol.* 262:R339–R345.

22. Helfman, D.M., Berthier, C., Grossman, J., et al. 1999. Nonmuscle tropomyosin-4 requires co-expression with other low molecular weight isoforms for binding to the thin filaments in adult cardiac myocytes. *J. Cell Sci.* 112:371–380.

23. Morkin, E. 1970. Postnatal muscle fiber assembly: localization of newly synthesized myofibrillar proteins. *Science* 167:1499–1501.

24. Lemanski, L.F., Mooseker, M.S., Peachey, L.D., and Iyengar, M.R. 1976. Studies of muscle proteins in embryonic myocardial cells of cardiac lethal mutant Mexican axolotls by use

of heavy meromyosin binding and sodium dodecyl sulfate polyacrylamide gel electrophoresis. *J. Cell Biol.* 68:375–388.

25. Liu, H. and Bretscher, A. 1989. Disruption of a single tropomyosin gene in yeast results in the disappearance of actin cables from the cytoskeleton. *Cell* 57:233–242.

26. Littlefield, R. and Fowler, V.M. 1998. Defining actin filament length in striated muscle: rulers and caps or dynamic stability? *Annu. Rev. Cell Dev. Biol.* 14:487–525.

27. Gregorio, C.C., Weber, A., Bondad, M., Pennise, C.R., and Fowler, V.M. 1995. Requirement of pointed end capping by tropomodulin to maintain actin filament length in embryonic chick cardiac myocytes. *Nature* 377:83–86.

28. Sussman, M.A., Baque, S., Uhm, C.S., et al. 1988. Altered expression of tropomodulin in cardiomyocytes disrupts the sarcomeric structure of myofibrils. *Circ. Res.* 82:94–105.

29. Tobacman, L.S. 1996. Thin filament mediated regulation of cardiac contraction. *Annu. Rev. Physiol.* 58:447–481.

30. Hibberd, M.G. and Jewell, B.R. 1982. Calcium dependent and length dependent force production in rat ventricular muscle. *J. Physiol. London* 329:527–540.

31. Trombitás, K., Baatsen, P.H.W.W., Lin, J.J.-C., Lemanski, L.F., and Pollack, G.H. 1990. Immunoelectron microscopic observations of tropomyosin localization in striated muscle. *J. Muscle Res. Cell Motil.* 11:445–452.

32. Wenderoth, M.P. and Eisenberg, B.R. 1987. Incorporation of nascent myosin heavy chains into thick filaments of cardiac myocytes in thyroid treated rabbits. *J. Cell Biol.* 105: 2771–2780.

33. Dome, J.S., Mittal, B., Pochapin, M.B., Sanger, J.M., and Sanger, J.W. 1988. Incorporation of fluorescently labeled actin and tropomyosin into muscle cells. *Cell Differ. Dev.* 23:37–52.

34. Watkins, H., Seidman, J.G., and Seidman, C.E. 1995. Familial hypertrophic cardiomyopathy: a genetic model of cardiac hypertrophy. *Hum. Mol. Genetics* 4:1721–1727.

35. Bonne, G., Carrier, L., Richard, P., Hainque, B., and Schwartz, K. 1998. Familial hypertrophic cardiomyopathy: from mutations to functional defects. *Circ. Res.* 83:580–593.

36. Mogensen, J., Klausen, I.C., Pedersen, AK., et al. 1999. α-cardiac actin is a novel disease gene in familial hypertrophic cardiomyopathy. *J. Clin. Invest.* 103:R39–R43.

37. Marian, A.J., Zhao, G.L., Seta, Y., Roberts, R., and Yu, Q.T. 1997. Expression of a mutant (Arg92Gln) human cardiac troponin T, known to cause hypertrophic cardiomyopathy, impairs adult cardiac myocyte contractility. *Circ. Res.* 81:76–85.

38. Yang, Q., Sanbe, A., Osinska, H., Hewett, T.E., Klevitsky, R., and Robbins, J. 1998. A mouse model of myosin binding protein C human familial hypertrophic cardiomyopathy. *J. Clin. Invest.* 102:1292–1300.

39. Marian, A.J., Yu, Q.-T., Mann, D.L., Graham, F.L., and Roberts, R. 1995. Expression of a mutation causing hypertrophic cardiomyopathy disrupts sarcomere assembly in adult feline cardiac myocytes. *Circ. Res.* 77:98–106.

40. Becker, K.D., Gottshall, K.R., Hickey, R., Perriard, J.-C., and Chien, K.R. 1997. Point mutations in human β cardiac myosin heavy chain have differential effects on sarcomeric structure and assembly: an ATP binding site change disrupts both thick and thin filaments whereas hypertrophic cardiomyopathy mutations display normal assembly. *J. Cell Biol.* 137:131–140.

41. Fatkin, D., Christe, M.E., Aristizabal, O., et al. 1999. Neonatal cardiomyopathy in mice homozygous for the Arg403Gln mutation in the alpha cardiac myosin heavy chain gene. *J. Clin. Invest.* 103:147–153.

42. Tardiff, J.C., Hewett, T.E., Palmer, BM., et al. 1999. Cardiac troponin T mutations result in allele-specific phenotypes in a mouse model for hypertrophic cardiomyopathy. *J. Clin. Invest.* 104:469–481.

CHAPTER **6**

Manipulation of Myofibrillogenesis in Whole Hearts

Robert W. Zajdel, Matthew D. McLean, Christopher R. Denz, Syamalima Dube, Larry F. Lemanski, and Dipak K. Dube

INTRODUCTION

Potentially powerful tools for treatment of acquired and inherited diseases of the myocardium are the introduction of replacement genes or targeted disruption of diseased proteins. Direct intramuscular injection (Ascadi et al., 1991) and viral vectors (Dunckley et al., 1992) have been previously used for the transfer of functional cDNA constructs into muscle cells, but these are dependent on possible low expression or negative responses to the viral vector. A promising alternative to these methods that currently is being extensively studied is the use of cationic liposomes. Benefits of cationic liposomes include the ability to protect nucleic acids from degradation and the efficient delivery of tagged oligonucleotides to the nucleus (Zelphati and Szoka, 1996). In this chapter, we describe different strategies for using cationic liposomes to analyze cardiac myofibril organization and formation in whole intact hearts. Results from these applications have yielded valuable information on myofibril initiation and segmental heart development.

Cardiomyocytes function in a highly integrative environment, and lack of contractions or function can alter the organization of the myofibrils. Whole heart transfection has the obvious advantage of maintaining intercellular connections and mechanical load compared to cultured cells. The advent of the confocal microscope has created the opportunity to examine cardiac myocytes in whole hearts. Whole hearts of embryos of the Mexican axolotl are quite small (one to several cells thick, depending upon the stage of the embryo), but they can be easily visualized by confocal microscopy. An added benefit of the Mexican axolotl as a heart model for myofibrillogenesis is a naturally occurring cardiac mutant that has no heartbeat in the ventricle, a lack of organized myofibrils, and reduced amounts of sarcomeric tropomyosin expression (Lemanski, 1973). This model is excellent for the examination of factors that initiate myofibril formation in the heart.

Throughout the chapter, we use tropomyosin as a focus in an examination of myofibril organization, primarily because of the lack of classical sarcomeric tropomyosin in the ventricle of mutant hearts. However, antisense and antibody methods used to block expression of tropomyosin can also be used on normal hearts, and all of these procedures could potentially be used to examine other specific proteins or isoforms. Myofibrillogenesis was studied with the lipofection of wild-type tropomyosin cDNA or protein into mutant hearts and transfection of Asp175Asn mutant tropomyosin cDNA into mutant and normal hearts.

PROTEIN INTRODUCTION INTO WHOLE EMBRYONIC HEARTS

Lipofection techniques may also be used for introduction of proteins and antibodies into cardiomyocytes. Microinjection of single cardiac cells with tagged proteins (Sanger et al., 1984; Imanaka-Yoshida et al., 1996) has been previously used for valuable insights into myofibrillogenesis. Protein delivery by cationic liposomes has been characterized in a small number of examples, but importantly the proteins were able to maintain functional activity. Enzymatic activity was maintained with cationic liposome delivery of purified prostatic acid phosphatase protein (100 kd) into prostate carcinoma cells (Lin et al., 1993). Glucocorticoid receptor protein, T7X556 (19 kd), activity was studied in cell lines after delivery by cationic liposomes (Debs et al., 1990). An acidic isoelectric point appears to be beneficial for the binding of a protein to the cationic liposomes; prostatic acid phosphatase protein (pI 5.0) has a PI similar to tropomyosin (Bailey, 1948).

Tropomyosin protein introduced into the cells of mutant whole hearts resulted in myofibrillogenesis and the formation of organized myofibrils (Zajdel et al., 1998). Examination of the kinetics of FITC protein incorporation into whole hearts demonstrated that mutant and normal hearts had FITC tropomyosin localized within the hearts within four hours after treatment with the liposome containing tropomyosin. At 24 hours, there was an increase in FITC-labeled tropomyosin within the treated hearts. Even at 4 hours, some of the FITC tropomyosin had been incorporated into nascent myofibrils in the mutant. These results showed that the FITC tropomyosin entered the hearts but is not completely incorporated into all myofibrils at this time. The natural reduction of tropomyosin in the mutant hearts (Starr et al., 1989) provided a clear backdrop for examination of introduction of exogenous tropomyosin.

In contrast, incorporation into a majority of myofibrils in normal hearts would be dependent on the degradation and turnover of the exogenous tropomyosin protein. It is unclear whether mutant hearts have a substantially different degradation/turnover rate of tropomyosin compared to normal hearts or whether the process is normal and the mutant degradation rate is dependent on the presence of mutant protein. Because mutant hearts lack α-tropomyosin in ventricles, they do not appear to have the ability to make new tropomyosin to replace the exogenous protein, whereas normal hearts would be able to make new tropomyosin. Previous studies have shown that fluorescently labeled exogenous tropomyosin was still faintly visible at four days in microinjected cells (Wehland and Weber, 1980), which suggests that tropomyosin has a low turnover rate within cells.

An alternative method to examine what effect blockage of a specific protein such as tropomyosin has on function is to introduce a specific antibody to that protein into normal cells. Lipofection of tropomyosin antibody (CH1) caused disruption of the myofibrils in the normal hearts after four days and beating was decreased at days 2 to 3. Control normal hearts treated with secondary antibody without previous liposome/antibody treatment did not have detectable fluorescence above background. Liposome treatment with antibodies to proteins not expressed in the hearts at this time did not disrupt myofibril organization compared to the tropomyosin antibody.

Introduction of a specific antibody to tropomyosin disrupts functional expression of the protein, which makes this technique useful for altering the expression of various proteins in normal as well as mutant hearts. In the absence of available "knockout" technology in amphibian systems, the use of antibody to curtail specific proteins or antisense oligonucleotide-mediated blocking of specific protein synthesis are valuable tools to examine structure–function relationships in cardiac myofibrillogenesis.

REPLACEMENT OF α-TROPOMYOSIN IN MUTANT HEARTS

Gene therapy is a potent method for the replacement of a missing protein in a diseased or deficient tissue. Transfection procedures using cationic liposomes were used to introduce an exogenous α-tropomyosin cDNA into cardiac mutant hearts (Figure 6.1a). These hearts had higher expression of tropomyosin, exhibited myofibrillogen-

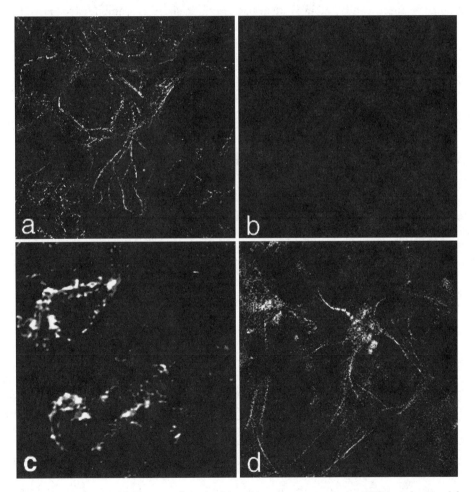

FIGURE 6.1. (a) Mutant heart transfected with wild-type tropomyosin cDNA and stained for CH1 tropomyosin antibody after five days. Tropomyosin protein has increased compared to mutant controls, and "nascent" organized myofibrils have formed. (b) Mutant control heart does not have organized myofibrils, and tropomyosin protein levels are reduced. (c) Mutant heart transfected with α-tropomyosin cDNA but stained with α-actinin antibody to detect organized myofibrils. α-actinin is localized to Z-bodies in nascent myofibrils after five days. (d) Transfection of wild-type tropomyosin cDNA into mutant hearts in the pericardial cavity resulted in the formation of well-organized myofibrils.

esis, and began contractions (Zajdel et al., 1998). Mutant control hearts do not have organized myofibrils and have a lower level of detectable tropomyosin protein (Figure 6.1b). Mutant hearts transfected with a reporter gene (*Lac z*) had β-galactosidase expression throughout all areas (i.e., ventricle, atrium, sinus venosus, and conus arteriosus), with no observed effect on myofibril formation. There was no effect with the β-galactosidase gene. Confirmation of the formation of organized myofibrils was obtained with staining of the wild-type tropomyosin transfected mutant heart ventricles with α-actinin antibody (Figure 6.1c). Transfection of β-tropomyosin cDNA did not result in the formation of organized myofibrils, although the expression of tropomyosin was increased compared to the mutant control hearts. These results strongly suggest that the lack of classical tropomyosin in mutant ventricles (Lemanski, 1973; Moore and Lemanski, 1982; Starr et al., 1989) results in a failure of organized myofibrils to form, probably due to a destabilization of the thin filaments. Significantly, actin is at near normal levels in mutant hearts (Starr et al., 1989) but mainly in amorphous collections instead of filaments. Therefore, a lack of or reduced expression of classical tropomyosin in the mutant hearts most likely leads to a destabilization of the thin filament with no well-formed myofibrils, as has been previously suggested by Lemanski et al. (1995).

TRANSFECTION OF LIVE EMBRYOS BY MICRODISSECTION INTO THE PERICARDIAL CAVITY

The chests of mutant embryos at stages 37 to 38 were opened microsurgically overlying the hearts similar to the method of Erginel-Unaltuna et al. (1995). The embryos with their hearts were then incubated with cationic liposomes containing an expression construct with wild-type murine tropomyosin cDNA or Asp175Asn mutant cDNA (Zajdel et al., unpublished results). Transfection of live mutant embryos with α-tropomyosin cDNA after 20 days resulted in the formation of organized myofibrils (Figure 6.1d). Survivability of the treated embryos was monitored by observation of active swimming. Remarkably, only rare embryos did not survive treatment, and healing occurred in most with recovery and closure of the pericardial cavity. Hearts transfected with Asp175Asn cDNA in live mutant embryos did not result in organized myofibrils, nor was there an initiation of beating.

The transfected embryos were also examined daily under a dissecting microscope for the presence of contracting hearts. Embryos at these stages have a thin semitransparent skin overlying the pericardial cavity that is conducive to monitoring. Synchronous contractions started in a significant number of mutant embryos transfected with wild-type tropomyosin cDNA, although beating was not maintained permanently. At approximately 20 days, beating in mutant hearts transfected with the wild-type tropomyosin cDNA became inconsistent, and stopped in some. Surgical procedures and retransfection with wild-type cDNA resulted in restarting of the contractions and reorganization of the myofibrillar structure. These retransfected hearts were arbitrarily maintained in culture up to 43 days when the experiment was terminated. Mutant embryos typically live for 20 days beyond the heartbeat initiation stage, when they develop ascites and subsequently hearts become thin-walled and distended. It was clear that mutant embryos transfected the wild-type tropomyosin cDNA survived for a longer period of time than the untreated mutant controls or the mutant embryos transfected with the Asp175Asn mutant tropomyosin cDNA. Mutant embryos transfected with wild-type murine α-tropomyosin cDNA also did not have ascites compared to control mutant embryos that were also surgically manipulated.

ASP175ASN MUTANT TROPOMYOSIN DOES NOT PROMOTE MYOFIBRIL FORMATION IN MUTANT HEARTS WITHOUT A MYOFIBRIL TEMPLATE

Transfection of Asp175Asn mutant tropomyosin cDNA into whole hearts provides a valuable opportunity to examine the mutant proteins in an environment without an existing myofibril template. The lack of organized myofibrils and classical α-tropomyosin in the ventricles of mutant hearts contrasts with the presence of organized myofibrils in transgenic models of familial hypertrophic cardiomyopathy (FHC). FHC has been characterized as a disease of the sarcomere with morphological changes, including myocyte hypertrophy, disarray, and fibrosis (Coviello et al., 1997). FHC involves seven sarcomeric genes with a heterogeneity that can result in similar pathophysiological changes. A proposed common pathway suggests that mutant genes contain missense mutations that are dominant and produce so-called poison polypeptides. These poison polypeptides incorporate into the sarcomere and change the function of the wild-type protein and the assembly of myofibrils (Bonne et al., 1998). Therefore, the importance of these mutant proteins to the stability of the sarcomere makes this model important to analyze whether mutant proteins can form organized myofibrils in isolation.

Two mutations, Asp175Asn and Glu180Gly, that have been identified in familial hypertrophic cardiomyopathy are near the Cys 190 region of tropomyosin (Thierfelder et al., 1994). Mutant protein expression of the Asp175Asn tropomyosin in the sarcomere of skeletal muscle has been confirmed and equal levels of mutant and wild-type tropomyosin observed (Bottinelli et al., 1998). However, this study examined the incorporation of the Asp175Asn mutant tropomyosin in heterozygous myofibers and in all wild-type myofibers. Significantly, they were unable to demonstrate the presence of Asp175Asn mutant tropomyosin in organized myofibers without the presence of at least some wild-type tropomyosin.

As previously described, transfection of exogenous wild-type tropomyosin cDNA resulted in myofibrillogenesis in the ventricle of mutant hearts. With this as a baseline control, mutant hearts transfected with Asp175Asn mutant tropomyosin cDNA did not form organized myofibrils, although the amount of tropomyosin protein detected by a secondary antibody was greatly increased (Figure 6.2a). Western blots of transfected mutant hearts also showed a substantial increase in tropomyosin protein expression. Transfection of Asp175Asn mutant tropomyosin cDNA into normal hearts resulted in a disruption of myofibril organization (Figure 6.2B). This result appears to support the hypothesis of a dominant-negative effect for the action of the mutant protein. Examination of heartbeat during in vitro experiments demonstrated a lack of beating ventricles in mutant hearts transfected with the Asp175Asn cDNA, in contrast to the initiation of contractions that can be seen in mutant hearts transfected with wild-type α-tropomyosin cDNA.

Because tropomyosin protein forms a dimer in an organized sarcomere, we also cotransfected mutant hearts with β-tropomyosin cDNA and Asp175Asn mutant cDNA. There was very little sarcomeric myofibril formation, although again tropomyosin levels were higher than in controls. Specific stages of myofibril forma-

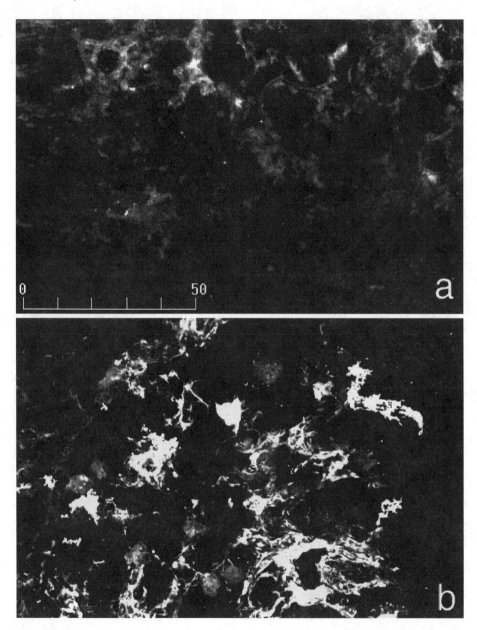

FIGURE 6.2. (a) Confocal laser scanning microscope Z-series of a mutant heart transfected with Asp175Asn mutant tropomyosin. Organized myofibrils have not formed, but tropomyosin protein levels have increased. (b) Transfection of normal hearts with Asp175Asn mutant tropomyosin demonstrated a disruption of myofibril organization.

tion were examined in the Asp175Asn mutant tropomyosin transfected hearts by using markers previously described for the premyofibril model of myofibrillogenesis (Rhee et al., 1994; Dabiri et al., 1997; LoRusso et al., 1997). Nonmuscle myosin IIB was detected in control and transfected mutant hearts, but it was not organized into nascent myofibrils. This protein was localized to dot-like structures similar to Z-bodies. Nonmuscle myosin IIB is a possible marker for pre- and nascent myofibrils. Zeugmatin staining of Asp175Asn transfected hearts did not demonstrate any nascent myofibrils as seen in normal controls. Zeugmatin is a marker for nascent and mature myofibrils, with Z-body/Z-line type staining typical of α-actinin. A commonly used marker for late myofibril formation is the presence of myosin binding protein C (MyBP-C). Mutant hearts, although expressing MyBP-C, did not show a striated pattern indicative of organized myofibrils. Although MyBP-C was present in the Asp175Asn transfected heart cells, it was not organized into myofibrils. These marker proteins suggest that mutant hearts are in a premyofibril stage of myofibrillogenesis and that the transfection with Asp175Asn tropomyosin cDNA into mutant hearts does not appear to advance the process. This is in contrast to the transfection of wild-type α-tropomyosin under control of an identical α-myosin heavy-chain promoter in which myofibrillogenesis was promoted such that nascent and some mature myofibrils form in mutant cardiomyocytes.

The contrast between the Asp175Asn tropomyosin mutant in FHC and the cardiac mutation in axolotls is interesting to compare. The cardiac mutation in axolotls results in a greatly decreased amount of tropomyosin protein, with a lack of organized myofibrils (Moore and Lemanski, 1982; Starr et al., 1989; Lemanski et al., 1995). There is also a dilation of the heart in these animals as they develop. It is not entirely clear that the cardiac mutation is solely associated with the tropomyosin genes, but studies in this laboratory have suggested that the reduced level of tropomyosin disrupts the stoichiometry of the contractile proteins and no organized myofibrils are formed. The mutation has been found to affect 25% of a heterozygous spawning, which suggests a simple Mendelian recessive (Humphrey, 1968, 1972). If an isoform of tropomyosin is directly involved, the cardiac mutation may involve a truncated protein instead of a poison polypeptide proposed for missense mutations such as Asp175Asn, with degradation of the protein so that it is practically undetectable in the ventricle.

TARGETED GENE DISRUPTION OF SPECIFIC TROPOMYOSIN ISOFORMS IN WHOLE HEARTS

Three isoforms of tropomyosin have been identified in the normal hearts of the Mexican axolotl (Luque et al., 1997). Two isoforms, Axolotl Tropomyosin Cardiac 1 (ATmC-1) and Axolotl Tropomyosin Cardiac 2 (ATmC-2), are putatively encoded by the α-gene (Luque et al., 1997). The third isoform, Axolotl Tropomyosin Cardiac 3 (ATmC-3), is encoded by a separate tropomyosin gene, most likely a TM4 type.

Several studies have used antisense oligonucleotides to examine the effects of blocking a specific gene on function. Romano and Runyan (1999) demonstrated inhibition of mesenchymal cell formation when slug antisense oligonucleotides were used. Incubation of stage 6 chick embryos with cXin antisense oligonucleotides resulted in abnormal cardiac morphogenesis and alteration of looping (Wang et al., 1999). Specific cardiac and skeletal isoforms for dihydropyridine receptors were compared by treatment with antisense oligonucleotides (Bulteau et al., 1998). Specific inhibition of GATA-4 expression by antisense transcripts disrupted expression of cardiac muscle genes, and beating did not develop (Grepin et al., 1997). Thus, antisense oligonucleotides are becoming increasingly important tools for blocking the function of specific genes for embryos or cultured cells. Targeted disruption of the axolotl tropomyosin isoforms facilitated the examination of two important factors in myofibril formation during heart development. First, smooth muscle isoforms of contractile proteins may be involved with the early formation or initiation of myofibrils; and second, during heart development, specific isoforms may be significant for specific heart segments such as the conus and ventricle.

A UNIQUE TROPOMYOSIN ISOFORM (ATMC-2) IN NORMAL AND MUTANT HEARTS

Analysis of the nucleotide sequence of the ATmC-2 isoform suggest that this tropomyosin is a unique striated muscle-type tropomyosin with a smooth muscle type exon II (Luque et al., 1997; Spinner et al., manuscript in preparation). RT-PCR of the ATmC-2 isoform in normal and mutant hearts did not reveal any difference in expression (Spinner et al., manuscript in preparation). Thus, it appears that this smooth muscle isoform is not directly involved in the cardiac mutation of the axolotl heart. A specific antibody to this isoform was raised and used to localize the protein in whole hearts by confocal laser scanning microscopy. The ATmC-2 isoform appeared at the Z-line area of the sarcomere in a periodic pattern (Figure 6.3a).

Assays on the functional importance of ATmC-2 were performed by lipofection of isoform-specific oligonucleotides and the anti-ATmC-2 antibody into normal whole hearts. Lipofection of isoform-specific ATmC-2 antisense oligonucleotide into normal hearts disrupted function and structure of myofibrils in both the conus and ventricle of embryonic whole hearts (Figure 6.3b). Those results were also reproduced with the lipofection of ATmC-2 isoform antibody. These studies demonstrate the necessity of the smooth muscle isoform for myofibrillogenesis in normal hearts. Binding of the protein or targeted disruption of transcription resulted in alteration of the sarcomeric structure of the myofibrils and disruption of heartbeat. Our results clearly show that a smooth muscle tropomyosin isoform (ATmC-2) was present with the α-tropomyosin isoform in developing myofibrils of the embryonic whole heart. These results corroborate previous findings for cultured cardiac myocytes (Greaser et al., 1989) and glycerinated cardiac muscle (Wang et al., 1990) in which muscle tropomyosin and nonmuscle tropomyosin isoforms were found in a sarcomeric pattern. In addition, our results strongly suggest a necessary role for smooth muscle tropomyosin in the formation of organized myofibrils.

DIFFERENTIAL EXPRESSION OF TROPOMYOSIN IN THE CONUS AND THE VENTRICLE

Targeted disruption of specific tropomyosin isoforms has revealed that the ATmC-3 tropomyosin isoform is necessary for function of the ventricle and that ATmC-1 is necessary for the conus. Early observations of mutant embryos have shown that the conus of mutant hearts can beat while the ventricles remain quiescent (Lemanski, 1973). Laser scanning confocal microscopy and electron microscopy have shown that tropomyosin levels are apparently normal in the conus and that organized myofibrils are ubiquitous (Figure 6.4a). Structure and function of the conus were disrupted with transfection of ATmC-1 antisense oligonucleotides (Figure 6.4b). Transfection of ATmC-3 antisense oligonucleotides did not significantly alter myofibril formation or tropomyosin protein levels in the conus (Figure 6.4c). Transfection of ATmC-3 antisense oligonucleotides replicated/mimicked the phenotype of the mutant hearts, which lack a beat, have disrupted myofibrils, and have a reduced expression of

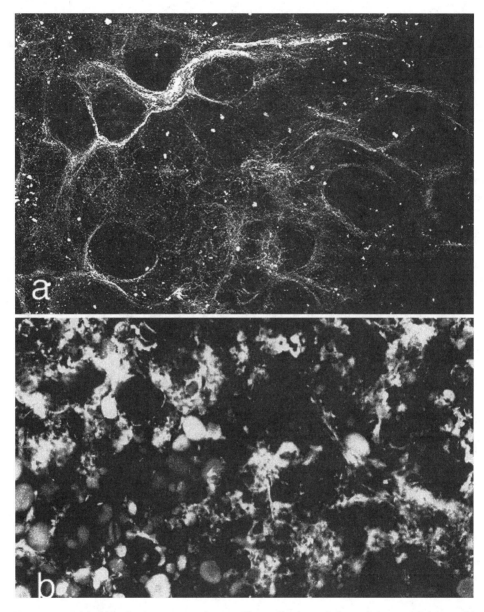

FIGURE 6.3. (a) ATmC-2 tropomyosin-specific antibody staining of a normal heart. (b) Normal heart transfected with ATmC-2 antisense oligonucleotide resulted in disruption of myofibril organization.

tropomyosin in the ventricle (Figure 6.4d). However, ATmC-3 sense oligonucleotides did not disrupt the ventricle, nor did ATmC-1 antisense oligonucleotides. These studies suggest a differential expression of tropomyosin isoforms in the conus and ventricle during development. The significance of the differential expression of tropomyosin in the conus versus the ventricle may be explained by the conus possibly having a different origin than the ventricle. The cardiac mutant hearts have a conus that beats and have organized myofibrils. This obvious difference may be explained

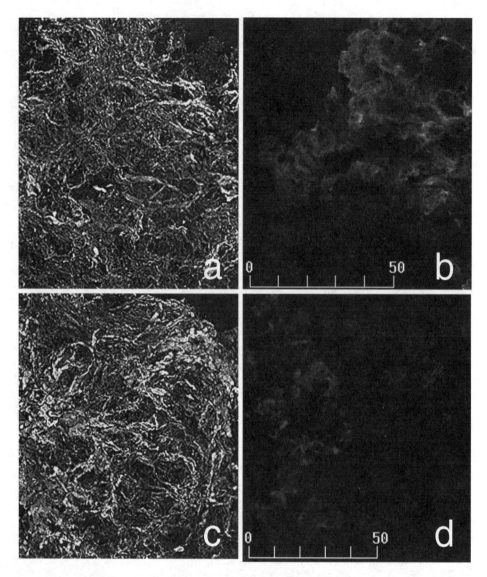

FIGURE 6.4. (a). Tropomyosin staining (CH1) of the conus of a mutant heart. Myofibrils are present, in contrast to the ventricle. (b) Transfection of ATmC-1 antisense oligonucleotide into the conus of a normal heart. (c) Transfection of ATmC-3 anti-sense oligonucleotide into the conus of a normal heart. (d) Transfection of ATmC-3 antisense oligonucleotide into the ventricle of a normal heart.

by different inducing signals or less of an inhibitory signal compared to the ventricle. It may also be explained by the possibility that the conus may arise from a late-appearing fifth heart segment of nonheart field origin (Markwald et al., 1999). Markwald and coworkers (1999) propose that the fifth heart segment is derived or recruited from extracardiac mesoderm and forms the conus, which integrates with the ventricular segments to form the outlets. A functional conus in cardiac mutant hearts may be derived from a nonheart field, which maintains the factors necessary for myofibrillogenesis. This implies that the heart fields that form the other heart segments may be deficient in an inducing factor necessary for differentiation of the cardiac myocytes.

CONCLUSIONS

An absence of organized sarcomeric myofibrils and defective myofibrillogenesis are hallmarks of the cardiac mutant phenotype in the Mexican axolotl. Remarkably, whole heart transfection of both mutant and normal hearts is possible while maintaining cellular connections. Initial experiments using this model have focused on the expression of tropomyosin isoforms in relation to initiation of organized myofibril formation and variable expression in heart segments.

1. Replacement of the missing tropomyosin in mutant ventricles was sufficient to complete the protein stoichiometry and promote organized myofibrils in vitro and in vivo.
2. Asp175Asn mutant tropomyosin involved with a type of FHC was insufficient to form organized myofibrils without a myofibril template and equal levels of wild-type tropomyosin.
3. Asp175Asn mutant tropomyosin disrupted myofibril organization in normal hearts following the dominant-negative model.
4. A smooth muscle tropomyosin is present in organized myofibrils, and targeted disruption negatively affects function.

Combined examination of the cardiac mutant heart and the targeted disruption of tropomyosin isoforms in normal hearts suggest that the conus and the ventricle have diverse developmental factors affecting myofibrillogenesis.

1. Mutant conus can contract normally and have organized myofibrils while the attached ventricle is quiescent.
2. Transfection of ATmC-3 tropomyosin antisense oligonucleotide disrupts myofibril formation in the ventricle replicating the mutant phenotype but not significantly in the conus.
3. Transfection of ATmC-1 antisense oligonucleotide disrupts myofibrils in the conus but not the ventricle.

The role of specific isoforms of contractile proteins is not clearly understood, but potentially valuable information concerning segmental expression of tropomyosin isoforms may lead to a better understanding of the formation of the conus from a nonheart field. Regulation of the conus appears to be different from the ventricle, and further delineation of possible factors is needed. A specific role in myofibril organization and formation for a smooth muscle isoform is another important area to be studied. Functional significance of a specific tropomyosin isoform ATmC-2 was examined by transfection of a specific antibody and isoform-specific antisense oligonucleotides into normal hearts. The importance of these procedures on whole hearts in organ culture lies in the ability to examine various aspects of myofibrillogenesis in a model approaching in vivo.

In vivo examination of mutant embryos that were transfected with wild-type α-tropomyosin cDNA demonstrated initiation of contractions and formation of organized myofibrils. This approach involved direct application of the liposome complexes into the pericardial cavity. Direct injection into the pericardial cavity may also be a possible method for delivery of the liposome/cDNA. Exogenous DNA was expressed for weeks to months following direct injection into skeletal muscle and myocardium (Ascadi et al., 1991; Jiao et al., 1992; Prentice et al., 1994).

The use of cationic liposomes in a clinical setting is dependent on overcoming possible problems inherent to the liposomes themselves. Delivery of cationic liposome/cDNA complexes by intracoronary perfusion or through the bloodstream may face potential problems of interactions with serum components. Serum components bound to the liposome complex may lead to an increase in particle diameter and entrapment in the capillary beds (Zelphati et al., 1998), decreased availability to the cells, or increased clearance by the reticuloendothelial system. In vivo use of a cationic liposome in a serum environment requires more extensive study for effective delivery of nucleic acids or proteins. It appears that the ability to deliver proteins into cells by forming a liposome/protein complex may also interfere with the ability to introduce cationic liposome complexes into the bloodstream. Our studies on direct exposure of liposomes/replacement cDNA and previous studies on direct injection into the myocardium would suggest the possibility of fewer complications and inhibitors compared to vascular delivery. The use of cationic liposomes has proven valuable for targeted disruption and replacement of the expression of specific proteins in whole hearts. Further study will determine whether cationic liposomes will be useful and safe clinically for gene or protein therapy.

ACKNOWLEDGMENTS

We are thankful to Ms. Susan T. Duhon of Indiana University Axolotl Colony for providing some of the axolotl embryos used in this study. This work was supported by an AHA Grant-in-Aid (New York Chapter) to D.K.D., an AHA Grant-in-Aid (New York Chapter) to D.K.D. and L.F.L., and NIH Grants HL58435 and HL 61246 to L.F.L. Dr. Robert W. Zajdel was an AHA (New York Affiliate) Postdoctoral Fellow.

REFERENCES

Ascadi, G., Dickson, G., Love, D.R., Jani, A., Walsh, F.S., Gurusinghe, A., Wolff, J.A., and Davies, K.E. 1991. Human dystrophin expression in *mdx mice* after intramuscular injection of DNA constructs. *Nature* 352:815–818.

Bailey, K. 1948. Tropomyosin: A new asymmetric protein component of the muscle fibril. *Biochem. J.* 43:271.

Bonne, G., Carrier, L., Richard, P., Hainque, B., and Schwartz, K. 1998. Familial hypertrophic cardiomyopathy: from mutations to functional defects. *Circ. Res.* 83:580–593.

Bottinelli, R., Coviello, D.A., Redwood, C.S., Pellegrino, M.A., Maron, B.J., Spirito, P., Watkins, H., and Reggiani, C. 1998. A mutant tropomyosin that causes hypertrophic cardiomyopathy is expressed in vivo and associated with an increased calcium sensitivity. *Circ. Res.* 82:106–115.

Bulteau, L., Raymond, G., and Cognard, C. 1998. Antisense oligonucleotides against "cardiac" and "skeletal" DHP-receptors reveal a dual role for the "skeletal" isoform in EC coupling of skeletal muscle cells in primary culture. *J. Cell Sci.* 111:2149–2158.

Coviello, D.A., Maron, B.J.M., Spirito, P., Watkins, H., Vosberg, H.P., Thierfelder, L., Schoen, F.J., Seidman, J.G., and Seidman, C.E. 1997. Clinical features of hypertrophic cardiomyopathy caused by mutation of a "hot spot" in the alpha-tropomyosin gene. *J. Am. Coll. Cardiol.* 29:635–640.

Dabiri, G.A., Turnacioglu, K.T., Sanger, J.M., and Sanger, J.W. 1997. Myofibrillogenesis visualized in living embryonic cardiomyocytes. *Proc. Natl. Acad. Sci. USA* 94:9493–9498.

Debs, R.J., Freedman, L.P., Edmunds, S., Gaensler, K.L., Duzgunes, N., and Yamamoto, K.R. 1990. Regulation of gene expression in vivo by liposome-mediated delivery of a purified transcription factor. *J. Biol. Chem.* 265:10189–10192.

Dunckley, M.G., Love, D.R., Davies, K.E., Walsh, F.S., Morris, G.E., and Dickson, G. 1992. Retroviral-mediated transfer of a dystrophin minigene into *mdx mouse* muscle in vitro. *FEBS Lett.* 296:128–134.

Erginel-Unaltuna, N., Dube, D.K., Robertson, D.R., and Lemanski, L.F. 1995. In vivo protein synthesis in developing hearts of normal and cardiac mutant axolotls, *Ambystoma mexicanum. Cell. Mol. Biol. Res.* 41:181–187.

Greaser, M.L., Handel, S.E., Wang, S.-M., Schultz, E., Bulinski, J.C., Lin, J.J.-C., and Lessard, J.L. 1989. Assembly of titin, myosin, actin, and tropomyosin into myofibrils in cultured chick cardiomyocytes. In: *Cellular and Molecular Biology of Muscle Development*, Eds. L.H. Kedes and F.E. Stockdale, Alan R. Liss, New York, pp. 247–257.

Grepin, C., Nemer, G., and Nemer, M. 1997. Enhanced cardiogenesis in embryonic stem cells overexpressing the GATA-4 transcription factor. *Development* 124:2387–2395.

Humphrey, R.R. 1968. A genetically determined absence of heart function in embryos of the Mexican axolotl (*Ambystoma mexicanum*). *Anat. Rec.* 162:475.

Humphrey, R.R. 1972. Genetic and experimental studies on a mutant gene (c) determining absence of heart action in embryos of the Mexican axolotl (*Ambystoma mexicanum*). *Dev. Biol.* 27:365–375.

Imanaka-Yoshida, K., Danoski, B.A., Sanger, J.M., and Sanger, J.W. 1996. Living adult rat cardiomyocytes in culture: evidence for dissociation of costameric distribution of vinculin from costameric distributions of attachments. *Cell Motil. Cytoskel.* 33:263–275.

Jiao, S., Williams, P., Berg, R.K., Hodgeman, B.A., Liu, L., Repetto, G., and Wolff, J.A. 1992. Direct gene transfer into nonhuman primate myofibers in vivo. *Hum. Gene Ther.* 3:21–33.

Lemanski, L. 1973. Morphology of developing heart in *cardiac* lethal mutant Mexican axolotl, *Ambystoma mexicanum. Dev. Biol.* 33:312–333.

Lemanski, L.F., LaFrance, S.M., Erigent-Unaltuna, N., Luque, E.A., Ward, S.M., Fransen, M.E., Mangiacapar, F.J., Nakatsugawa, M., Lemanski, S.L., Capone, R.B., Goggins, K.J., Nash, B.P., Bhatia, R., Dube, A., Gaur, A., Zajdel, R.W., Zhu, Y., Spinner, B.J., Pietras, K.M., Lemanski, S.F., Kovacs, C.V., and Dube, D.K. 1995. The cardiac mutant gene c in Axolotls: Cellular, developmental, and molecular studies. *Cell. Mol. Biol. Res.* 41:293–305.

Lin, M.-F., DaVolio, J., and Garcis, R. 1993. Cationic liposome-mediated incorporation of prostatic acid phosphatase protein into human prostate carcinoma cells. *Biochem. Biophys. Res. Commun.* 192:413–419.

LoRusso, S.M., Rhee, D., Sanger, J.M., and Sanger, J.W. 1997. Premyofibrils in spreading adult cardiomyocytes in tissue culture: evidence for reexpression of the embryonic program for myofibrillogenesis in adult cells. *Cell Motil. Cytoskel.* 37:183–198.

Luque, E.A., Spinner, B.J., Dube, S., Dube, D.K., and Lemanski, L.F. 1997. Differential expression of a novel isoform of alpha-tropomyosin in cardiac and skeletal muscle of the Mexican axolotl (*Ambystoma mexicanum*). *Gene* 185:175–185.

Markwald, R.R., Trusk, T., and Moreno-Rodriquez, R. 1998. Formation and septation of the tubular heart: integrating the dynamics of morphology with emerging molecular concepts. In: *Living Morphogenesis of the Heart*, Eds. M.V. De la Cruz and R.R. Markwald, Birkhauser, Boston.

Moore, P.B. and Lemanski, L.F. 1982. Quantitation of tropomyosin by radioimmunoassay in developing hearts of *cardiac* mutant axolotls, *Ambystoma mexicanum. J. Muscle Res. Cell Motil.* 3:161–167.

Prentice, H., Kloner, R.A., Prigozy, T., Christensen, T., Newman, L., Li, Y., and Kedes, L. 1994. Tissue-restricted gene expression assayed by direct DNA injection into cardiac and skeletal muscle. *J. Mol. Cell. Cardiol.* 26:1393–1401.

Rhee, D., Sanger, J.M., and Sanger, J.W. 1994. The premyofibril: evidence for its role in myofibrillogenesis. *Cell Motil. Cytoskel.* 28:1–24.

Romano, L.A. and Runyan, R.B. 1999. Slug is a mediator of epithelial-mesenchymal cell transformation in the developing chicken heart. *Dev. Biol.* 212:243–254.

Sanger, J.W., Mittal, B., and Sanger, J.M. 1984. Formation of myofibrils in spreading chick cardiac myocytes. *Cell Motil. Cytoskel.* 4:405–416.

Spinner, B.J., Zajdel, R.W., Mehta, S., Choudhury, A., Lemanski, L.F., and Dube, D.K. 2000. A novel isoform of alpha-tropomyosin, ATmC-2, is essential for structure and function in the Mexican axolotl. (submitted).

Starr, C., Diaz, J.G., and Lemanski, L.F. 1989. Analysis of actin and tropomyosin in hearts of cardiac mutant axolotls by two-dimensional gel electrophoresis, western blots, and immunofluorescent microscopy. *J. Morphol.* 210:1–10.

Theirfelder, L.H., Watkins, C., MacRae, C., Lamas, R., McKenna, W., Vosberg, H.-P., Seidman, J.G., and Seidman, C.E. 1994. Alpha-tropomyosin and cardiac troponin-T mutations cause familial hypertrophic cardiomyopathy: disease of the sarcomere. *Cell* 77:701–712.

Wang, S.-M., Wang, S.-H., Lin, J.L.-C., and Lin, J.J.-C. 1990. Striated muscle tropomyosin-enriched microfilaments of developing muscles of chicken embryos. *J. Muscle Res. Cell Motil.* 11:191–202.

Wang, D.Z., Reiter, R.S., Lin, J.L., Wang, Q., Williams, H.S., Krob, S.L., Schultheiss, T.M., Evans, S., and Lin, J.J. 1999. Requirement of a novel gene, Xin, in cardiac morphogenesis. *Development* 126:1281–1294.

Wehland, J. and Weber, K. 1980. Distribution of fluorescently labeled actin and tropomyosin after microinjection in living tissue culture cells as observed with TV image intensification. *Exp. Cell Res.* 127:397–408.

Zajdel, R.W., McLean, M.D., Lemanski, S.L., Muthuchamy, M., Wieczorek, D.F., Lemanski, L.F., and Dube, D.K. 1998. Ectopic expression of tropomyosin promotes myofibrillogenesis in mutant axolotl hearts. *Dev. Dyn.* 213:412–420.

Zajdel, R.W., Dube, D.K., and Lemanski, L.F. 1999. The cardiac mutant Mexican axolotl is a unique animal model for evaluation of cardiac myofibrillogenesis. *Exp. Cell Res.* 248:557–566.

Zajdel, R.W., McLean, M.D., Muthuchamy, M., Mehta, S., Lemanski, L.F., Wieczorek, D.F., and Dube, D.K. 2000a. Mutation of Asprtic acid[175] of alpha-tropomyosin prohibits myofibril formation in the mutant axolotl heart. (submitted).

Zajdel, R.W., Choudhury, A., McLean, M.D., Spinner, B.J., Mehta, S., and Dube, D.K. 2000b. Differential expression of tropomyosin isoforms during development of the conus and ventricle (submitted).

Zelphati, O. and Szoka, Jr., F.C. 1996. Mechanism of oligonucleotide release from cationic liposomes. *Proc. Natl. Acad. Sci. USA* 93:11493–11498.

Zelphati, O., Uyechi, L.S., Barron, L.G., and Szoka, Jr., F.C. 1998. Effect of serum components on the physico-chemical properties of cationic lipid/oligonucleotide complexes and on their interactions with cells. *Biochim. Biophys. Acta* 1390:119–133.

PART III

Regulation of Expression of Myofibrillar Proteins

Molecular Regulation of Cardiac Myofibrillogenesis: Roles of Serum Response Factor, Nkx2-5, and GATA-4

Robert J. Schwartz, Jorge Sepulveda, and Narasimhaswamy S. Belaguli

A central theme of this chapter will be to review some of the evidence supporting serum response factor's (SRF) biological role in driving cardiac gene expression of contractile protein genes leading to cardiac myofibrillogenesis. Specification of the cardiac lineage and expression of cardiac-specified genes requires combinatorial interactions between transcription factors found to be enriched during the emergence of cardiac-progenitor cells. The homologous recombinant knockout of the murine SRF gene locus supports the observation that SRF is absolutely required for the appearance of cardiac mesoderm during mouse embryogenesis. In addition, analysis of SRF null mice revealed a severe block to the activation of SRF-regulated immediate early genes, as well as inhibited expression of the cardiac, skeletal, and smooth muscle α-actin genes.[2] Thus, analysis of mouse SRF null mutants indicates that SRF is an essential regulator of mammalian mesoderm formation and places SRF at a very high point in the regulatory hierarchy for all types of muscle cells in their commitment, differentiation, and expression of genes leading to myofibrillogenesis. Consistent with this view, we have strong evidence that SRF acts as a myogenic restricted platform to interact with other regulatory proteins and ultimately alter the regulation of muscle-specific gene programs.[7,22,25] GATA-4, Nkx2-5, and SRF are among the earliest known molecular markers of the cardiac myocyte lineage and are first expressed in the anterior lateral plate mesoderm.[26,27,47,59,62,69] In this chapter, we will also focus on mechanisms that allow SRF to interact with accessory factors. For example, we have shown that SRF facilitates binding of the murine homeobox transcription factor, Nkx2-5, to serum response elements resulting in the activation of the endogenous cardiac α-actin gene.[22-25] In addition, we and others[32,66,95] have found that Nkx2-5 can associate with the dual zinc finger transcription factor, GATA-4, to activate a variety of cardiac-specified genes. Likewise, we found that SRF recruits GATA factors to activate serum response element (SRE) containing myogenic and nonmyogenic promoters. SRF may play a leading role in the commitment of cardiac progenitors by virtue of its obligatory requirement for mesoderm formation and by its ability to make specific protein–protein associations with other early cardiac enriched transcription factors.

SRF AND MADS BOX PROTEINS

Serum response factor (SRF), a 67 kd DNA binding protein, first cloned by Dr. Richard Treisman and colleagues[84] from a HeLa cDNA library, was generally presumed to be a ubiquitous transcription factor. SRF monomers homodimerize and symmetrically contact DNA binding sites with a consensus sequence CC (A/T)$_6$GG. SRF is a member of an ancient DNA binding protein family, which shares a highly conserved DNA binding/dimerization domain of 90 amino acids, termed the MADS box. SRF, yeast transcription factors MCM1 and ARG80, and several plant proteins, such as Deficiens, all have a related ancient evolutionarily conserved MADS box with similar DNA sequence binding specificity.[67,84,97] In addition, SRF-related proteins (RSRF/MEF-2) constitute a subfamily of the MADS box family of transcription factors.[67,89] (reviewed in 12) MEF-2 factors contain the MADS box and an adjacent MEF-2 box.[12] MEF-2 proteins bind to MEF-2 sites, CTA(A/T)4TAG, which can be found in the regulatory regions of both nonmuscle- and muscle-specific genes.[12,89] Despite their similarities, MADS box proteins have evolved to perform important biological functions such as specification of mating type in yeast, homeotic activities in plants, pharyngeal muscle specification in *C. elegans*, pulmonary development in *Drosophila*, and elaboration of mesoderm structures in vertebrates. Interestingly, the overall structural divergence of SRF proteins among evolutionarily distant species of animals appears to be related to differences in their spatial expression patterns. For example, pruned, the SRF homolog in *Drosophila*, is the most divergent member, with sequence conservation limited only to the MADS box domain.[46] *Drosophila* SRF expression is maternally expressed and enriched in somatic muscle and early heart cell precursors and later is strongly expressed and localized to the tracheal system.[1,44]

SRF EMBRYONIC EXPRESSION

In vertebrates, SRF expression is restricted to tissues of mesoderm and neuroecto-derm origins.[2,5,27] During chicken embryogenesis and the progression of gastrulation, strongly localized SRF mRNA expression was observed in an asymmetric pattern at Hensen's node, along the primitive streak, the neural groove, lateral plate mesoderm, the precardiac splanchic mesoderm, the myocardium, and the myotomal portion of the somites.[27] We have shown that SRF is expressed at high levels in a symmetrically split crescent capping the extreme anterior and lateral parts of the embryo, appearing like the split cardiac progenitor cell populations described from fate mapping experiments.[27] The initial expression pattern resolves into a complete crescent and undergoes changes consistent with morphogenesis of the linear and S-shaped heart tube (see Figure 7.1). In the mouse, we showed that the highest SRF mRNA levels were seen in adult skeletal and cardiac muscles. During mouse embryonic development, SRF transcripts were found to be enriched in smooth muscle media of the vessels, the myocardium of the heart, and the myotome portion of somites.[5] The homologous recombinant knockout of the murine SRF gene has demonstrated a severe block for mesoderm formation during mouse embryogenesis.[2] These very early lethal SRF-deficient embryos, which appear to have normal cell replication, also have a severe gastrulation defect that is lethal in early embryonic stages. They consist of poorly folded ectoderm and endoderm cell layers that do not form a primitive streak or any detectable mesoderm and fail to express the very early developmental marker genes Bra (T), Bmp 2/4, and *Shh*.[2]

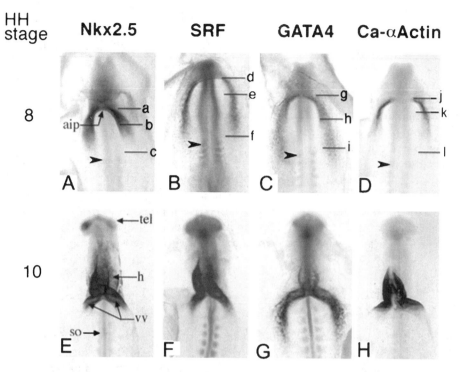

FIGURE 7.1. Expression of SRF, Nkx2-5, GATA-4, and cardiac α-actin in the developing chicken embryo. Sections of chicken embryos at H-H stages 8 and 10 were hybridized with digoxigenin labeled antisense probes specific for chicken SRF, cNkx2-5, cGATA-4, and cardiac α-actin (α-CA). In the developing chicken embryo, by H-H stage 8, cNkx2.5, cSRF, cGATA-4 and cardiac α-actin have similar patterns of expression, with the anterior intestinal portal (AIP) marking the apex of expression. However, SRF and cardiac α-actin are not expressed in the medial AIP. Thus, SRF expression appears to mark the anterior border of cardiac α-actin expression. cNkx2.5 is expressed throughout the mesoderm and endoderm of the AIP, whereas cGATA-4 is expressed only in the endoderm. The first somite marks the posterior border of cSRF and cGATA-4 expression, but cNkx2.5 expression never extends to this level. Thus, cNkx2.5 expression appears to mark the posterior border of cardiac α-actin expression. At H-H stage 10, the heart tube is fused at the midline and cNkx2.5, cSRF, and cardiac α-actin are expressed throughout the single straight heart tube, whereas cGATA-4 expression is primarily in the sinus venosus. All four genes are expressed in the myocardium. The fact that the appearance of α-cardiac actin coincides exactly with the overlapping coexpression of Nkx2-5, SRF, and GATA-4 in the developing heart is consistent with a cooperative role between the three proteins in the regulation of the cardiac α-actin gene.

SIGNALING VIA MEMBERS OF THE TGF-β FAMILY IS IMPORTANT FOR MESODERM FORMATION AND HEART DEVELOPMENT

BMPs are centrally involved in establishing early developmental pathways leading to cardiac mesoderm in higher vertebrates (reviewed in Hogan, reference 49). BMP-4 knockout mice showed little or no mesoderm differentiation and did not express the mesoderm marker, the *Brachyury* (T) gene, and disruption of BMP signals leads to direct neural induction.[49] Studies indicate that mouse mutants lacking BMP-2 form a disorganized heart that does not undergo looping morphogenesis.[106] Furthermore, a significant portion of BMP-2 null mutant embryos do not express *Nkx2-5*, the earliest known vertebrate cardiac lineage marker.[106] Schultheiss et al.[94] showed that ectopic application of BMP-2, -4, and -7 to regions of chick embryos not usually specified to become heart tissue allowed for the induction of Nkx2-5 and GATA-4. Application of noggin, an antagonist of BMPs, to chick embryos completely inhibits differentiation of the precardiac mesoderm.[94] To elicit beating heart cells, Lough et al.[70] found that treatment of noncardiac mesoderm with BMPs alone was not sufficient but required FGF-4 in addition to support cardiogenesis. Lough et al.[70] reported that the combined treatment with BMP-2 and FGF-4 was required to induce the appearance of Nkx2-5 and SRF in naive mesoderm; neither Nkx2-5 nor SRF could be induced by either growth factor alone. Indeed, this observation was confirmed by Dr. Matt Barron, a postdoctoral fellow in our laboratory, in experiments in which SRF transcripts were induced in the anterior medial region of a stage 8 chick embryo with beads loaded with BMP-2 and FGF-4 (data not shown).

SRF, AN OBLIGATORY MYOGENIC TRANSCRIPTION FACTOR

The regulatory regions of a number of muscle-specific genes such as skeletal,[64,65,75] cardiac,[17,22-25,75-77,93] smooth muscle α-actin,[18,20,47,72] and other myogenic-specified genes contain multiple SREs that are required for promoter activity and depend upon SRF for activity. Mutations that prevent SRF binding severely impair the expression of *c-fos,* as well as these muscle-restricted promoters. In addition, high levels of SRF expression and increased SRF mass appear to coincide with the expression of cardiac, skeletal, and smooth muscle α-actins[5,27] as early markers for terminal striated and smooth muscle differentiation. SRF also transactivates α-actin gene transcription under conditions that block myogenic differentiation.[68] SRF expression is actually suppressed in replicating myoblasts, but SRF mRNA and protein levels increase approximately 40-fold during the progression from replicating primary myoblast cultures to nonreplicating postdifferentiated myotubes.[27] SRF protein levels precede the induction of skeletal α-actin and smooth muscle α-actin expression during myogenic differentiation.[18,27] Vandromme et al.[103] and Soulez et al.[98] demonstrated that the microinjection of SRF antibodies and SRF antisense oligonucleotides prevented the myoblast–myotube progression, the expression of the myogenic factor myogenin, and terminal differentiation, thus underscoring the obligate role for SRF in skeletal muscle differentiation. We showed that an SRF dominant-negative mutant (SRFpm1), which has three point mutations in its DNA binding domain, converted basic amino acids to neutral-charge amino acids at positions aa143 arginine to isoleucine, aa145 lysine to alanine, and aa146 lysine to glycine. SRFpm1 retains dimerization properties and blocks transcriptional activation of the skeletal α-actin gene.[27] We observed that stable transfections of the Sol 8 and C2C12 myogenic cell lines with SRFpm1 allowed for cell replication but inhibited myoblast fusion and postreplicative myogenic differentiation.[27] Taken together, these observations clearly support a tissue-restricted pattern of SRF gene activity and an obligatory role for SRF during striated and smooth muscle differentiation.

SRF ALTERNATIVE SPLICING

We have shown that murine SRF primary RNA transcripts are alternatively spliced at the fifth exon, deleting approximately one-third of the C-terminal activation domain, described as SRFΔ5.[6] Among the different muscle types examined, visceral smooth muscles have a very low ratio of SRFΔ5/wild-type SRF. Increased levels of SRFΔ5 correlate well with reduced smooth muscle contractile gene activity within the elastic aortic arch,[6] suggesting important biological roles for differential expression of SRFΔ5 variant relative to wild-type SRF. SRFΔ5 forms DNA binding competent homodimers and heterodimers. SRFΔ5 acts as a naturally occurring dominant-negative regulatory mutant that blocks SRF-dependent skeletal, cardiac, and smooth α-actins, SM22 α, and SRF promoter luciferase reporter activities.[6] These results indicate that the absence of exon 5 might be bypassed through recruitment of transcription factors such as ATF6 that interact with extra-exon 5 regions in the transcription activation domain. The novel alternatively spliced isoform of SRF, SRFΔ5, might serve an important regulatory role in modulating the differentiation of contractile protein gene expression. We have observed SRFΔ5 transcripts during embryonic stem-cell induced differentiation into committed cardiac myocytes, but we do not yet know whether alternatively spliced SRF species play a regulatory role during cardiogenesis and the initiation of heart failure.

MADS BOX, A NOVEL DNA BINDING MOTIF

SRF is a key regulator of immediate early gene expression, which frequently results in mitogenesis, and also of terminal differentiation. How can SRF play two apparently opposite roles? The answer is related in part to the current view in which SRF serves as a versatile protein that binds to its cognate sites in a multitude of promoters and serves as a docking surface for the binding of many different accessory proteins that confer specific functional abilities to the promoter.[25,41,42] The recent elucidation of the x-ray crystal structure of the SRF core bound to DNA[88] provides an explanation for mutually inclusive binding of coaccessory factors to a single SRE. As shown schematically in Figure 7.2, the coiled coil formed by the MADS box αI helices (aa153 to aa179) lies parallel and on top of a narrow DNA major groove making contacts with the phosphate backbone on an SRE half-site. In addition, the unstructured N-terminal extension from the αI helix (aa132 to 152aa) makes critical base contacts in the minor groove. Dimerization of the MADS box occurs above the αI helix by a structure composed of two β-sheets in the monomer that interact with the same unit in its partner. A second αII helix in the C-terminal portion of the MADS box, stacked above these β-sheets, completes this stratified structure. As we will point out in this review, proteins touching the MADS box at either the first or second α-coil may play important roles in determining whether gene activity is directed toward growth or cell differentiation.

SRF ACCESSORY PROTEIN INTERACTIONS

Studies indicate the versatile nature of SRF, which can act as a platform to interact with other regulatory proteins and ultimately the regulation of specific gene programs. Most SRF accessory proteins identified are coregulators of c-fos induction

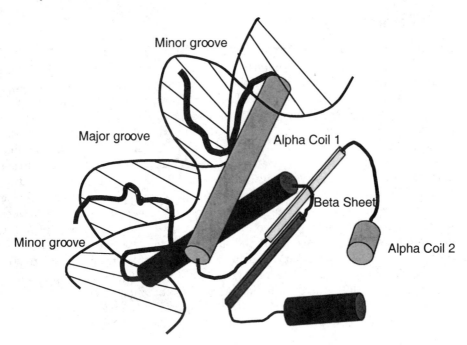

FIGURE 7.2. Schematic diagram of SRF MADS box binding to its DNA binding element. This figure is adapted from L. Pellegrini, S. Tan, and T.J. Richmond,[88] who described the X-ray crystallographic structure of serum response factor core bound to DNA. The paired monomers form a coiled coil by the MADS box αI helices (aa153 to aa179) which lay on top of a narrow DNA major groove making contacts with the phosphate backbone on an SRE half-site, while an unstructured N-terminal extension from the αI helix (aa132 to aa152) makes critical base contacts on paired Gs in the minor groove. Dimerization of the MADS box occurs above the αI helix by a structure composed of two β-sheets in the monomer that interact with the same unit in its partner. A second αII helix in the C-terminal portion of the MADS box, stacked above these β-sheets, completes this stratified structure.

and act as endpoints of signal transduction cascades, often leading to mitogenesis. For example, the Ets-related ternary complex factor, Elk-1, interacts with the SRF bound to the c-fos SRE and with the SRE in a cooperative manner.[96] Phosphorylation of Elk-1 by mitogen-responsive JNK and ERK groups of mitogen-activated protein (MAP) kinases causes increased DNA binding, ternary complex formation, and transcriptional activation.[104] Activated Elk-1 touches the αII coil in the MADS box.[88,96] Conversely, the mesoderm-restricted paired homeobox protein Phox-1/MHOX/ Prx-1 binds to SRF in a competitive manner with Elk-1, inhibiting ternary complex formation and induction of c-fos by mitogens.[68] However, in HeLa cells, Phox-1 imparts serum-responsive transcriptional activity to SRF binding sites[42] in a model of synergistic cooperation requiring the stabilization of the SRF–Phox-1 complexes by the Inr binding protein, TFII-I.[41] Interestingly, Phox-1 stimulates transcriptional activity of SRF at some muscle-specific promoters, possibly by stabilizing the binding of SRF to these lower-affinity SREs.[41]

In contrast to the c-fos gene, which contains a single high-affinity binding site for SRF in its promoter, many muscle-specific genes, including skeletal, cardiac, SM α-actin, SM MHC, and SM22 contain two or more low-affinity SREs that bind SRF in a cooperative manner (see reference 64). The precise role of Prx-1/Phox-1 and its relative Prx-2/S8 in myogenesis is not clear because they fail to activate α-cardiac actin in cooperation with SRF while they promote transcription of smooth muscle actin and muscle creatine kinase.[29,47] It is possible that these homeobox genes play a complex role in either inducing or repressing certain SRE-responsive genes, depending on the cellular and signaling environment.

TINMAN AND NKX2-5/CSX HOMEODOMAIN GENES EXPRESSED IN THE HEART

Studies on *Drosophila* have shown that *tinman* (msh2, NK4) is initially expressed in the presumptive mesoderm at the cellular blastoderm stage and continues to be expressed in all mesodermal cells during germ-band elongation. Later, expression becomes restricted to the developing dorsal vessel, the insect equivalent of the vertebrate heart, and visceral mesoderm.[3,14] Mutations in the *tinman* gene do not affect mesoderm invagination or dorsal spreading but result in loss of heart formation in the *Drosophila* embryo. In addition, *tinman* is known to regulate NK3 (*bagpipe*) expression in the visceral mesoderm,[3] related to Nkx3-1[10] and the expression of the *Drosophila* MEF-2 factor, dMEF-2 gene, required for muscle differentiation and late cardiogenesis.[38] These observations suggest that *tinman* may be involved in cardiac mesoderm patterning and makes it a likely inducer of cardiac mesoderm.

The vertebrate *tinman* relative, Nkx2-5, also appears to be instrumental in the patterning of the vertebrate embryonic heart. Nkx2-5 is expressed in early cardiac progenitor cells, prior to cardiogenic differentiation, and continues through adulthood.[9,59,69] Nkx2-5 genes identified in other vertebrates such as zebrafish,[26,63] *Xenopus*,[101] and chickens[92] are highly conserved and exhibit expression patterns that demarcate the heart field.[26] In the mouse embryo, expression of Nkx2-5 is first noted around the time of the precardiac mesoderm, which undergoes a transformation from mesenchyme to a cuboidal epithelium, the first physical sign of the committed state.[59,69] Nkx2-5 seems to be insufficient to specify myocardial cell fate determination by itself, but appears to be sufficient to control assymetric expression of eHand.[8] High-level ectopic overexpression does cause low-level expression of cardiac-specific genes; these cells do not beat, suggesting the need for at least one more signal.[25,26,63]

Embryos homozygous for the disrupted Nkx-2-5 allele displayed heart morphogenetic defects at embryonic day 8.5.[71] A beating linear heart tube developed, but looping morphogenesis, a critical determinant of heart formation, was not initiated in Nkx2-5 null mice. Although these mice die at 9 days of embryonic development, the expression of several cardiogenic-restricted genes marking the differentiation of early cardiac myocytes did not appear to be affected. However, expression of ventricular myosin light chain 2 (MLC-2v), ANF, SM22/transgelin, CARP, and eHand was abolished or spatially deranged. These genes are thus candidates for direct or indirect targets of Nkx2-5 in cardiogenesis. The Cardiac Ankyrin Repeat Protein (CARP) is a cardiac-restricted transcription factor involved in the expression of the ventricle-specific myosin light chain 2 (MLC-2v) gene.[108] It is interesting that the expression of both CARP and MLC-2v is reduced in Nkx2-5 null mice,[108] thus suggesting one of the pathways.

THE NKX CODE

The existence of at least nine Nkx-2 family members, their overlapping DNA binding specificity, and most importantly their partially overlapping patterns of expression raise the possibility of an Nkx code. The term Hox code means that a particular combination of Nkx genes is functionally active in a region and thereby specifies the developmental fate of this region (see reference 92). During early heart formation, Nkx2-3, 2-5, 2-7 and 2-8 are expressed in partially overlapping domains in the lateral plate mesoderm and pharyngeal endoderm.[16,34,86,92] However, after the linear heart tube has undergone the looping process, Nkx2-3, 2-5, 2-6, and 2-8 are expressed in distinct and partially overlapping domains. In the pharyngeal region, Nkx2-3 and 2-5 are expressed in broad, partially overlapping domains primarily within the ectoderm and endoderm.[19] By contrast, Nkx2-1 and Nkx-2.8 expression is spatially more restricted; transcripts are found at the site at which the anlagen for thyroid, thymus, parathyroid, and lung reside. These data are consistent with a model called "The Nkx code" in which each Nkx member performs a unique temporally and spatially restricted function in the developing embryonic heart and pharyngeal region. It is possible that the position and identity of these organ rudiments are determined by the combinatorial expression of Nkx genes.

DOWNSTREAM TARGETS OF NKX2-5

The three-dimensional structure of the 60 amino acid Nk homeodomain is comprised of three α helices, in which helix II and helix III form a helix-turn-helix motif. DNA binding assays shown in Chen and Schwartz[24,25] indicated that Nkx2-5 bound as a monomer to various DNA targets, probably using the recognition helix (helix III), which fits across the AT-rich major groove of the DNA binding site.[31,43] DNA binding assays have also revealed that both Nkx2-1, also named thyroid transcription factor,[31,43] and Nkx2-5 prefer DNA sequences containing 5'-TNAAGTG-3'[24,33,91] named NKEs. However, Nkx-2 factors also bind with weaker avidity to a second subset of sequences that contain the 5'-TTAATT-3' motif, recognized by Hox proteins.[24,33] Tyrosine 54, normally lying on the external face of the third helix in the unbound state, becomes positioned to make crucial contacts with the 5'AAG 3' core of the binding site as the helix lengthens upon DNA binding.[45] NKEs are present as multiple sites in lung Clara cell-specific protein (mCC10). Ray et al.[91] showed that Nkx-2-1 and Nkx-2-5 bound the same NKEs required for mCC10 gene promoter activity. Given the abundance and relatively low specificity of the NKEs, it is reasonable to postulate that tissue and developmental specificity is enhanced by interactions of the DNA-bound Nkx-2 factors with other proteins, in the context of particular promoter structures.

SREs of the cardiac α-actin promoter served as both high-affinity (SRE2 and SRE3) and intermediate-strength binding targets (SRE1 and SRE4) of Nkx2-5.[22,25] Durocher et al.[33] showed that two NKE sites on the proximal region of the cardiac atrial natriuretic factor (ANF) promoter directed high levels of cardiac-specific promoter activity. The ANF NKE is composed of two near-consensus NK2 binding sites that are each able to bind purified Nkx2-5. The NKE is sufficient to confer cardiac-cell–specific activity to a minimal TATA-containing promoter and is required for Nkx2-5 activation of the ANF promoter in heterologous cells. Interestingly, in primary cardiomyocyte cultures, the NKE contributes to ANF promoter activity in a chamber- and developmental-stage–specific manner, suggesting that Nkx2-5 and/or other related cardiac proteins may play a role in chamber specification.[33] Gajewski et al.[38] demonstrated that two NKE promoter sites direct *D-mef2* expression in response to *tinman*.

Nkx2-5 serves as a modest transcription activator in transfection assays when analyzed with reporter genes carrying multimerized NKE binding sites. However, deletion of the Nkx-2-5 C-terminal inhibitory domain stimulated transcriptional activity over 50-fold.[24] Both the inhibitory domain and the highly charged activation domain of Nkx2-5 are enriched in alanine and proline. These findings suggest that potential hydrophobic interactions between the inhibitory and activation domains might block access of transcription initiation factors to the highly charged moiety. It is likely that interactions with other coaccessory factors might cause conformational changes in the Nkx2-5 protein to cause it to become a more effective transcriptional activator. In fact, Nkx2-5 does not appear to work in isolation but forms combinatorial binding complexes with other transcription factors, such as SRF.

INTERACTIONS BETWEEN SRF AND NKX2-5

As mentioned earlier, SRF is highly enriched in embryonic cardiac progenitors and plays a mandatory role in myogenic differentiation.[27] Therefore, studies were initiated to identify possible cardiac-restricted accessory proteins that might be involved in cardiac differentiation. Nkx2-5 was a strong candidate, given the paradigm of Phox-1/SRF interaction in c-fos activation and the role of Nkx2-5 in cardiogenesis. Transient transfection of Nkx2-5 and SRF resulted in about a 20-fold activation of the cardiac α-actin promoter in fibroblasts, while stable transfection of both factors resulted in expression of the endogenous cardiac α-actin gene.[35–38] We asked how this mechanism was mediated. In Figure 7.3, we showed that DNA binding to multiple SREs is a primary feature of Nkx2-5, and that Nkx2-5 and SRF together elicited robust activation of the cardiac α-actin promoter in 10T1/2 fibroblasts. Analysis of the α-cardiac actin promoter mutants in cotransfection experiments indicated that maximal transcriptional activity required multiple intact upstream SREs. The deletion mutant, del-100, which removed three of the four SREs, caused a 70% reduction in overall cotransfection activity. The proximal SRE1 appears to play the central role in regulating SRF/Nkx-2.5-dependent promoter activity, as demonstrated by the complete loss of promoter function by the mutant SRE1M, as shown in Figure 7.3. Therefore, even though Nkx2-5 can bind weakly to some SREs, we found that activation of a minimal promoter consisting of a single SRF binding site was dependent upon increasing the cellular levels of SRF.[25] When Nkx2-5 binding activity was blocked by a point mutation in the third helix of the homeodomain, SRF was still capable of recruiting mutated Nkx2-5 to the cardiac α-actin promoter.[25] Investigation of protein–protein interactions demonstrated that Nkx2-5 could bind to SRF in the absence of DNA as soluble protein complexes isolated from cardiac myocyte nuclei. In addition, Nkx2-5 and SRF complexes could be detected as coassociated binding complexes on the proximal SRE1.[25]

Recruitment of Nkx2-5 to an SRE was dependent upon SRF DNA binding activity and could be blocked by the dominant-negative SRFpm1 mutant, which dimerizes with wild-type SRF monomers but cannot itself bind to DNA. In addition, Nkx-2.5 protein and SRF interact directly in the absence of the SRE. A short 30 amino acid peptide (amino acids 142 to 171), which encompasses the basic region of

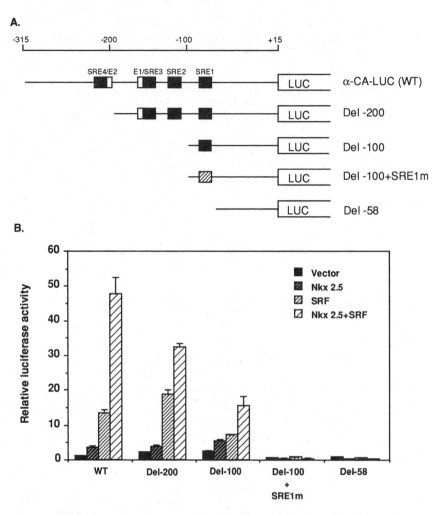

FIGURE 7.3. Multiple intact SREs are required for the coactivation of cardiac actin promoter by Nkx2.5 and SRF. In panel A, deletion and site-directed mutants of the avian cardiac actin promoter were shown above. Dependence upon a single intact SRE for transfactor activation was evaluated by site-directed mutagenesis of the proximal SRE1 (Del-100+ SRE1m). SRE1 was converted from (−86) CCAAATAAGG (−76) to CCAAAGATCT. In panel B, transfection analysis of various reporter constructs in the presence of Nkx2.5 and SRF. Single SRE is sufficient for coactivation by Nkx2.5 and SRF, although the promoter activity is lower than that of the entire promoter.

SRF in the αI coil of the MADS box, is sufficient to mediate protein–protein contacts with the Nkx-2.5. The N-terminus/helix I and helix II regions of the Nkx2-5 homeodomain interact with the MADS box, as modeled in Figure 7.4.[25] The significance for Nkx-2.5 and SRF interaction is demonstrated by cotransfection experiments, in which Nkx-2.5 and SRF transactivate promoters consisting of either the SRF or Nkx-2.5 binding site. The fact that Phox-1/SRF[41,42] and Nkx2-5/SRF interactions all required the amino terminal arm/helix 1/helix 2 region of the homeodomain lead us to propose a model that amino terminus and helix 3 are responsible for homeodomain–DNA interactions. Helix 1/2 is probably mediating

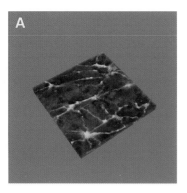

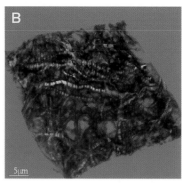

FIGURE 3.1. Three-dimensional representation of confocal microscope data sets from whole mount preparations of embryonic chicken heart at the 8 somite (A) and the 12 somite (B) stages, stained with polyclonal antibodies to MyBP-C (myosin binding protein C, green) and with rhodamine-phalloidin to visualize F-actin (red). In the heart before the start of beating, actin runs in filaments along the cell membranes, and A-band components such as MyBP-C are distributed in a diffuse fashion throughout the cytoplasm (A). Only a couple of hours later, the actin filaments stretch through the cytoplasm and clear double bands can be observed for MyBP-C (B).

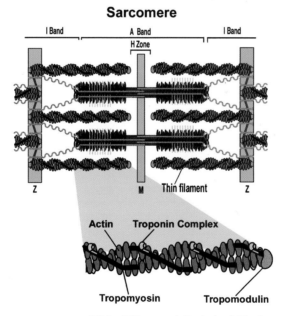

FIGURE 4.1. Schematic of the known organization of the major components of a cardiac sarcomere (a single contractile unit). Inset is of the pointed end of a thin filament at a higher magnification. The boundaries of the sarcomere are described by the Z-lines. The actin filaments are made up of actin monomers (red). The barbed ends are capped by capZ (pink) and embedded in the Z-line. The pointed ends are capped by tropomodulin (turquoise). Tropomyosin dimers bind head-to-tail, forming two polymers that wrap around each thin filament (black lines). One troponin complex (yellow) binds to each tropomyosin dimer. The thic filaments are made up of bipolar myosin filaments (dark green) and myosin binding protein C (light green). The N-terminus of the giant protein, titin (orange), is crosslinked to the thin filaments in the Z-disk by α-actinin (purple) and interacts with Tcap (green). Note that, for simplification, one tropomodulin molecule is depicted per thin filament. Whether there are one or two tropomodulins per thin filament awaits further clarification.

Sarcomere

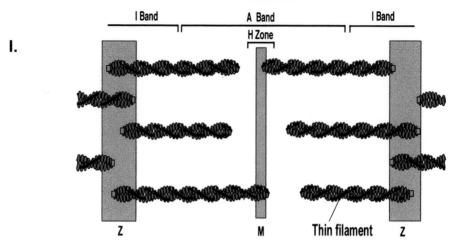

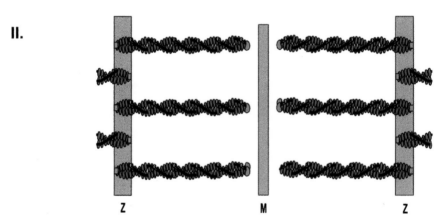

FIGURE 4.3. A model for cardiac thin filament assembly. (I) The thin filament proteins, actin (red), tropomyosin (black lines), and the troponin complex (yellow) are assembled. The barbed end is capped by capZ (pink). The thin filaments are of variable lengths, and the pointed ends are uncapped. (II) The thin filaments that are too long have been disassembled, and those that are too short have been lengthened. Most of the thin filaments are capped by one or two tropomodulin molecules, which maintains their lengths. For simplicity, the myosin-containing thick filaments, Z-filaments, and titin filaments have been omitted.

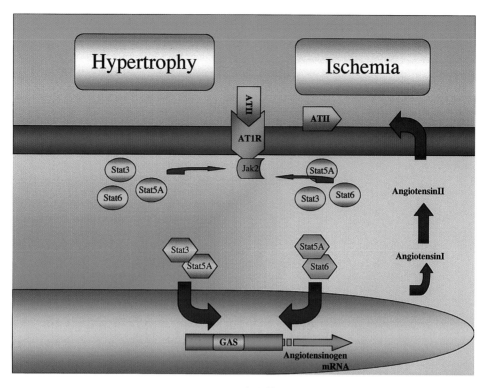

FIGURE 9.1. Diagrammatic representation of differential mobilization of STAT factors responding to signals associated with myocardial hypertrophy and ischemic injury.

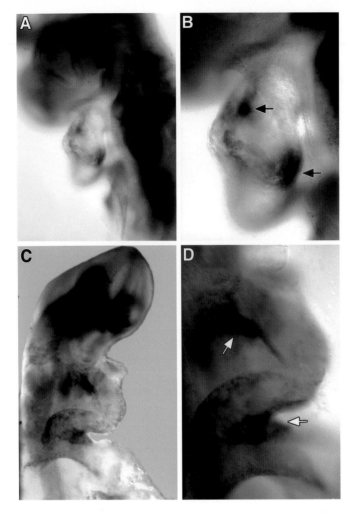

FIGURE 11.9. WNT2 and WNT14 RNA expression in the tubular heart of the chick. Whole mount in situ hybridization used antisense RNA probes to detect either WNT2 mRNA in H-H stage 15 chick embryos (A,B) or WNT14 mRNA in H-H stage 12 chick embryos (C,D). Low (A) and high (B) magnification views of embryonic WNT2 distribution show that expression of this molecule in the developing heart localizes exclusively to the outflow tract and atrioventricular junction (arrows). WNT14 expression, as shown at low (C) and high (D) magnifications, is also restricted to the outflow tract and atrioventricular segments (arrows), but at an earlier stage than WNT2. Because the primary heart tube is not fully formed at this earlier stage, the region-specific pattern of WNT14 expression may indicate that it plays a role either in establishing segmental identity and/or segment formation. The later expression of WNT2 suggests that it may be involved in the remodeling of these respective heart segments.

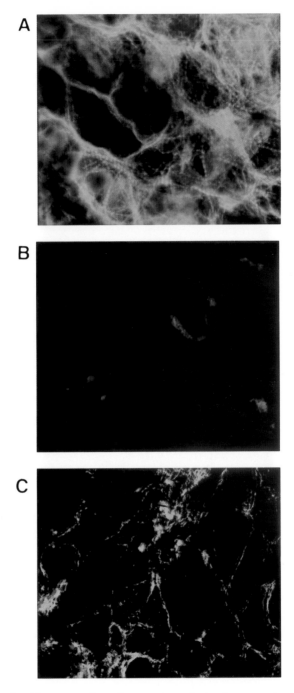

FIGURE 12.1. Confocal laser scanning microscopic analysis of sarcomeric tropomyosin distribution in embryonic axolotl hearts. Normal heart cultured in Steinberg's solution has well-organized sarcomeric myofibrils (upper, A). *Cardiac* mutant heart cultured in Steinberg's solution does not beat and has almost no staining for tropomyosin (middle, B). *Cardiac* mutant heart incubated with RNA synthesized from Clone #4 by T7 polymerase exhibits rhythmic contractions and enhanced labeling for sarcomeric tropomyosin (lower, C).

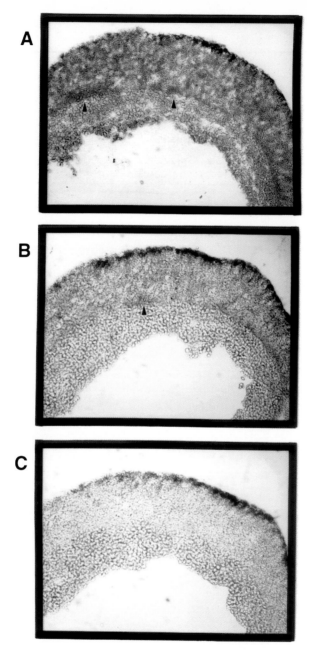

FIGURE 12.4. In situ hybridization analysis of Clone #4 RNA in axolotl embryos. Paraffin sections of the axolotl embryos were treated with digoxygenin-labeled probes to Clone #4. Intense staining was observed in sections from stage 15 embryos stained with the antisense probes (A). The staining with the sense probes was significantly lower (B), and staining was not observed with the no-probe control (C).

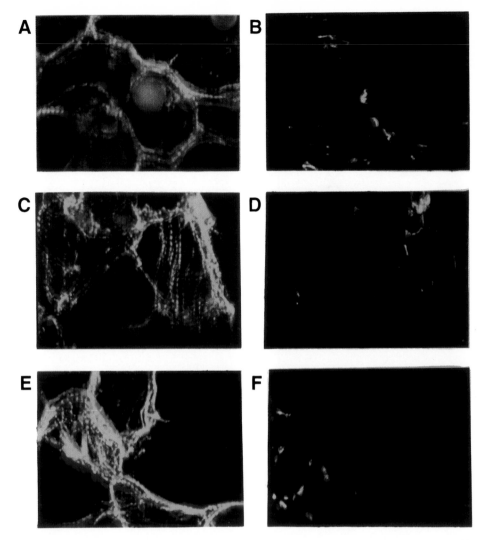

FIGURE 12.6. Effects of Clone #4 RNA and the mutant Clone #4 RNA on promoting myofibrillogenesis in mutant axolotl hearts. Confocal laser scanning microscope analysis of tropomyosin immunostaining of axoltl hearts after two days of lipofectin-mediated transfection followed by 3 days of incubation in normal Steinberg's solution. (A) Normal heart in Steinberg–lipofectin solution. Myofibrils containing tropomyosin are abundant. (B) Mutant heart in Steinberg–lipofectin solution. Myofibrils containing tropomyosin are few. (C) Mutant heart transfected with T7 sense RNA (nt 166) from Clone #4. Myofibrils containing tropomyosin are present, evidence for the bioactivity of the RNA. (D) Mutant heart transfected with Sp6 antisense RNA from Clone #4. Myofibrils containing tropomyosin are few, indicating no significant correction of the mutant heart cells by this RNA. (E) Mutant heart transfected with T7 sense RNA (nt 142) from mutant Clone #4, which has a deletion of 24 base pairs from the 3′ end. Myofibrils containing tropomyosin are present, suggesting that the bioactivity of Clone #4 RNA has not been reduced due to the deletion. (F) Mutant heart transfected with T7 sense RNA (nt 90) from mutant Clone #4, which has a deletion of 76 base pairs from the 3′ end. Myofibrils containing tropomyosin are few, indicating that the bioactivity of Clone #4 RNA has been modified due to the deletion.

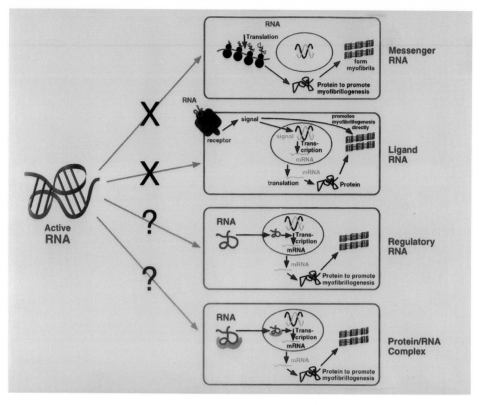

FIGURE 12.8. Model of hypothetical mechanisms by which Clone #4 RNA could rescue mutant hearts.

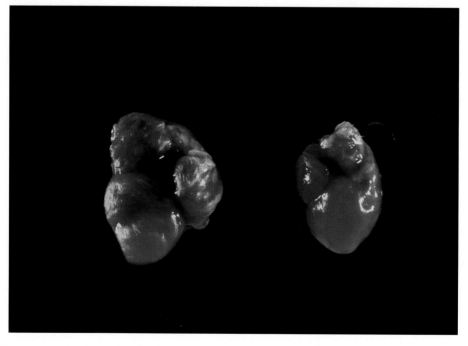

FIGURE 13.5A. Control and α-TM FHC 180 mouse hearts at four months of age. Panel A shows the whole hearts of control (right) and FHC mutant (left) mice. Notice the enlarged atria and left ventricle.

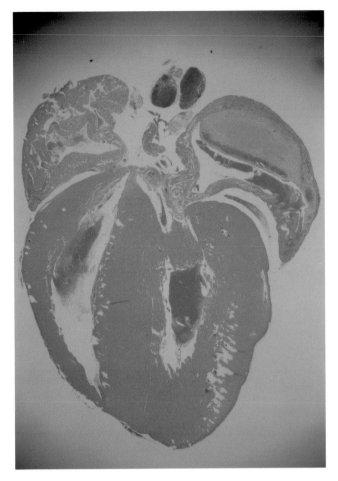

FIGURE 13.5B. Control and α-TM FHC 180 mouse hearts at four months of age. Panel B shows a cross section of a four-month-old α-TM FHC 180 heart. Note the thickening of the interventricular septum and the left ventricular wall. Also, both atria exhibit large centrally located thrombi that have formed within them.

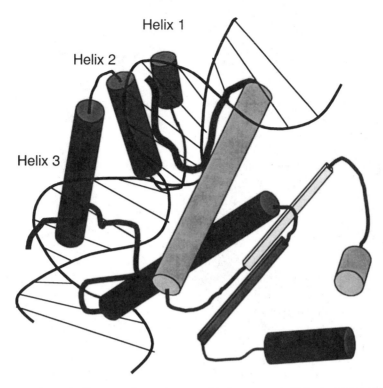

FIGURE 7.4. Schematic diagram of recruitment of Nkx2-5 to an SRE via SRF DNA binding activity. Nkx2.5 protein and SRF interact directly in the absence of the SRE. A short 30 amino acid peptide (aa142 to aa171), which encompasses the basic region of SRF in the α1 coil of the MADS box, is sufficient to mediate protein–protein contact with the Nkx-2.5. The N-terminus/helix I and helix II regions of the Nkx2-5 homeodomain interact with the MADS box. The fact that Phox-1 SRF[43,44] and Nkx2-5/SRF[25] interactions all required the amino terminal arm/helix 1/helix 2 region of the homeodomain lead us to propose a model that helices 1/2 are responsible for homeodomain-SRF interactions.[25]

protein–protein interactions of HOX genes among themselves and with other protein factors, which might be an essential requirement for regulating their specificity of action as activators, coactivators, or repressors.

Under appropriate extracellular signals in the HeLa cell background, the formation of the SRF/Phox-1/Elk-1 tertiary complex on the c-fos SRE can transactivate the c-fos promoter. In contrast, Phox-1 and Elk-1 cannot activate the cardiac α-actin promoter.[25] Possibly, during early cardiogenesis, paired-like homeodomain genes such as Prx-1/Phox-1/MHox or Prx-2/S8[68] form nonproductive inhibitory complexes with SRF on the α-actin promoter SREs. The high level of S8 expression in the endocardial cushions, in fact, correlates well with regions of the heart that do not express Nkx2-5, α-actin genes, and other contractile genes. Thus, SRF/MHox complexes may serve a repressive role to block contractile activity in regions of the heart that will form septations and valves. Possibly, Nkx2-5 might actually compete, through its SRF-interactive subdomains, with Phox-1 for SRF MADS box binding. The outcome of Nkx2-5 and SRF interactions may simply preclude Phox-1 or Elk-1 from the complex and result in activation of the cardiac α-actin promoter in cardiac myocytes. Conversely, the Nkx2-5/SRF complexes that are activated in nonreplicating cardiac myocytes might serve to repress the c-fos promoter through the formation of nonproductive complexes on its SRE. Consistent with this idea, Nkx2-5 blocked the serum-inducible expression of a minimal c-fos SRE-containing promoter in transient transfection assays (Belaguli and Schwartz, unpublished observation).

Therefore, in cardiac myocyte precursors, cardiac α-actin gene activation may require increased levels of SRF in combination with the coappearance of Nkx2-5 to foster cooperative transactivation complex formation. We have shown that the increase in nuclear localized SRF/Nkx2-5 complexes competed off negative-acting factors such as YY1, allowing for the cooperative SRF binding and saturation of the multiple SREs with positive-acting SRF-complexes that activated the cardiac α-actin promoter.[23]

GATA FACTORS

The zinc finger containing GATA factors[41,53,84,86–89] is also high in the hierarchical order of regulatory factors that might specify the cardiac cell lineage. The GATA family has been subdivided, with GATA-1/-2/-3 being linked to hematopoiesis, whereas GATA-4/-5/-6 are thought to be involved with cardiac, gut, and blood vessel formation (reviewed in reference 87). Each of the six GATA proteins contains a highly conserved DNA binding domain consisting of two C4 zinc fingers of the motif Cys-X2-Cys-X17-Cys-X2-Cys.[39,62] These two zinc fingers have been shown to direct binding to the DNA sequence element (A/T)GATA(A/G),[58,74] although the carboxy-terminal zinc finger is sufficient for site-specific binding.[105] Examination of the DNA binding site specificities of all six GATA factors indicated that they are capable of binding to the same target sequence, thus suggesting their ability to substitute for one another in cells where they are coexpressed. The GATA-4/-5/-6 subfamily of transcription factors are expressed in an overlapping pattern in the extraembryonic endoderm, precardiac mesoderm, embryonic and adult heart, and gut epithe-lium.[48,53,55,62,81,82] However, whereas GATA-4 and GATA-6 are developmentally co-expressed in the precardiac mesoderm and embryonic heart,[87,89] GATA-5 has a temporally and spatially distinct pattern of expression from that of GATA-4 and GATA-6 during embryonic cardiac development[82] (see also Figure 7.1). In addition, murine GATA-4/-5/-6 proteins are each expressed in a unique cell-lineage–restricted pattern in tissues including the lung, bladder, and vascular smooth muscle cells.[81–83] Taken together, these data strongly suggest that whereas GATA-4 and -6 may serve partially or completely redundant functions in the developing vertebrate heart, each member of the GATA-4/-5/-6 subfamily of transcription factors performs a unique function during vertebrate development.

GATA-4 has been found to be expressed in a developmentally and lineage-specific pattern within the cardiac mesoderm and is coexpressed with Nkx2-5 and SRF in the nascent myocardial cells. Experiments have shown that GATA-4 regulates expression of cardiac-specific genes, such as cardiac troponin C[49] and cardiac α-myosin heavy chain.[79] For example, injections of GATA-4 DNA into Xenopus oocytes resulted in premature expression of cardiac-specific myofibrillar proteins.[53] Forced expression of antisense DNA for GATA-4 blocked expression of retinoic acid-inducible, cardiac-specific genes in pluripotent P19 embryonal carcinoma cells.[39] GATA-4 null mice display a severe defect in formation of the cardiac tube, required for the migration and folding morphogenesis of the precardiogenic splanchnic mesoderm.[60,78] Rather normal expression of cardiac-specific genes was observed in these homozygous GATA-4 knockout embryos. Taken together, these results indicate that GATA-4 is not essential for terminal differentiation of cardiomyocytes and suggest that additional GATA binding proteins known to be in cardiac tissue, such as GATA-5 or GATA-6, may compensate for a lack of GATA-4, probably reflecting redundancy of some functions in the GATA-4/-5/-6 subfamily.

GATA-4 AND NKX2-5 INTERACTIONS

It has been shown that Nkx2-5 and GATA-4 are capable of coactivating the ANF promoter through binding to their adjacent DNA binding sites.[32] Binding of both Nkx2-5 and GATA-4 was required for synergistic coactivation.[32] We demonstrated that physical and functional interactions between Nkx2-5 and GATA-4 result in dramatic transcriptional activation of target promoters that contain only NKEs. Because these two transcription factors are coexpressed in the precardiac mesoderm, it is likely that this interaction plays a significant role in regulating transcriptional activity during early cardiogenesis in the embryo.

Our work examined the functional and physical interactions between two cardiac-restricted transcription factors, Nkx2-5 and GATA-4, which are coexpressed during the earliest stages of cardiogenesis. Cotransfection experiments in CV1 fibroblasts showed dramatic activation of cardiac-specific promoters only when Nkx2-5 and GATA-4 were cotransfected, in contrast to when the transcription factors were transfected alone. This synergistic activation attained powerful levels, 500-fold over baseline, when a promoter with the trimeric Nkx binding site A20[25] was used in cotransfection assays, as compared to the 15- to 20-fold activation observed with the cardiac α-actin promoter. The synergistic activation levels may be amplified by the absence of inhibitory sequences in the A20[25] minimal promoter, in contrast to the more complex cardiac α-actin and α-MHC promoters. Alternatively, the high level of activation may simply reflect the increased affinity of Nkx2-5 for the A20 site in comparison with the lower-affinity NKE sites present in the cardiac α-actin and α-MHC promoters.

Transfection experiments with deletion mutants show that at least the homeobox of Nkx2-5 and the second zinc finger of GATA-4 are required for potent transcriptional coactivation through an Nkx2-5 binding site.[95] Moreover, a mutant containing only the second zinc finger of GATA-4 and its C-terminal extension is sufficient for 200-fold coactivation with Nkx2-5. On the other hand, the N-terminal activation domain of Nkx2-5 is responsible for most of the transcriptional activity. However, the isolated homeodomain is still able to cooperate with GATA-4 to produce a modest activation at the A20(3) site. In contrast, SRF was unable to synergistically activate the same promoter with a truncated Nkx2-5 mutant containing only the homeobox.[95] This discrepancy may relate to differences in the function of the activation domains of the SRF and GATA-4 when complexed with the Nkx2-5 homeodomain. Physical mapping by pulldown assays with purified proteins confirmed that the minimal interacting domains consist of the C-terminal zinc finger of GATA-4 and the homeodomain of Nkx2-5.

Transfection data revealed that the effect of GATA-4 on full-length Nkx2-5 is very similar to the changes observed with C-terminal–deleted Nkx2-5 mutants. Because the C-terminal region has been previously shown to act as a transcriptional inhibitor,[95] these data suggest that GATA-4 may cause a conformational change in

Nkx2-5 that relieves the negative constraints exerted by the inhibitory Nkx2-5 C-terminal domain. What is the nature of this putative conformational change? We favor a model, shown in Sepulveda et al.,[95] in which activation of Nkx2-5 is through its homeodomain's interaction with GATA-4, which then causes a change in the physical ordering of Nkx2-5. Perhaps changes in Nkx2-5 protein shape elicited by the touching of the homeodomain with GATA-4's second zinc finger displace the inhibitory C-terminal domain, which then further improves DNA binding activity and reveals the N-terminal activation domain. In additional support of this model, GATA-4 facilitates Nkx2-5 DNA binding activity but is unable to improve the binding efficiency of a mutant that contains only the Nkx2-5 homeodomain. Thus, GATA-4 interaction with Nkx2-5 might evoke strong cotransactivation by removing physical impediments that occlude the Nkx2-5 N-terminal activation domain, by increasing DNA binding affinity, and/or by enhancing interactions with ancillary proteins. The interaction of GATA-4 with Nkx2-5 to induce a conformational change that increases the transcriptional activity of Nkx2-5 might be similar to the effects of Extradenticle and Pbx1 on the DNA binding affinity and transcriptional activity of engrailed and Hox proteins.[57]

Studies by Durocher et al.[32] and Lee et al.[66] have shown that the atrial natriuretic factor (ANF) promoter is also activated in synergy with Nkx2-5 and GATA-4 through their binding to separate DNA elements. They have shown that GATA-5 but not GATA-6 can substitute for GATA-4 for interaction with Nkx2-5. Further determination of the precise structural interactions among the various Nkx and GATA proteins and their biological significance will be important because a combination of factors might reveal novel coactivating and corepressive pairs, especially in light of the increasing number of proteins that contain both zinc finger and homeobox motifs.[50] For example, *Drosophila* ZFH1, a factor downstream of *tinman*, is also required for insect dorsal vessel formation, and its vertebrate homolog, ZEB, is expressed during the early stage of mammalian cardiogenesis.[61]

The biological significance of the GATA-4/Nkx2-5 interaction is underscored by studies in zebrafish showing that the cardiogenic field corresponds to the zone of overlap of the larger anterior GATA-4 expression field and the posterior Nkx2-5 expression area in the lateral plate mesoderm.[26] The posterior half of the Nkx2-5 field, adjacent to the notochord, does not include cardiac progenitors, and the posterior Nkx2-5-expressing cells do not contribute to the heart, even after ablation of the normal cardiogenic region. The cells that can acquire a cardiac cell fate after injury to the normal progenitors reside near the prechordal plate, but anterior to the Nkx2-5-expressing domain, in the anterior region of the GATA-4 expression field. Normally, they give rise to head mesenchyme. These data, together with the in situ hybridization studies performed in chicken embryos, further support the possibility of a necessary interaction between GATA-4 and Nkx2-5 to trigger heart development.

GATA-4 INTERACTS WITH SRF

Because Nkx2-5 interacts with GATA-4 and with SRF, we reasoned that it was also likely that SRF and GATA-4 may function as coaccessory factors. To address this issue, luciferase reporter constructs for cardiac, skeletal, and smooth-muscle–restricted, SRE-dependent promoters such as cardiac α-actin, skeletal α-actin, smooth muscle α-actin, and SM22 α and ubiquitously expressed c-fos promoter were tested in cotransfection assays. Cotransfection of these reporter constructs into CV1 fibroblasts along with an expression vector encoding SRF elicited modest activation (Figure 7.5A). Similarly, expression of GATA-4 with these reporters resulted in weak activation. However, coexpression of both GATA-4 and SRF, from transfected CMV-driven plasmid expression vectors, resulted in robust activation of both muscle-restricted and ubiquitous SRE-dependent promoters, as shown in Figure 7.5.

Regulatory domains of GATA-4 and SRF were evaluated by deletion, and point-mutational analysis of GATA-4 revealed the second zinc finger and the immediate C-terminal basic region to be essential for coactivation, as shown by Belaguli et al.[7] and in Figure 7.6. In an earlier study,[81] by using deletion mutants of GATA-4, two transcriptional domains were mapped to the N-terminus of GATA-4. These activation domains were effective when fused to a heterologous DNA binding domain.[81]

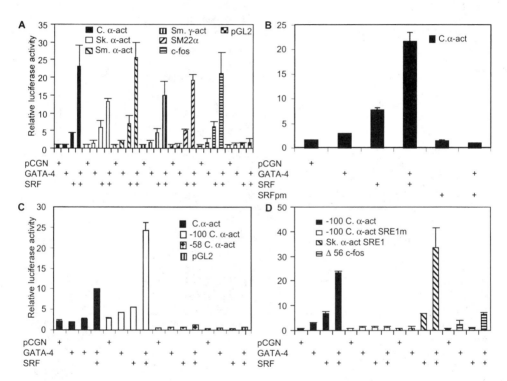

FIGURE 7.5. SRF and GATA-4 synergistically activate cardiac α-actin promoter and other SRE-containing promoters. Subconfluent CV1 cells were transfected with 1 μg of numerous myogenic and nonmyogenic promoter luciferase reporters indicated in panel A along with 150 ng of expression vector for SRF alone or in combination with 400 ng of GATA-4 (panel A). For panel B, 150 ng of a DNA binding mutant of SRF (SRFPpm) was used in addition to wild-type SRF and GATA-4. For panel C, wild-type and deletion mutants of cardiac α-actin promoter and the control pGL2 basic luciferase reporters were used. For panel D, a deletion mutant of cardiac α-actin containing a single wild-type or mutated SRE1 and a truncated c-fos minimal promoter (Δ56 c-fos) with or without skeletal α-actin SRE1 cloned upstream was used in the cotransfection assay. The total amount of DNA was adjusted to 2 μg by balancing with the pCGN empty vector. *Cells* were harvested 48 hours posttranscription, and the luciferase activity was measured. Results shown are mean ± standard error of the mean for three duplicate experiments (panels B and C) and two duplicate experiments (panels A and D).

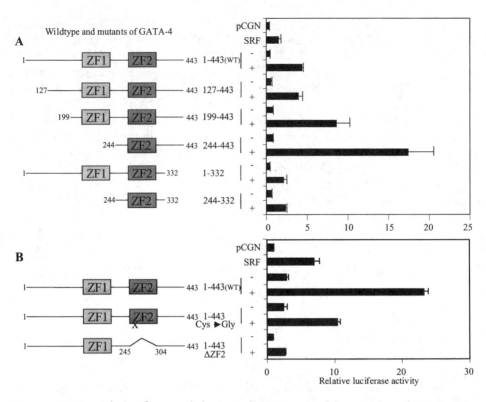

FIGURE 7.6. Second zinc finger and the immediate C-terminal basic region of GATA-4 are essential for synergistic activation of cardiac α-actin promoter. Subconfluent CV1 cells were transfected with 1 μg of cardiac α-actin luciferase reporter along with 400 ng of wild-type and various deletion mutants of GATA-4 either alone or in combination with 150 ng of SRF (panel A). The total amount of DNA was adjusted to 2 μg by balancing with the pCGN empty vector. For panel B, –100 cardiac α-actin promoter containing the proximal SRE was used as the reporter. Cells were harvested 48 hours posttranscription, and the luciferase activity was measured. Results shown are mean ± standard error of the mean for three duplicate experiments (panel A) and two duplicate experiments for panel B. GATA-4 domains retained in each deletion mutant are diagrammatically represented on the left. ZF1 and ZF2 refer to the N- and C-terminal zinc fingers, respectively. Single amino acid mutation in ZF2 (cysteine 273 to glycine) abolished GATA-4 DNA binding activity and is indicated by "X." This figure is adapted from Belaguli et al.[7]

In addition to the N-terminal activation domains, the C-terminus was also necessary for the transcriptional activity of GATA-4. However, this domain was transcriptionally inert in the context of a heterologous DNA binding domain, indicating indirect participation. It has been shown that acetylation of lysine residues located in the basic region C-terminal to the second zinc finger and the inter–zinc-finger linker region of GATA-1 results in enhanced DNA binding and transcriptional activity.[15] Several of these lysine residues are conserved between GATA-1 and GATA-4. Enhancement of transcriptional activation of a mutated GATA-4 containing the second zinc finger, inter–zinc-finger linker region, and the C-terminus by SRF may suggest that SRF, which binds CBP/p300,[90] may facilitate the access of GATA-4 to these transacetylating activities by interacting with and subtly altering the conformation of GATA-4.

Deletion of N-terminal activation domains of GATA-4 located between amino acids 1 to 74 and 130 to 177 did not affect the ability of GATA-4 to coactivate with SRF, suggesting that the activation domain of SRF can compensate for lack of activation domains on GATA-4 (Figure 7.6). Interestingly, deletion of the second N-terminal activation domains and the first zinc finger of GATA-4 increased the ability of GATA-4 to synergize with SRF, suggesting that these domains interfere with the interaction of SRF and GATA-4. This interference could be mediated by the binding of other proteins to these domains of GATA-4, which might preclude efficient interaction of SRF with the second zinc finger of GATA-4. This notion is supported by reports describing interaction of a variety of cofactors with the amino finger of GATA proteins. Multizinc-finger coactivator proteins such as FOG-1 and FOG-2 modulate the transcriptional activity of GATA-1 and GATA-4 by interacting with the first zinc finger.[36,73,99,100,101,102,104] In a similar manner, *Drosophila* GATA protein, pannier, interacts with a zinc finger protein called U-shaped (Ush)[30,46] that negatively regulates the transcriptional activity of pannier toward the expression of proneural basic HLH proteins, achete, and scute.

The C-terminal activation domains of both SRF and GATA-4 were required for the coactivation of the cardiac α-actin promoter because deletion of the C-terminal activation domain of SRF or GATA-4 abolished coactivation. The coactivation of cardiac α-actin promoter by SRF and GATA-4 is mediated through SRE1 because deletion and specific point mutations of SRE1 reduced the basal activity of the promoter and eliminated the synergistic activation. Coactivation of the cardiac α-actin promoter was strictly dependent on SRF binding to SRE1, as deletion or point mutations that abolish DNA binding of SRF also abrogated synergistic activation. The coactivation appears to be independent of GATA-4 DNA binding because no GATA binding site is detected in the minimal fragment of cardiac α-actin promoter (−100) that was responsive to SRF–GATA-4 combination. Because GATA factors are known to bind divergent GATA sites,[58,74] we performed a gel-shift analysis of potential GATA sites present in the cardiac α-actin promoter to rule out direct DNA binding of GATA-4 in the context of the entire plasmid. None of these sites were bound by GATA-4.[95] Further, skeletal α-actin SRE1 cloned upstream of the c-fos minimal promoter was sufficient to confer synergistic activation by SRF and GATA-4. These results and our earlier report demonstrating the absence of functional cryptic GATA sites in the luciferase vector strongly suggest that GATA-4 is recruited to the cardiac α-actin promoter by SRF independent of GATA-4 binding to DNA. Our claim is further supported by the ability of related GATA proteins such as GATA-1 to activate transcription independent of DNA binding.[28,85]

Transcriptional activation of GATA binding-site–dependent genes such as cTnC, ANF, BNP, and troponin I required the N-terminal activation domains of GATA-4. Further, GATA-5 and GATA-6, which share extensive homology within the N-terminal activation domains, were capable of activating these genes and substituting for GATA-4.[82] Interestingly, GATA site-independent coactivation of NKE-driven reporters by Nkx-2.5 and GATA-4 was independent of N- and C-terminal activation domains of GATA-4.[95] In contrast, synergistic activation of the ANF promoter, which contains binding sites for both GATA-4 and Nkx-2.5, by combinations of GATA-4 and Nkx-2.5 required both the N- and C-terminal activation domains of GATA-4.[32,66] However, the C-terminal activation domain of GATA-4 was essential for the coactivation of ANF promoter by GATA-4 and GATA-6.[21] These results indicate that differential utilization of GATA-4's activation domains may depend on the promoter context and other interactive proteins. In support of this hypothesis, tran-

scriptional activity of GATA-1 and GATA-4 was dependent on both the target promoters and their interaction with cofactors FOG-1 and FOG-2.

GATA-4 synergistically activated various muscle-restricted promoters, which are
expressed in differentiated muscle types. Other cardiovascular tissue-enriched GATA
factors such as GATA-5 and GATA-6, which have a distinct yet overlapping expression pattern with GATA-4, were capable of interacting with SRF and substitute for
GATA-4 in coactivation assays. These results suggest that the pairing of SRF with
different GATA factors confers muscle subtype specificity (such as cardiac versus
skeletal versus smooth). Additional degrees of muscle subtype specificity could be
conferred by interaction of the SRF–GATA-complex with tissue–restricted factors
such as Nkx-2.5 and MyoD. Our unpublished results show that the cardiac tissue-
restricted homeoprotein, Nkx-2.5, combinatorially interacts with both SRF and
GATA-4 to strongly activate cardiac α-actin promoter (Sepulveda et al., manuscript
under preparation). In addition to cardiac and smooth-muscle–restricted promoters,
skeletal α-actin promoter and the ubiquitous c-fos promoter, which are normally
upregulated during cardiac hypertrophic response, were also coactivated by SRF and
GATA-4. Given the role of GATA-4 in mediating cardiac hypertrophy,[42] the interaction of SRF with GATA-4 may have a functional role in physiological hypertrophic
response.

Pulldown assays with bacterially expressed GST-SRF and in vitro translated
GATA-4, as well as with protein A-GATA-4 and protein A-SRF, indicated that these
two factors interact in solution and in mammalian cells (see Figure 7.7). By analogy
with the Nkx2-5–GATA-4 interaction[32,66,69,95] and the SRF–Nkx2-5 synergy reported

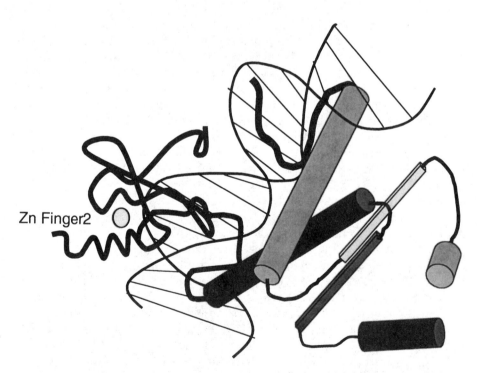

FIGURE 7.7. The second zinc finger of GATA-4 is sufficient for binding to SRF. Schematic
diagram of binding interaction between the C-terminal zinc finger (aa244 to aa331) as the
minimal required region for binding with the α1 coil of the SRF MADS box.

by Chen and Schwartz,[25] the interaction between SRF and GATA-4 required the conserved DNA binding domains of both proteins. More specifically, the C-terminal zinc finger of GATA-4 and the 142 to 171 region (N-terminal half of helix 1) of the MADS box were minimally required. This region of SRF is also minimally required to interact with Nkx2-5,[11] and includes the N-terminal extension of the MADS box that wraps around the DNA to interact with the minor groove of the SRE.

SRF increases the rate of preinitiation complex assembly at the target promoter,[107] in part by interacting with the Rap74 subunit of TFIIF.[54] Little is currently understood about the molecular mechanisms by which GATA-4 activates transcription. One possible mechanism by which SRF and GATA-4 interaction results in increased transcriptional activity relates to the ability of SRF to recruit the coactivator and protein acetylases CBP/p300[4,90] and SRC-1.[56] GATA-1 also binds CBP/p300[13] and undergoes a conformational change after acetylation by CBP and p300 that correlates with activation.[52] It is possible that synergistic activation results from a cooperative recruitment of the holoenzyme by SRF (through TFIIF) and of CBP/p300 by the SRF–GATA-4 complex.

GATA proteins have been reported to interact with a multitude of transcription factors, but this is the first demonstration of interaction between a GATA protein and an MADS protein. Several functional interactions of SRF with the zinc finger protein Sp1 have been described,[11,93] but physical association of the two proteins has not been demonstrated. MEF-2C, a member of the MADS box family, activates the expression of several muscle-specific genes either directly by binding to the regulatory regions of the target genes or indirectly by interacting with other muscle-restricted factors such as MyoD and Myogenin (reviewed in references 12 and 79). Reciprocal recruitment of SRF and GATA-4 to the promoter independent of DNA binding by either SRF or GATA-4 is analogous to the cross recruitment between SRF or MEF-2C and myogenic bHLH proteins.[40,79] The reciprocal recruitment between SRF and GATA-4 would expand the spectrum of genes regulated by either of these factors, while conferring an additional level of specificity. Our results demonstrating interaction between the MADS box proteins such as SRF and Mef-2C with the zinc finger protein, GATA-4, underscore the ability of these proteins to interact in a combinatorial manner to drive myogenic gene-expression programs.

REFERENCES

1. Affolter, M., Montagne, J., Walldorf, U., Groppe, J., Kloter, U., LaRosa, M., and Gehring, W.J. 1994. The *Drosophila* SRF homolog is expressed in a subset of tracheal cells and maps within a genomic region required for tracheal development. *Development* 120:743–753.

2. Arsenian, S., Weinhold, B., Oelgeschlager, M., Ruther, U., and Nordheim, A. 1998. Serum response factor is essential for mesoderm formation during mouse embryogenesis. *EMBO J.* 17:6289–6299.

3. Azpiazu, N. and Frasch, M. 1993. *Tinman* and *bagpipe*: two homeobox genes that determine cell fates in the dorsal mesoderm of *Drosophila*. *Genes Dev.* 7:1325–1340.

4. Bannister, A.J. and Kouzarides, T. 1996. The CBP co-activator is a histone acetyltransferase. *Nature* 384:641–643.

5. Belaguli, N.S., Schildmeyer, L.A., and Schwartz, R.J. 1997. Organization and myogenic restricted expression of the murine serum response factor gene: a role for autoregulation. *J. Biol. Chem.* 272:18222–18231.

6. Belaguli, N.S., Zhou, W., Trinh, T.-H.T., Majesky, M., and Schwartz, R.J. 1999. Dominant negative murine serum response factor: Alternative splicing within the activation

domain inhibits transactivation of serum response factor binding targets. *Mol. Cell. Biol.* 19:4582–4591.

7. Belaguli, N.S., Sepulveda, J.L., Nigam, V., Charron, F., Nemer, M., and Schwartz, R.J. 2000. Cardiac tissue enriched factors serum response factor and GATA-4 are mutual coregulators. *Mol. Cell. Biol.* 20:7550—7558.

8. Biben, C. and Harvey, R.P. 1997. Homeodomain factor Nkx2-5 controls left/right asymmetric expression of bHLH gene eHand during murine heart development. *Genes Dev.* 11:1357–1369.

9. Biben, C., Palmer, S., Elliott, D.A., and Harvey, R.P. 1997. Homeobox genes and heart development. *Cold Spring Harb. Symp. Quant. Biol.* 62:395–403.

10. Bieberich, C.J., Fujita, K., He, W.W., and Jay, G. 1996. Prostate-specific and androgen-dependent expression of a novel homeobox gene. *J. Biol. Chem.* 271:31779–31782.

11. Biesiada, E., Hamamori, Y., Kedes, L., and Sartorelli, V. 1999. Myogenic basic helix-loop-helix proteins and Sp1 interact as components of a multiprotein transcriptional complex required for activity of the human cardiac alpha-actin promoter. *Mol. Cell. Biol.* 19:2577–2584.

12. Black, B.L. and Olson, E.N. 1998. Transcriptional control of muscle development by myocyte enhancer factor-2 (MEF2) proteins. *Annu. Rev. Cell Dev. Biol.* 14:167–196.

13. Blobel, G.A., Nakajima, T., Eckner, R., Montminy, M., and Orkin, S.H. 1998. CREB-binding protein cooperates with transcription factor GATA-1 and is required for erythroid differentiation. *Proc. Natl. Acad. Sci. USA* 95:2061–2066.

14. Bodmer, R. 1993. The gene *tinman* is required for specification of the heart and visceral muscles in *Drosophila*. *Development* 118:719–729.

15. Boyes, J., Byfield, P., Nakatani, Y., and Ogryzko, V. 1998. Regulation of activity of the transcription factor GATA-1 by acetylation. *Nature* 396:594–598.

16. Brand, T., Andree, B., Schneider, A., Buchberger, A., and Arnold, H.H. 1997. Chicken NKx-8, a novel homeobox gene expressed during early heart and foregut development. *Mech. Dev.* 64:53–59.

17. Boxer, L.M., Prywes, R., Roeder, R.G., and Kedes, L. 1989. The sarcomeric actin CArG-binding factor is indistinguishable from the c-fos serum response factor. *Mol. Cell. Biol.* 9:515–522.

18. Browning, C.L., Culberson, D.E., Aragon, I.V., Fillmore, R.A., Croissant, J.D., Schwartz, R.J., and Zimmer, W.E. 1997. The developmentally regulated expression of serum response factor plays a key role in the control of smooth muscle specific genes. *Dev. Biol.* 194:18–37.

19. Buchberger, A., Pabst, O., Brand, T., Seidl, K., and Arnold, H.H. 1996. Chick NKx-2.3 represents a novel family member of vertebrate homologues to the *Drosophila* homeobox gene tinman: differential expression of cNKx-2.3 and cNKx-2.5 during heart and gut development. *Mech. Dev.* 56:151–163.

20. Bushel, P., Kim, J.H., Chang, W., Catino, J.J., Ruley, H.E., and Kumar, C.C. 1995. Two serum response elements mediate transcriptional repression of human smooth muscle alpha-actin promoter in ras-transformed cells. *Oncogene* 10:1361–1370.

21. Charron, F., Paradis, P., Bronchain, O., Nemer, G., and Nemer, M. 1999. Cooperative interaction between GATA-4 and GATA-6 regulates myocardial gene expression. *Mol. Cell. Biol.* 19:4355–4365.

22. Chen, C.Y., Croissant, J., Majesky, M., Topouzis, S., McQuinn, T., Frankovsky, M.J., and Schwartz, R.J. 1996. Activation of the cardiac alpha-actin promoter depends upon serum response factor, Tinman homologue, Nkx-2.5, and intact serum response elements. *Dev. Genet.* 19:119–130.

23. Chen, C.Y. and Schwartz, R.J. 1997. Competition between negative acting YY1 versus positive acting serum response factor and *tinman* homologue Nkx-2.5 regulates cardiac a-actin promoter activity. *Mol. Endo.* 11:812–822.

24. Chen, C.Y. and Schwartz, R.J. 1995. Identification of novel DNA binding targets and regulatory domains of a murine tinman homeodomain factor, nkx-2.5. *J. Biol. Chem.* 270:15628–15633.

25. Chen, C.Y. and Schwartz, R.J. 1996. Recruitment of the tinman homolog Nkx-2.5 by serum response factor activates cardiac alpha-actin gene transcription. *Mol. Cell. Biol.* 16:6372–6384.

26. Chen, J.N. and Fishman, M.C. 1996. Zebrafish tinman homolog demarcates the heart field and initiates myocardial differentiation. *Development* 122:3809–3816.

27. Croissant, J.D., Kim, J.-H., Eichele, G., Goering, L., Lough, J., Prywes, R., and Schwartz, R.J. 1996. Avian serum response factor expression restricted primarily to muscle cell lineages is required for α-actin gene transcription. *Dev. Biol.* 177:250–264.

28. Crossley, M., Merika, M., and Orkin, S.H. 1995. Self-association of the erythroid transcription factor GATA-1 mediated by its zinc finger domains. *Mol. Cell. Biol.* 15:2448–2456.

29. Cserjesi, P., Lilly, B., Hinkley, C., Perry, M., and Olson, E.N. 1994. Homeodomain protein MHox and MADS protein myocyte enhancer-binding factor-2 converge on a common element in the muscle creatine kinase enhancer. *J. Biol. Chem.* 269:16740–16745.

30. Cubadda, Y., Heitzler, P., Ray, R.P., Bourouis, M., Ramain, P., Gelbart, W., Simpson, P., and Haenlin, M. 1997. U-shaped encodes a zinc finger protein that regulates the proneural genes achaete and scute during the formation of bristles in *Drosophila*. *Genes Dev.* 11:3083–3095.

31. Damante, G., Fabbro, D., Pellizzari, L., Civitareale, D., Guazzi, S., Polycarpou-Schwartz, M., Cauci, S., Quadrifoglio, F., Formisano, S., and Di Lauro, R. 1994. Sequence-specific DNA recognition by the thyroid transcription factor-1 homeodomain. *Nucl. Acids Res.* 22:3075–3083.

32. Durocher, D., Charron, F., Warren, R., Schwartz, R.J., and Nemer, M. 1997. The cardiac transcription factors Nkx2-5 and GATA-4 are mutual cofactors. *EMBO J.* 16:5687–5696.

33. Durocher, D., Chen, C.Y., Ardati, A., Schwartz, R.J., and Nemer, M. 1996. The atrial natriuretic factor promoter is a downstream target for Nkx-2.5 in the myocardium. *Mol. Cell. Biol.* 16:4648–4655.

34. Evans, S.M., Yan, W., Murillo, M.P., Ponce, J., and Papalopulu, N. 1995. Tinman, a *Drosophila* homeobox gene required for heart and visceral mesoderm specification, may be represented by a family of genes in vertebrates: XNkx-2.3, a second vertebrate homologue of tinman. *Development* 121:3889–3899.

35. Fischer, K.D., Haese, A., and Nowock, J. 1993. Cooperation of GATA-1 and Sp1 can result in synergistic transcriptional activation or interference. *J. Biol. Chem.* 268:23915–23923.

36. Fox, A.H., Liew, C., Holmes, M., Kowalski, K., Mackay, J., and Crossley, M. 1999. Transcriptional cofactors of the FOG family interact with GATA proteins by means of multiple zinc fingers. *EMBO J.* 18:2812–2822.

37. Frasch, M. 1995. Induction of visceral and cardiac mesoderm by ectodermal Dpp in the early *Drosophila* embryo. *Nature* 374:464–467.

38. Gajewski, K., Kim, Y., Lee, Y.M., Olson, E.N., and Schultz, R.A. 1997. *D-mef2* is a target for *Tinman* activation during *Drosophila* heart development. *EMBO J.* 16:515–522.

39. Grepin, C., Robitaille, L., Antakly, T., and Nemer, M. 1995. Inhibition of transcription factor GATA-4 expression blocks in vitro cardiac muscle differentiation. *Mol. Cell. Biol.* 15:4095–4102.

40. Groisman, R., Masutani, H., Leibovitch, M.P., Robin, P., Soudant, I., Trouche, D., and Harel-Bellan, A. 1996. Physical interaction between the mitogen-responsive serum response factor and myogenic basic-helix-loop-helix proteins. *J. Biol. Chem.* 271:5258–5264.

41. Grueneberg, D.A., Henry, R.W., Brauer, A., Novina, C.D., Cheriyath, V., Roy, A.L., and Gilman, M. 1997. A multifunctional DNA-binding protein that promotes the formation of serum response factor/homeodomain complexes: identity to TFII-I. *Genes Dev.* 11:2482–2493.

42. Grueneberg, D.A., Natesan, S., Alexandre, C., and Gilman, M.Z. 1992. Human and *Drosophila* homeodomain proteins that enhance the DNA-binding activity of serum response factor. *Science* 257:1089–1095.

43. Guazzi, S., Price, M., De Felice, M., Damante, G., Mattei, M.G., and Di Lauro, R. 1990. Thyroid nuclear factor 1 (TTF-1) contains a homeodomain and displays a novel DNA binding specificity. *EMBO J.* 9:3631–3639.

44. Guillemin, K., Groppe, J., Ducker, K., Treisman, R., Hafen, E., Affolter, M., and Krasnow, M.A. 1996. The pruned gene encodes the *Drosophila* serum response factor and regulates cytoplasmic outgrowth during terminal branching of the tracheal system. *Development* 122:1353–1362.

45. Harvey, R.P. 1996. NK-2 homeobox genes and heart development. *Dev. Biol.* 178:203–216.

46. Haenlin, M., Cubadda, Y., Blondeau, P., Heitzler, P., Lutz, P., Simpson, P., and Ramain, P. 1997. Transcriptional activity of pannier is regulated negatively by heterodimerization of the GATA DNA-binding domain with a cofactor encoded by the u-shaped gene of *Drosophila*. *Genes Dev.* 11:3096–3108.

47. Hautmann, M.B., Madsen, C.S., Mack, C.P., and Owens, G.K. 1998. Substitution of the degenerate smooth muscle (SM) alpha-actin CC(A/T-rich)6GG elements with c-fos serum response elements results in increased basal expression but relaxed SM cell specificity and reduced angiotensin II inducibility. *J. Biol. Chem.* 273:8398–8406.

48. Heikinheimo, M., Scandrett, J.M., and Wilson, D.B. 1994. Localization of transcription factor GATA-4 to regions of the mouse embryo involved in cardiac development. *Dev. Biol.* 164:361–373.

49. Hogan, B.L. 1996. Bone morphogenetic proteins in development. *Curr. Opin. Genet. Dev.* 6:432–438.

50. Ido, A., Miura, Y., and Tamaoki, T. 1994. Activation of ATBF1, a multiple-homeodomain zinc-finger gene, during neuronal differentiation of murine embryonal carcinoma cells. *Dev. Biol.* 163:184–187.

51. Ip, H.S., Wilson, D.B., Heikinheimo, M., Tang, Z., Ting, C.N., Simon, M.C., Leiden, J.M., and Parmacek, M.S. 1994. The GATA-4 transcription factor transactivates the cardiac muscle-specific troponin C promoter-enhancer in nonmuscle cells. *Mol. Cell. Biol.* 14:7517–7526.

52. Hung, H.-L., Lau, J., Kim, A.Y., Weiss, M.J., and Blobel, G.A. 1999. CREB-binding protein acetylates hematopoietic transcription factor GATA-1 at functionally important sites. *Mol. Cell. Biol.* 19:3496–3505.

53. Jiang, Y. and Evans, T. 1996. The *Xenopus* GATA-4/5/6 genes are associated with cardiac specification and can regulate cardiac-specific transcription during embryogenesis. *Dev. Biol.* 174:258–270.

54. Joliot, V., Demma, M., and Prywes, R. 1995. Interaction with RAP74 subunit of TFIIF is required for transcriptional activation by serum response factor. *Nature* 373:632–635.

55. Kelley, C., Blumberg, H., Zon, L.I., and Evans, T. 1993. GATA-4 is a novel transcription factor expressed in endocardium of the developing heart. *Development* 118:817–827.

56. Kim, J.-H., Kim, H.-J., and Lee, W.J. 1998. Steroid receptor coactivator-1 interacts with serum response factor and coactivates serum response element-mediated transactivations. *J. Biol. Chem.* 273:28564–28567.

57. Knoepfler, P.S. and Kamps, M.P. 1995. The pentapeptide motif of Hox proteins is required for cooperative DNA binding with Pbx1, physically contacts Pbx1, and enhances DNA binding by Pbx1. *Mol. Cell. Biol.* 15:5811–5819.

58. Ko, J.L. and Engel, J.D. 1993. DNA-binding specificities of the GATA transcription factor family. *Mol. Cell. Biol.* 13:4011–4022.

59. Komuro, I. and Izumo, S. 1993. Csx: a murine homeobox-containing gene specifically expressed in the developing heart. *Proc. Natl. Acad. Sci. USA* 90:8145–8149.

60. Kuo, C.T., Morrisey, E.E., Anandappa, R., Sigrist, K., Lu, M.M., Parmacek, M.S., Soudais, C., and Leiden, J.M. 1997. GATA4 transcription factor is required for ventral morphogenesis and heart tube formation. *Genes Dev.* 11:1048–1106.

61. Lai, Z.-C., Rushton, E., Bate, M., and Rubin, G. 1993. Loss of function of the *Drosophila zfh-1* gene results in abnormal development of mesodermally derived tissues. *Proc. Natl. Acad. Sci. USA* 90:4122–4126.

62. Laverriere, A.C., MacNeill, C., Mueller, C., Poelmann, R.E., Burch, J.B., and Evans, T.G. 1994. GATA-4/5/6, a subfamily of three transcription factors transcribed in developing heart and gut. *J. Biol. Chem.* 269:23177–23184.

63. Lee, K.H., Xu, Q., and Breitbart, R.E. 1996. A new tinman-related gene, nkx2.7, anticipates the expression of nkx2.5 and nkx2.3 in zebrafish heart and pharyngeal endoderm. *Dev. Biol.* 180:722–731.

64. Lee, T.C., Chow, K.L., Fang, P., and Schwartz, R.J. 1991. Activation of skeletal alpha-actin gene transcription: the cooperative formation of serum response factor-binding complexes over positive *cis*-acting promoter serum response elements displaces a negative-acting nuclear factor enriched in replicating myoblasts and nonmyogenic cells. *Mol. Cell. Biol.* 11:5090–5100.

65. Lee, T.C., Shi, Y., and Schwartz, R.J. 1992. Displacement of BrdUrd-induced YY1 by serum response factor activates skeletal α-actin transcription in embryonic myoblasts. *Proc. Natl. Acad. Sci. USA* 91:9814–9818.

66. Lee, Y., Shioi, T., Kasahara, H., Jobe, S.M., Wiese, R.J., Markham, B.E., and Izumo, S. 1998. The cardiac tissue-restricted homeobox protein Csx/Nkx2.5 physically associates with the zinc finger protein GATA4 and cooperatively activates atrial natriuretic factor gene expression. *Mol. Cell. Biol.* 18:3120–3129.

67. Leifer, D., Krainic, D., Yu, Y.T., McDermott, J., Breibart, R.E., Heng, J., Neve, R.L., Kosofsky, B., Nadal-Ginard, B., and Lipton, S.A. 1993. MEF2C, a MADS/MEF2-family transcription factor expressed in a laminar distribution in cerebral cortex. *Proc. Natl. Acad. Sci. USA* 90:1546–1560.

68. Leussink, B., Brouwer, A., el Khattabi, M., Poelmann, R.E., Gittenberger-de Groot, A.C., and Meijlink, F. 1995. Expression patterns of the paired-related homeobox genes MHox/Prx1 and S8/Prx2 suggest roles in development of the heart and the forebrain. *Mech. Dev.* 52:51–64.

69. Lints, T.J., Parsons, L.M., Hartley, L., Lyons, I., and Harvey, R.P. 1993. Nkx-2.5: a novel murine homeobox gene expressed in early heart progenitor cells and their myogenic descendants. *Development* 119:969.

70. Lough, J., Barron, M., Brogley, M., Sugi, Y., Bolender, D.L., and Zhu, X. 1996. Combined BMP-2 and FGF-4, but neither factor alone, induces cardiogenesis in non-precardiac embryonic mesoderm. *Dev. Biol.* 178:198–202.

71. Lyons, I., Parsons, L.M., Hartley, L., Li, R., Andrews, J.E., Robb, L., and Harvey, R.P. 1995. Myogenic and morphogenetic defects in the heart tubes of murine embryos lacking the homeobox gene Nkx2-5. *Genes Dev.* 9:1654–1666.

72. Li, L., Liu, Z.-C., Mercer, B., Overbeek, P., and Olson, E.N. 1997. Evidence for serum response factor-mediated regulatory networks governing SM22α transcription in smooth, skeletal, and cardiac muscle cells. *Dev. Biol.* 187:311–321.

73. Lu, J.-R., McKinsey, T.A., Xu, H., Wang, D.-Z., Richardson, J.A., and Olson, E.N. 1999. FOG-2, a heart- and brain-enriched cofactor for GATA transcription factors. *Mol. Cell. Biol.* 19:4495–4502.

74. Merika, M. and Orkin, S.H. 1993. DNA binding specificity of the GATA family transcription factors. *Mol. Cell. Biol.* 13:3999–4010.

75. Minty, A. and Kedes, L. 1986. Upstream regions of the human cardiac actin gene that modulate its transcription in muscle cells: presence of an evolutionarily conserved repeated motif. *Mol. Cell. Biol.* 6:2125–2136.

76. Mohun, T.J., Taylor, M.W., Garrett, N., and Gurden, J.B. 1989. The CArG promoter sequence is necessary for muscle-specific transcription of the cardiac actin gene in *Xenopus* embryos. *EMBO J.* 8:1153–1161.

77. Mohun, T.J., Chambers, A.E., Towers, N., and Taylor, M.V. 1991. Expression of genes encoding the transcription factor SRF during early development of *Xenopus* laevis: identification of a CArG box binding activity as SRF. *EMBO J.* 10:933–940.

78. Molkentin, J., Lin, Q., Duncan, S.A., and Olson, E.N. 1997. Requirement of the transcription factor GATA4 for heart tube formation and ventral morphogenesis. *Genes Dev.* 11:1061–1072.

79. Molkentin, J.D., Black, B.L., Martin, J.F., and Olson, E.N. 1995. Cooperative activation of muscle gene expression by MEF2 and myogenic bHLH proteins. *Cell* 83:1125–1136.

80. Molkentin, J.D., Lu, J., Antons, C.L., Markham, B., Richardson, J., Robbins, J., Grant, S.R., and Olson, E.N. 1999. A calcineurin-dependent transcriptional pathway for cardiac hypertrophy. *Cell* 93:1–20.

81. Morrisey, E.E., Ip, H.S., Tang, Z., and Parmacek, M.S. 1997. GATA-4 activates transcription via two novel domains that are conserved within the GATA-4/5/6 subfamily. *J. Biol. Chem.* 272:8515–8524.

82. Morrisey, E.E., Ip, H.H., Tang, Z., Lu, M.M., and Parmacek, M.S. 1997. GATA-5: A transcriptional activator expressed in a novel temporally and spatially-restricted pattern during embryonic development. *Dev. Biol.* 183:21–36.

83. Morrisey, E.E., Ip, H.S., Lu, M.M., and Parmacek, M.S. 1996. GATA-6: a zinc finger transcription factor that is expressed in multiple cell lineages derived from lateral mesoderm. *Dev. Biol.* 177:309–322.

84. Norman, C., Runswick, M., Pollock, R., and Treisman, R. 1988. Isolation and properties of cDNA clones encoding SRF, a transcription factor that binds to the c-fos serum response element. *Cell* 55:989–1003.

85. Osada, H., Grutz, G., Axelson, H., Forster, A., and Rabbitts, T.H. 1995. Association of erythroid transcription factors: complexes involving the LIM protein RBTN2 and the zinc finger protein GATA1. *Proc. Natl. Acad. Sci. USA* 92:9585–9589.

86. Pabst, O., Schneider, A., Brand, T., and Arnold, H.H. 1997. The mouse Nkx2-3 homeodomain gene is expressed in gut mesenchyme during pre- and postnatal mouse development. *Dev. Dyn.* 209:29–35.

87. Parmacek, M.S. and Leiden, J.M. 1999. GATA transcription factors and cardiac development. In: *Heart Development*, Eds. R. Harvey and N. Rosenthal, Academic Press, San Diego, pp. 291–306.

88. Pellegrini, L., Tan, S., and Richmond, T.J. 1995. Structure of serum response factor core bound to DNA. *Nature* 376:490–498.

89. Pollock, R. and Treisman, R. 1991. Human SRF-related proteins: DNA-binding properties and potential regulatory targets. *Genes Dev.* 5:2327–2341.

90. Ramirez, S., Ali, S.A.S., Robin, P., Trouche, D., and Harel-Bellan, A. 1997. The CREB-binding protein (CBP) cooperates with the serum response factor for transactivation of the c-fos serum response element. *J. Biol. Chem.* 272:31016–31021.

91. Ray, M.K., Chen, C.Y., Schwartz, R.J., and DeMayo, F.J. 1996. Transcriptional regulation of a mouse Clara cell-specific protein (mCC10) gene by the NKx transcription factor family members thyroid transcription factor 1 and cardiac muscle-specific homeobox protein (CSX). *Mol. Cell. Biol.* 16:2056–2064.

92. Reecy, J.M., Yamada, M., Cummings, K., Sosic, D., Chen, C.Y., Eichele, G., Olson, E.N., and Schwartz, R.J. 1997. Chicken Nkx-2.8: a novel homeobox gene expressed in early heart progenitor cells and pharyngeal pouch −2 and −3 endoderm. *Dev. Biol.* 188:295–311.

93. Sartorelli, V., Webster, K.A., and Kedes, L. 1990. Muscle-specific expression of the cardiac alpha-actin gene requires MyoD1, CArG-box binding factor, and Sp1. *Genes Dev.* 4:1811–1822.

94. Schultheiss, T.M., Burch, J.B., and Lassar, A.B. 1997. A role for bone morphogenetic proteins in the induction of cardiac myogenesis. *Genes Dev.* 11:451–462.

95. Sepulveda, J.L., Belaguli, N., Nigam, V., Chen, C.Y., Nemer, M., and Schwartz, R.J. 1998. GATA-4 and Nkx-2.5 coactivate Nkx-2 DNA binding targets: role for regulating early cardiac gene expression. *Mol. Cell. Biol.* 18:3405–3415.

96. Shore, P. and Sharrocks, A.D. 1994. The transcription factors Elk-1 and serum response factor interact by direct protein-protein contacts mediated by a short region of Elk-1. *Mol. Cell. Biol.* 14:3283–3291.

97. Sommer, H., Beltran, J.P., Huijser, P., Pape, H., Lonnig, W.E., Saedler, H., and Schwarz-Sommer, Z. 1990. Deficiens, a homeotic gene involved in the control of flower morphogenesis in *Antirrhinum majus*: the protein shows homology to transcription factors. *EMBO J.* 9:605–613.

98. Soulez, M., Rouviere, C.G., Chafey, P., Hentzen, D., Vandromme, M., Lautredou, N., Lamb, N., Kahn, A., and Tuil, D. 1996. Growth and differentiation of C2 myogenic cells are dependent on serum response factor. *Mol. Cell. Biol.* 16:6065–6074.

99. Svensson, E.C., Tufts, R.L., Polk, C.E., and Leiden, J.M. 1999. Molecular cloning of FOG-2: a modulator of transcription factor GATA-4 in cardiomyocytes. *Proc. Natl. Acad. Sci. USA* 96:956–961.

100. Tevosian, S.G., Deconinck, A.E., Cantor, A.B., Rieff, H.I., Fujiwara, Y., Corfas, G., and Orkin, S.H. 1999. FOG-2: a novel GATA-family cofactor related to multitype zinc-finger proteins friend of GATA-1 and U-shaped. *Proc. Natl. Acad. Sci. USA* 96:950–955.

101. Tonissen, K.F., Drysdale, T.A., Lints, T.J., Harvey, R.P., and Krieg, P.A. 1994. XNkx-2.5, a *Xenopus* gene related to Nkx-2.5 and tinman: evidence for a conserved role in cardiac development. *Dev. Biol.* 162:325–328.

102. Tsang, A.P., Visvader, J.E., Turner, C.A., Fujiwara, Y., Yu, C., Weiss, M.J., Crossley, M., and Orkin, S.H. 1997. FOG, a multitype zinc finger protein, acts as a cofactor for transcription factor GATA-1 in erythroid and megakaryocytic differentiation. *Cell* 90:109–119.

103. Vandromme, M., Gauthier-Rouviere, C., Carnac, G., Lamb, N., and Fernandez, A. 1992. Serum response factor p67SRF is expressed and required during myogenic differentiation of both mouse C2 and rat L6 muscle cell lines. *J. Cell Biol.* 118:1489–1500.

104. Whitmarsh, A.J., Shore, P., Sharrocks, A.D., and Davis, R.J. 1995. Integration of MAP kinase signal transduction pathways at the serum response element. *Science* 269:403–407.

105. Yang, H.Y. and Evans, T. 1992. Distinct roles for the two cGATA-1 finger domains. *Mol. Cell. Biol.* 12:4562–4570.

106. Zhang, H. and Bradley, A. 1996. Mice deficient for BMP2 are nonviable and have defects in amnion/chorion and cardiac development. *Development* 122:2977–2986.

107. Zhu, C., Johansen, F.E., and Prywes, R. 1997. Interaction of ATF6 and serum response factor. *Mol. Cell. Biol.* 17:4957–4966.

108. Zou, Y., Evans, S., Chen, J., Kuo, H.C., Harvey, R.P., and Chien, K.R. 1997. CARP, a cardiac ankyrin repeat protein, is downstream in the Nkx2-5 homeobox gene pathway. *Development* 124:793–804.

CHAPTER 8

Regulation and Organization of Human Troponin Genes

Paul J.R. Barton, Kimberley A. Dellow, Pankaj K. Bhavsar, Martin E. Cullen, Antony J. Mullen, and Nigel J. Brand

INTRODUCTION

In spite of considerable advances, understanding the mechanisms that regulate correct temporal and spatial gene expression during development remains one of the major challenges of molecular biology. During cardiac development, intricate patterns of gene expression underlie the processes of precursor cell specification and differentiation. Subsequent alterations in expression can be identified that correlate both with the formation of distinct cell types, including ventricular, atrial, and conduction system myocytes, and with the general process of myocyte maturation. One approach to identifying mechanisms that regulate these processes is through the dissection of specific regulatory pathways required for expression of particular genes, and this has led to the identification of a number of important transcriptional pathways. We have chosen to investigate the regulation of expression of troponin genes in the human heart as a route to identifying such pathways.

The troponin complex is located on the thin filament of the sarcomere and is composed of three subunits: troponin C (TnC), the calcium binding subunit; troponin T (TnT), which is involved in the attachment of the complex to tropomyosin; and troponin I (TnI), the inhibitory subunit. Together, these form the calcium-sensitive molecular switch that regulates striated muscle contraction. Multiple isoforms have been identified for each of these subunits, each of which are expressed with a distinct pattern of tissue specificity and developmental regulation (Schiaffino and Reggiani, 1996). Three troponin I genes have been identified in vertebrates, and in the adult these are expressed in cardiac muscle, slow-twitch skeletal muscle, and fast-twitch skeletal muscle, respectively. However, during human cardiac development, the predominant isoform in the fetal myocardium is slow skeletal troponin I. Subsequently, there is an isoform switch, such that in the adult heart cardiac troponin I is the only isoform detected in the bulk of the myocardium (Sasse et al., 1993), although expression of the slow skeletal isoform persists in myocytes of the conduction system (Schiaffino et al., 1993). Evidence from analysis of mRNA and protein levels in humans and from transgenic mice indicates that this developmental regulation occurs at the level of gene transcription.

The human cardiac troponin I gene therefore offers a model of cardiac-specific and developmentally regulated expression. Moreover, the strict cardiac-specificity of this gene contrasts with all other cardiac contractile protein genes, which are also expressed in skeletal muscle, at least during development. This unique pattern of expression is especially intriguing, as we have recently shown that the cardiac troponin I gene is located just 2.6 kb upstream of the slow skeletal troponin T gene, which is expressed in slow skeletal muscle fibers but not in the heart (Barton et al., 1999). The differential regulation of these two closely linked troponin genes is therefore of interest in the broad context of locus control and chromatin remodeling, as well as in terms of transcriptional regulation driven through proximal promoter elements. Troponin genes are also of importance in familial hypertrophic cardiomyopathy (FHC), an autosomal dominant disease characterized by left ventricular hypertrophy, myocyte disarray, and risk of sudden death. Mutations in troponin T are a common cause of this myocardial disease (Thierfelder et al., 1994). Mutations in the cardiac troponin I gene have also been detected in a Japanese cohort of FHC families (Kimura et al., 1997), although these appear to be rare and have not been found in European families.

THE HUMAN CARDIAC TROPONIN I GENE PROMOTER

In order to determine the factors required for its expression, we isolated and characterized the human cardiac troponin I gene (Bhavsar et al., 1996). Sequence analysis of the proximal 5′-flanking sequence reveals the presence of a number of potentially important *cis*-acting elements. These include putative Nkx, E box, Sp1/3, CACC box, GATA, A/T-rich (MEF-2/TATA), and Initiator elements. In addition, eleven copies of a 36 to 38 bp, chromosome-19–specific minisatellite sequence are located close to the promoter. Experiments in which constructs containing varying lengths of flanking DNA were transfected into neonatal rat cardiac myocytes in culture suggest the presence of both positive and negative regulatory regions within the first 6.5 kb of the flanking region but also reveal that 98 bp of proximal promoter sequence is sufficient to drive a significant level of transcription (Bhavsar et al., 2000). Within this compact region, there is significant homology among rat, mouse, and human genes, allowing identification of several conserved putative *cis*-acting elements (Figure 8.1). In all three genes, the site of initiation of transcription is located within an Initiator element more commonly associated with genes that lack a TATA box. Two consensus (WGATAR) GATA elements are present, one overlapping the Initiator element and the other further upstream, both in reverse orientation. Also present is a conserved A/T-rich region, which contains a potential binding site for members of the widely expressed MEF-2 family of transcription factors and a nonconsensus TATA box. Finally, a CACC box, similar in sequence to that described in the myoglobin gene promoter and the cardiac troponin C enhancer, is located at the distal end of this region in the human gene and overlaps a consensus Sp1 element. This overlapping arrangement of CACC box and Sp1 elements is not present in the rodent proximal promoters at this position, although both contain a more proximal and shorter CACC box. The functional importance of the human promoter elements and the various factors that they bind is described later.

GATA and A/T-Rich Elements Bind Multiple Factors and Are Required for Maximal Expression

Electrophoretic mobility shift assays (EMSA) and mutational analysis were used to determine the importance of the identified elements within the proximal promoter. The two GATA elements contained within the first 98 bp of promoter have been

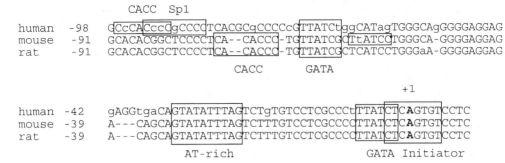

FIGURE 8.1. The proximal promoter regions of the human rat and mouse cardiac troponin I genes. The site of initiation of transcription (bold A) lies within an Initiator element (Inr), which is overlapped by the proximal GATA element. Note that in the case of the distal GATA element only the human sequence complies with the "WGATAR" consensus. A C-rich element comprised of overlapping CACC box and Sp1 elements is located toward the distal end of the human sequence. In the rodent genes, a related CACC box sequence is located more proximally. Sequences derived from Ausoni et al. (1994), Bhavsar et al. (1996), and Murphy et al. (1997). Redrawn from Bhavsar et al. (2000).

shown, by EMSA analysis, to bind both GATA-4 and GATA-6 (data not shown). A characteristic band is observed, most of which can be removed by addition of GATA-4 antibody, indicating that GATA-4 is the principal binding activity. The residual band can be eliminated by further addition of GATA-6 antibody. Site-directed mutagenesis of either of these GATA elements results in a reduction in promoter activity, with the mutation of the distal element resulting in the greatest effect, with 90% loss of activity compared to 50% loss of activity with mutation of the proximal element. This contrasts with the mouse gene, where mutation of either the proximal or distal GATA element results in a similar significant reduction in activity (Di Lisi et al., 1998).

GATA factors are known to be involved in the regulation of a number of cardiac genes (Charron and Nemer, 1999). GATA-4, GATA-5, and GATA-6 are detected in the precardiac mesoderm from embryonic day 8 (ED 8) in the mouse (Heikinheimo et al., 1994), and GATA-4 and GATA-6 are expressed throughout embryonic and fetal cardiogenesis. GATA-5 has a more transient expression pattern, initially detectable in the primitive heart, becoming restricted to the atrial endocardium by ED 12.5 and no longer detected by ED 16.5. In vitro experiments using the pluripotent embryonal carcinoma cell line P19 showed that GATA-4 overexpression increases the number of differentiated cardiac myocytes, whereas inhibition of GATA-4 blocks myocyte differentiation (Grepin et al., 1997). Injection of GATA-4 mRNA into *Xenopus* oocytes stimulates cardiac α-actin and α-myosin heavy-chain gene expression (Jiang and Evans, 1996), suggesting that cardiac-restricted GATA factors may play a role in cardiac myocyte differentiation.

Targeted disruption of GATA genes has identified more complex roles for these factors in cardiogenesis. GATA-4–deficient mice died between ED 8.5 and ED 10.5, with disruptions to body patterning and severe folding defects (Kuo et al., 1997; Molkentin et al., 1997). Cardiac progenitors and differentiated cardiac myocytes are able to form in these embryos but are unable to migrate to the ventral midline to form the heart tube. This suggests that GATA-4 is not required for the determination and differentiation of cardiac myocytes per se but may be part of the morphogenic signaling mechanisms involved in embryo folding and the migration of the cardiac progenitors. GATA-4 knockout mice also have upregulated levels of GATA-6, which may account for the presence of differentiated cardiac myocytes. In contrast, a GATA-6 homozygous knockout is lethal by ED 6.5 to 7.5 due to severe disruption in the formation of the mesoderm and endoderm (Morrisey et al., 1998), suggesting a role for these GATA factors in early embryo patterning and potentially in cell lineage determination, as seen with GATA-1, -2, and -3. The role of GATA factors in the formation of the heart and the specification and differentiation of the cardiac myocyte is therefore complex. This may be explained in part by data suggesting that GATA-4 and GATA-6 together can differentially regulate various groups of genes (Charron et al., 1999). GATA factors also undergo posttranslational modification, including acetylation and phosphorylation, and can interact with a number of other factors, including FOG-2, NFAT, Nkx2.5, and SRF (Charron and Nemer, 1999), the relative influence of each of which may vary during development. It therefore seems likely that a combination of these regulatory pathways leads to the GATA-dependent regulation of cardiac troponin I gene expression and to changes in expression during development and maturation.

The A/T-rich element in the cardiac troponin I gene lies in a region fully conserved between the human and rodent gene promoters. Many genes expressed in the heart contain A/T-rich elements within their regulatory elements that have been shown to be required for expression. A number of factors from cardiac tissue have been identified binding these regions, such as MEF-2 (Brand, 1997), HF1b (Navankasattusas et al., 1994), and Oct-1 (Lakich et al., 1998). Both Oct-1 and MEF-2 have been identified binding to the human cardiac troponin I A/T-rich element from neonatal rat cardiac myocyte nuclear extracts. This element has also been shown to bind the recombinant human TATA box binding protein (TBP). The ability of the A/T-rich element to bind MEF-2 proteins, Oct-1 proteins, and TBP indicates a potential mechanism in the regulation for this gene. Steric arguments suggest it is unlikely that these factors can bind simultaneously to the A/T-rich element in the cardiac troponin I genes, and higher-order complexes are not seen on EMSA. It is therefore unclear which factors bind to this element in vivo, and it is possible that different factors bind at different developmental stages. Such a mechanism has been proposed for the skeletal muscle myosin heavy-chain 2B (MHC 2B) gene promoter (Lakich et al., 1998), which contains two A/T-rich elements. MHC 2B is expressed in adult skeletal muscle, but during development both A/T-rich elements are occupied by Oct-1, which acts to inhibit expression. In the adult, Oct-1 is displaced by MEF-2 at both sites, leading to derepression and transcriptional activation. It is possible that similar mechanisms act on this element in the cardiac troponin I gene. Surprisingly, and in spite of the general importance of the factors that are able to bind to it, mutation of the human cardiac troponin I A/T-rich element in a way that ablates all detectable binding activity only results in a 60% reduction in promoter activity (Bhavsar et al., 2000), suggesting that the element, although necessary for full activity is not essential for promoter function.

Multiple Factors Bind the Human CACC/Sp Element

The human cardiac troponin I gene contains a C-rich element within the 98-bp proximal promoter region, which comprises a CACC box (CCCACCCC) overlapping a consensus Sp1 element (CCCCGCCCC). This element is not conserved in the rodent promoters, although there are three more distal GA repeat sequences that have been shown to bind Sp1 family members in the mouse (Di Lisi et al., 1998). Both rodent genes also contain a more distal inverted CACC box (GGGGGTGGG) not found in the human promoter and a proximally located CACC box (CACACCC) similar to that required for the regulation of globin genes.

EMSA analysis of the human cardiac troponin I C-rich element has identified the binding of at least four factors (see Figure 8.2). Two of these bind preferentially to the Sp1 element and have been shown by supershift assay to be Sp1 and Sp3. The Sp family of transcription factors are widely expressed and are involved in the regulation of cell growth and differentiation and have been shown to play a role in the regulation of skeletal myogenesis (Vinals et al., 1997). Their role in the heart is unclear, although there is evidence for their involvement both in the regulation of genes during fetal heart development and in response to hypertrophic stimuli. The two other factors identified binding the C-rich element require an intact CACC box for binding. These factors do not appear to correspond to any previously described CACC box binding factor on the basis of their binding sequence requirements or in terms of their tissue distribution. Moreover, these factors are important for activity of the troponin I gene promoter, as mutations that disrupt their binding result in up to 90% loss of promoter activity (Dellow et al., 2001), and they appear to be cardiac-restricted in their expression (see Figure 8.2b). We therefore believe that these represent previously unidentified factors, which we have named HCB1 and HCB2 (for *h*eart *C*ACC *b*ox factors 1 and 2).

CACC box binding factors have been extensively studied in other systems, but their role in cardiac gene expression has been less well investigated. All known bona fide CACC box binding factors identified to date are zinc finger proteins. These include the Krüppel-like factor (KLF) family of ubiquitous and tissue-restricted factors, which belong to the Sp1/KLF superfamily of C_2H_2 zinc finger factors. These recognize various CACC box (CCCACCC) and CACC box-like (CACACC) elements. Many Krüppel-like factors have been identified, including the erythroid EKLF (Miller and Bieker, 1993), gut-enriched GKLF (Shields et al., 1996), lung LKLF (Anderson et al., 1995), and FKLF-2, which is highly expressed in bone marrow and striated muscle (Asano et al., 2000). KLF factors are involved in a wide variety of functions, including regulating cell growth and differentiation. EKLF, which is probably the best characterized to date, binds the CACC box (CACACC) found in the β-globin gene regulatory region and has been shown to be central to erythropoiesis (Nuez et al., 1995; Perkins et al., 1995).

CACC box elements have been identified in a number of genes expressed in striated muscle, including those encoding the slow and fast skeletal muscle troponin I isoforms (Nakayama et al., 1996), slow skeletal muscle myosin light chain 2 (Esser et al., 1999), slow skeletal/cardiac troponin C (Parmacek et al., 1994), cardiac α-actin (Biesiada et al., 1999), and myoglobin (Bassel-Duby et al., 1993). Although the function of these CACC boxes has been correlated with binding of the ubiquitously expressed Sp1, identification of muscle-specific CACC box binding factors has been limited to date. Two factors which were originally identified in extracts from skeletal muscle cell lines by their interactions with the myoglobin CACC box are myocyte nuclear factor (MNF) and CBF40 (Bassel-Duby et al., 1992, 1994). Although MNF was originally identified through its ability to bind the myoglobin CACC box, it was subsequently found to have a greater affinity for an A/T-rich element and belongs to the winged-helix family of factors (Yang et al., 1997).

The HCB1 and HCB2 factors we have identified appear to represent novel cardiac-restricted factors. They are unlikely to be one of the known CACC box binding

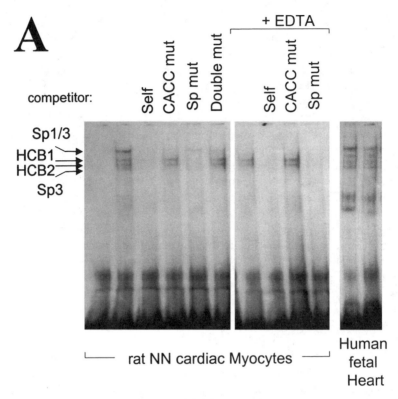

FIGURE 8.2. The C-rich element binds multiple factors, including novel factors HCB1 and HCB2. EMSA analysis using nuclear extracts of neonatal rat cardiac myocytes and human fetal cardiac muscle with an oligonucleotide cassette containing the C-rich element (CCCACCC CGCCCC) with or without a variety of competitor oligonucleotide cassettes. (A) The C-rich element binds at least four factors. These are competed by addition of unlabeled oligonucleotide (self). Mutation of the CACC box region (CCCACCC → TTCACCC) in the competitor (CACC mut) removes its ability to compete for binding of HCB1 and HCB2, but leaves it able to compete for Sp1/3. Mutation of the Sp1 region (CCCCGCCCC → CCC CGTTCC) of the competitor (Sp mut) results in reduced ability to bind Sp1 and Sp3, but leaves it able to compete for HCB1 and HCB2. Mutation of the central region (CCCACCC CGCCCC → CCCGGTACGCCCC) of the competitor (double mut) removes its ability to compete for any of the four complexes. Addition of EDTA removes binding of Sp1 and Sp3 but does not affect binding of HCB1 or HCB2 or the influence of competitors on the binding of these factors. A similar pattern of binding is seen with human fetal heart extracts as with the neonatal rat cardiac myocytes, confirming the presence of HCB1 and HCB2 in the human heart.

factors on at least three counts. First, all bona fide CACC box binding factors characterized to date have been shown to contain zinc finger DNA binding domains. These include members of the Sp1/KLF superfamily, the human T-cell receptor factor htβ, and the related rat ZF89 and human BERF-1 (Passantino et al., 1998), as well as the immediate early gene zinc finger factor Egr-1, which recognizes a GC-rich sequence but, as with Sp1 and Sp3, also has a low affinity for CACC box sequences (Rafty and Khachigian, 1998). Addition of chelating agents such as 1,10-

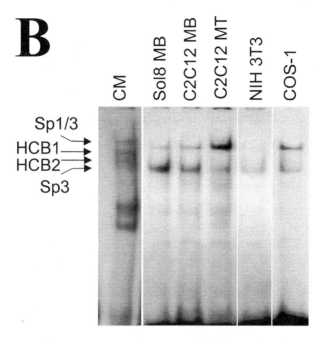

FIGURE 8.2, *Continued.* (B) EMSA analysis using an oligonucleotide cassette containing the C-rich element (CCCACCCCGCCCC) with nuclear extracts of neonatal rat cardiac myocytes (CM) compared with skeletal muscle cell lines Sol 8 and C2C12 (as myoblasts, MB, or myotubes, MT), NIH3T3 fibroblasts, and COS-1. Complexes containing Sp1 and Sp3 are readily detectable in all extracts tested, but HCB1 and HCB2 are only detected in cardiac myocytes.

phenanthroline and EDTA disrupts binding of such factors yet has no apparent effect on binding of HCB1 or HCB2 (see Figure 8.2A) under conditions where Sp1/Sp3 are clearly disrupted. Second, the cell line distribution of many of the known CACC box binding factors is different from that of HCB1 and HCB2. In particular, MNF, CBF40, and Egr-1 have all been detected by EMSA in the skeletal muscle cell lines Sol 8 and C2C12 cell lines (Bassel-Duby et al., 1992, 1994; Tounay and Benezra, 1996), whereas HCB1 and HCB2 have not. Nor are they detected in COS or NIH3T3 cells. Third, binding of HCB1 and HCB2 is unaffected by the addition of competitor oligonucleotides corresponding to the known binding sites for other CACC box binding factors. The CACC-like element in the human cardiac troponin I gene therefore appears to identify a potentially novel family of CACC box binding factors required for maximal activity of the promoter and appear to be distinct from previously described CACC box binding factors.

In summary, the human cardiac troponin I gene contains a compact promoter region that is sufficient for expression in cardiac myocytes and that binds multiple factors including MEF-2, Oct-1, GATA-4, Sp1, Sp3 and two novel factors that we have called HCB1 and HCB2 (Figure 8.3). Of these, GATA-4 (binding to the distal of two GATA elements in this region) and HCB1/2 appear to play the greatest role in transcriptional activation.

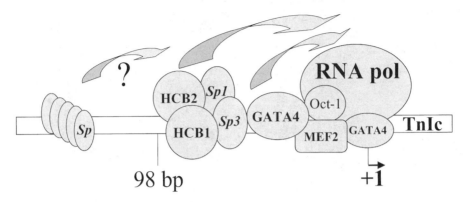

FIGURE 8.3. Model of transcription factor interactions on the human cardiac troponin I gene promoter. The 98 bp of the proximal promoter region are sufficient to drive the human cardiac troponin I promoter in cardiac myocytes. Within this region, multiple factors combine to drive transcription. Functional analysis suggests that of these GATA-4, bound to the distal element, and HCB1 and HCB2, bound to the CACC box region within the C-rich element, are the most significant. Sp factors may also influence transcriptional activity through binding to both the C-rich element within the 98-bp region and to multiple binding sites present within upstream repeat sequences. The roles of MEF-2, Oct-1, and the proximal GATA element are less clear.

GENES FOR TROPONIN I AND TROPONIN T ARE ORGANIZED IN PARALOGOUS PAIRS

Isoforms of troponin are encoded by multigene families, which are subject to both transcriptional regulation and, in some cases, tissue-specific and developmentally regulated alternative splicing. In total, eight genes have been identified in the human genome that encode the components of the troponin complex found in cardiac and skeletal muscle. Multigene families such as this may be dispersed throughout the genome or clustered at particular sites. In some cases, clustering plays an important role in regulation, as in the well-characterized developmental regulation of globin gene transcription and the temporal–spatial regulation of the HOX homeobox gene clusters. We therefore investigated the genomic organization of the troponin genes in humans (Barton et al., 1997) and in mice (Barton et al., 2000) and identified an unexpected pattern of linkage whereby the six genes encoding cardiac, slow skeletal, and fast skeletal isoforms of troponin I and troponin T are organized at three genomic loci, each containing a paralogous troponin I–troponin T gene pair (Figure 8.4). Although all sarcomeric proteins, including isoforms of actin, myosin light chain, and tropomyosin, are encoded by multigene families, only the myosin heavy-chain genes have been previously identified as being linked. These are grouped at two loci, one containing the α- and β-cardiac genes, the other containing the skeletal muscle genes and probably originated through tandem gene duplication. Studies have shown that although the organization of myosin genes does not relate to their temporal or spatial pattern of expression, as in the globin and homeobox gene clusters, it is highly conserved between the mouse and the human (Weiss et al., 1999).

The linkage pattern of the troponin I and troponin T genes is different from that seen for myosin because it occurs between members of two different gene families—

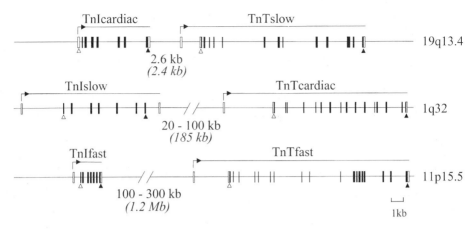

FIGURE 8.4. Overall organization of human troponin I and troponin T genes. The organization of the six human troponin I and troponin T genes is illustrated showing relative exon distribution and the intergene distances in both man and mouse (italics in parentheses) (Barton et al., 2000). Chromosomal location of human genes is shown on the right (Barton et al., 1997). Data derived from the following references: cardiac troponin I—slow skeletal troponin T locus (Barton et al., 1999), the slow skeletal troponin I gene (Corin et al., 1994), cardiac troponin T gene (Farza et al., 1998), fast skeletal troponin I gene (Mullen and Barton, 2000), and fast skeletal troponin T gene (EMBL/Genbank AF026276). Open and filled triangles point to initiation and stop codons, respectively. Arrows denote direction of transcription. Note that although the locations of the slow skeletal muscle and fast skeletal muscle troponin I genes relative to the respective troponin T genes are known, their orientation is assumed.

although troponin I and troponin T proteins are associated in the troponin complex, they are not related in sequence or structure. The pairing is also intriguing because there is no obvious correlation with expression pattern. For example, it is striking that the cardiac troponin I and slow skeletal troponin T genes are separated by only 2.6 kb in humans whereas their expression is mutually exclusive. The other troponin gene pairs show common sites of expression in either embryo, fetus, or adult (Schiaffino et al., 1993), but cardiac troponin I and slow skeletal troponin T are restricted to cardiac and skeletal muscle, respectively. These genes are similarly close in the mouse (where they are 2.4 kb apart), whereas there is significant variation in intron sizes between the two species, suggesting that the conserved distance separating the genes is of importance. It has been proposed that elements required for expression of the mouse slow skeletal troponin T gene may be embedded in the cardiac troponin I gene (Huang and Jin, 1999), which might indeed act to limit any evolutionary drift in their physical separation. Consistent with this idea, there is little sequence similarity between human and mouse intergene regions, outside the first 100 bp of proximal promoter of the slow skeletal troponin T promoter (data not shown).

In contrast to the close linkage described earlier, the troponin I and troponin T genes that are coexpressed in fast skeletal muscle are separated by about 300 kb in humans and about 1.2 Mb in the mouse. Not surprisingly, more recent data have established that unrelated genes are located between them. For example, the lymphocyte-specific protein (*Lsp1*) gene is just 4.3 kb upstream of the fast skeletal troponin T gene in the mouse (Misener et al., 1998). By analogy with the organization of the cardiac troponin I and slow skeletal troponin T genes, this would place *Lsp1* between the two fast skeletal troponin genes. The fast skeletal troponin T and *Lsp1* genes are transcribed in the same direction with the 5′ end of the troponin T gene toward the telomere. *Lsp1* is not expressed in skeletal muscle and, as with the closely linked cardiac troponin I and slow skeletal troponin T genes, this raises issues as to the regulation of closely linked but independent transcriptional units.

Their overall organization suggests that the troponin I and troponin T genes probably arose by successive rounds of duplication of an ancestral gene pair as part of the general process of chromosomal duplication during vertebrate evolution. In agreement with this viewpoint, *Drosophila* has single troponin I (*wingsup*) and troponin T (*heldup*) genes, both of which are located on chromosome X, separated by about 4.5 Mb of intervening sequence. The exact events that generated the current vertebrate gene organization remain unclear, but one possibility is that the ancestral locus contained loosely linked genes, which were subsequently brought together by some form of deletion. The arrangement of troponin I and troponin T genes in *Drosophila* is consistent with this idea, as is the more evolutionarily recent specialization of cardiac muscle, which would suggest that the more tightly linked pairs, both of which contain a cardiac gene, arose later in evolution. An alternative is that tightly linked genes in the ancestral locus became increasingly separated. Ultimately, it is conceivable that there was a single ancestral gene, which combined the functions of troponin I and troponin T in a single protein. This is plausible, given both the head-to-tail organization of their genes and considering that the proteins are intimately associated within the troponin protein complex. In this respect, it is of interest that caldesmon has been proposed as an allosteric regulator of actin–myosin interaction in smooth muscle and, although unrelated in sequence, combines functional features of both troponin I and troponin T (Marston, 1995). If such a combined ancestral gene existed, separation of two distinct transcriptional units would have most likely preceded duplication of the locus and the subsequent divergence of the independent patterns of expression seen in the current gene pairs. Genes encoding the third protein of the complex, troponin C, are not linked either to the troponin I–troponin T gene pairs or to each other and appear therefore to have evolved independently. Moreover, only two troponin C genes have been identified. One is expressed in fast skeletal muscle, the other in both slow skeletal and cardiac muscles. Whatever the precise nature of the evolutionary events that gave rise to the troponin I and troponin T genes, their organization raises questions as to their regulation. It is unclear, for example, how high-level, independent transcription is achieved by the closely linked cardiac troponin I and slow skeletal troponin T genes, or whether the coexpressed fast skeletal troponin I and troponin T genes share any common regulatory elements in spite of the distance between them. Further experiments are required to examine the precise expression patterns and regulatory mechanisms of these genes and to determine what part, if any, physical linkage plays in their regulation.

ACKNOWLEDGMENTS

This work was supported by the British Heart Foundation (FS297, PG/98194, and PG/99007).

REFERENCES

Anderson, K.P., Kern, C.B., Crable, S.C., and Lingrel, J.B. 1995. Isolation of a gene encoding a functional zinc finger protein homologous to erythroid Kruppel-like factor: identification of a new multigene family. *Mol. Cell. Biol.* 15:5858–5965.

Asano, H., Li, X.S., and Stamatoyannopoulos, G. 2000. FKLF-2: a novel Kruppel-like transcription factor that activates globin and other erythroid lineage genes. *Blood* 95:3578–3584.

Ausoni, S., Campione, M., Picard, A., Moretti, P., Vitadello, M., De Nardi, C., and Schiaffino, S. 1994. Structure and regulation of the mouse cardiac troponin I gene. *J. Biol. Chem.* 269:339–346.

Barton, P.J.R., Cullen, M.E., Townsend, P.J., Brand, N.J., Mullen, A.J., Norman, D.A.M., Bhavsar, P.K., and Yacoub, M.H. 1999. Close physical linkage of troponin genes: organization and complete sequence of the locus encoding cardiac troponin I and slow skeletal troponin T. *Genomics* 57:102–109.

Barton, P.J.R., Mullen, A.J., Dhoot, G.K., Cullen, M.E., Simon-Chazottes, D., and Guénet, J.-L. 2000. Genes encoding troponin I and troponin T are organised as three paralogous pairs in the mouse genome. *Mamm. Genome* 11:926–929.

Barton, P.J.R., Townsend, P.J., Brand, N.J., and Yacoub, M.H. 1997. Localization of the fast skeletal muscle troponin I gene (TNNI2) to 11p15.5: Genes for troponin I and T are organised in pairs. *Ann. Hum. Genetics* 61:519–523.

Bassel-Duby, R., Grohe, C.M., Jessen, M.E., Parsons, W.J., Richardson, J.A., Chao, R., Grayson, J., Ring, W.S., and Williams, R.S. 1993. Sequence elements required for transcriptional activity of the human myoglobin promoter in intact myocardium. *Circ. Res.* 73:360–366.

Bassel-Duby, R., Hernandez, M.D., Gonzalez, M.A., Krueger, J.K., and Williams, R.S. 1992. A 40-kilodalton protein binds specifically to an upstream sequence element essential for muscle-specific transcription of the human myoglobin promoter. *Mol. Cell. Biol.* 12:5024–5032.

Bassel-Duby, R., Hernandez, M.D., Yang, Q., Rochelle, J.M., Seldin, M.F., and Williams, R.S. 1994. Myocyte nuclear factor, a novel winged-helix transcription factor under both developmental and neural regulation in striated myocytes. *Mol. Cell. Biol.* 14:4596–4605.

Bhavsar, P.K., Brand, N.J., Yacoub, M.H., and Barton, P.J.R. 1996. Isolation and characterisation of the human cardiac troponin I gene (*TNNI3*). *Genomics* 35:11–23.

Bhavsar, P.K., Dellow, K.A., Yacoub, M.H., Brand, N.J., and Barton, P.J.R. 2000. Identification of *cis*-acting DNA elements required for expression of the human cardiac troponin I gene promoter. *J. Mol. Cell. Cardiol.* 32:95–108.

Biesiada, E., Hamamori, Y., Kedes, L., and Sartorelli, V. 1999. Myogenic basic helix-loop-helix proteins and Sp1 interact as components of a multiprotein transcriptional complex required for activity of the human cardiac alpha-actin promoter. *Mol. Cell. Biol.* 19:2577–2584.

Brand, N.J. 1997. Myocyte enhancer factor 2 (MEF2). *Int. J. Biochem. Cell. Biol.* 29:1467–1470.

Charron, F. and Nemer, M. 1999. GATA transcription factors and cardiac development. *Semin. Cell Dev. Biol.* 10:85–91.

Charron, F., Paradis, P., Bronchain, O., Nemer, G., and Nemer, M. 1999. Cooperative interaction between GATA-4 and GATA-6 regulates myocardial gene expression. *Mol. Cell. Cardiol.* 19:4355–4365.

Corin, S.J., Juhasz, O., Zhu, L., Conley, P., Kedes, L., and Wade, R. 1994. Structure and expression of the human slow twitch skeletal muscle troponin I gene. *J. Biol. Chem.* 269:10651–10659.

Dellow, K.A., Bhavsar, P.K., Brand N.J., and Barton, P.J.R. 2001. Identification of novel, cardiac-restricted transcription factors binding to a CACC-box within the human cardiac troponin I promoter. *Cardiovasc. Res.* 50:24–32.

Di Lisi, R., Millino, C., Calabria, E., Altruda, F., Schiaffino, S., and Ausoni, S. 1998. Combinatorial *cis*-acting elements control tissue-specific activation of the cardiac troponin I gene in vitro and in vivo. *J. Biol. Chem.* 273:25371–25380.

Esser, K., Nelson, T., Lupa-Kimball, V., and Blough, E. 1999. The CACC box and myocyte enhancer factor-2 sites within the myosin light chain 2 slow promoter cooperate in regulating nerve-specific transcription in skeletal muscle. *J. Biol. Chem.* 274:12095–12102.

Farza, H., Townsend, P.J., Carrier, L., Barton, P.J.R., Mesnard, L., Bahrend, E., Forissier, J.F., Fiszman, M.Y., Yacoub, M.H., and Schwartz, K. 1998. Genomic organisation, alternative splicing and polymorphisms of the human cardiac troponin T gene. *J. Mol. Cell. Cardiol.* 30:1247–1253.

Grepin, C., Nemer, G., and Nemer, M. 1997. Enhanced cardiogenesis in embryonic stem cells overexpressing the GATA-4 transcription factor. *Development* 124:2387–2397.

Heikinheimo, M., Scandrett, J.M., and Wilson, D.B. 1994. Localization of transcription factor GATA-4 to regions of the mouse embryo involved in cardiac development. *Dev. Biol.* 164:361–373.

Huang, Q.-Q. and Jin, J.P. 1999. Preserved close linkage between the genes encoding troponin I and troponin T, reflecting an evolution of adapter proteins coupling the Ca2+ signaling of contractility. *J. Mol. Evol.* 49:780–788.

Jiang, Y. and Evans, T. 1996. The *Xenopus* GATA-4/5/6 genes are associated with cardiac specification and can regulate cardiac-specific transcription during embryogenesis. *Dev. Biol.* 174:258–270.

Kimura, A., Harada, H., Park, J.-K., Nishi, H., Satoh, M., Takahashi, M., Hiroi, S., Sasaoka, T., Ohbuchi, N., Nakamura, T., Koyanagi, T., Hwang, T.-H., Choo, J.-A., Chung, K.-S., Hasegawa, A., Nagai, R., Okazaki, O., Nakamura, H., Matsuzaki, M., Sakamoto, T., Toshima, H., Koga, Y., Imaizumi, T., and Sasazuki, T. 1997. Mutations in the cardiac troponin I gene associated with hypertrophic cardiomyopathy. *Nat. Genetics* 16:379–382.

Kuo, C.T., Morrisey, E.E., Anandappa, R., Sigrist, K., Lu, M.M., Parmacek, M.S., Soundais, C., and Leiden, J.M.. 1997. GATA4 transcription factor is required for ventral morphogenesis and heart tube formation. *Gene Dev.* 11:1048–1060.

Lakich, M.M., Diagana, T.T., North, D.L., and Whalen, R.G. 1998. MEF-2 and Oct-1 bind to two homologous promoter sequence elements and participate in the expression of a skeletal muscle-specific gene. *J. Biol. Chem.* 273:15217–15226.

Marston, S. 1995. Ca^{2+}-dependent protein switches in actomyosin based contractile systems. *Int. J. Biochem. Cell. Biol.* 27:97–108.

Miller, I.J. and Bieker, J.J. 1993. A novel, erythroid cell-specific murine transcription factor that binds to the CACCC element and is related to the Kruppel family of nuclear proteins. *Mol. Cell. Biol.* 13:2776–2786.

Misener, V.L., Wielowieyski, A., Brennan, L.A., Beebakhee, G., and Jongstra, J. 1998. The mouse Lsp1 and Tnnt3 genes are 4.3 kb apart on distal mouse chromosome 7. *Mamm. Genome* 9:846–848.

Molkentin, J.D., Lin, Q., Duncan, S.A., and Olson, E.N. 1997. Requirement of the transcription factor GATA4 for heart tube formation and ventral morphogenesis. *Gene Dev.* 11:1061–1072.

Morrisey, E.E., Tang, Z., Sigrist, K., Lu, M.M., Jiang, F., Ip, H.S., and Parmacek, M.S. 1998. GATA6 regulates HMF4 and is required for differentiation of visceral endoderm in the mouse embryo. *Genes Dev.* 1210:3579–3590.

Mullen, A.J. and Barton, P.J.R. 2000. Structural characterization of the human fast skeletal muscle troponin I gene (*TNNI2*). *Gene* 242:313–320.

Murphy, A.M., Thompson, W.R., Peng, L.F., and Jones II, L. 1997. Regulation of the rat cardiac troponin I gene by the transcription factor GATA-4. *Biochem. J.* 322:393–401.

Nakayama, M., Stauffer, J., Cheng, J., Bannerjee-Basu, S., Wawrousek, E., and Buonanno, A. 1996. Common core sequences are found in skeletal muscle slow- and fast-fiber-type-specific regulatory elements. *Mol. Cell. Biol.* 16:2408–2417.

Navankasattusas, S., Sawadogo, M., van Bilsen, M., Dang, C.V., and Chien, K.R. 1994. The basic helix-loop-helix protein upstream stimulating factor regulates the cardiac ventricular myosin light-chain 2 gene via independent *cis* regulatory elements. *Mol. Cell. Biol.* 14: 7331–7339.

Nuez, B., Michalovich, D., Bygrave, A., Ploemacher, R., and Grosveld, F. 1995. Defective haematopoiesis in fetal liver resulting from inactivation of the EKLF gene. *Nature* 375: 316–318.

Parmacek, M.S., Ip, H.S., Jung, F., Shen, T., Martin, J.F., Vora, A.J., Olson, E.N., and Leiden, J.M. 1994. A novel myogenic regulatory circuit controls slow/cardiac troponin C gene transcription in skeletal muscle. *Mol. Cell. Biol.* 14:1870–1885.

Passantino, R., Antona, V., Barbieri, G., Rubino, P., Melchionna, R., Cossu, G., Feo, S., and Giallongo, A. 1998. Negative regulation of β enolase gene transcription in embryonic muscle is dependent upon a zinc finger factor that binds to the G-rich box within the muscle-specific enhancer. *J. Biol. Chem.* 273:484–494.

Perkins, A.C., Sharpe, A.H., and Orkin, S.H. 1995. Lethal beta-thalassaemia in mice lacking the erythroid CACCC-transcription factor EKLF. *Nature* 375:318–322.

Rafty, L.A. and Khachigian, L.M. 1998. Zinc finger transcription factors mediated high constitutive platelet-derived growth factor-B expression in smooth muscle cells derived from aortae of newborn rats. *J. Biol. Chem.* 273:5758–5764.

Sasse, S., Brand, N.J., Kyprianou, P., Dhoot, G.K., Wade, R., Arai, M., Periasamy, M., Yacoub, M.H., and Barton, P.J.R. 1993. Troponin I gene expression during human cardiac development and in end-stage heart failure. *Circ. Res.* 72:932–938.

Schiaffino, S., Gorza, L., and Ausoni, S. 1993. Troponin isoform switching in the developing heart and its functional consequences. *Trends Cardiovasc. Med.* 3:12–17.

Schiaffino, S. and Reggiani, C. 1996. Molecular diversity of myofibrillar proteins: Gene regulation and functional significance. *Physiol. Rev.* 76:371–423.

Shields, J.M., Christy, R.J., and Yang, V.W. 1996. Identification and characterisation of a gene encoding a gut-enriched Kruppel-like factor expressed during growth arrest. *J. Biol. Chem.* 271:20009–20017.

Thierfelder, L., Watkins, H., MacRae, C., Lamas, R., McKenna, W., Vosberg, H.P., Seidman, J.G., and Seidman, C.E. 1994. Alpha-tropomyosin and cardiac troponin T mutations cause familial hypertrophic cardiomyopathy: a disease of the sarcomere. *Cell* 77:701–712.

Tounay, O. and Benezra, R. 1996. Transcription of the dominant-negative helix-loop-helix protein Id1 is regulated by a protein complex containing the immediate-early response gene Egr-1. *Mol. Cell. Biol.* 16:2418–2430.

Vinals, F., Fandos, C., Santalucia, T., Ferre, J., Testar, X., Palacin, M., and Zorzano, A. 1997. Myogenesis and MyoD down-regulate Sp-1: a mechanism for the repression of Glut 1 during muscle cell differentiation. *J. Biol. Chem.* 272:12913–12921.

Weiss, A., McDonough, D., Wertman, B., Acakpo-Satchivi, L., Montgomery, K., Kucherlapati, R., Leinwand, L., and Krauter, K. 1999. Organization of human and mouse skeletal myosin heavy chain gene clusters is highly conserved. *Proc. Natl. Acad. Sci. USA* 96:2958–2963.

Yang, Q., Bassel-Duby, R., and Williams, R.S. 1997. Transient expression of a winged-helix protein, MNF-beta, during myogenesis. *Mol. Cell. Biol.* 17:5236–5243.

CHAPTER 9

Signal Transduction in Myofibrillogenesis, Cell Growth, and Hypertrophy

M.A.Q. Siddiqui, Michael Wagner, Eduardo Mascareno, and Saiyid Shafiq

In that segment of the adult population at risk for hypertension and cardiac hypertrophy, hemodynamic pressure overload of the heart can lead to its enlargement and functional demise. Such enlargement results from an adaptive response on the part of the heart to alleviate pressure overload by providing increased contraction for more forceful pumping of blood. Cardiac hypertrophy also occurs during the normal development of the heart. As the developing cardiovascular system of the fetus places increased hemodynamic demands upon the fetal heart, the heart responds by undergoing hypertrophic growth to increase its size and contractility. Interestingly, many aspects of pathological adult cardiac hypertrophy appear to be similar to those of fetal cardiac hypertrophy, leading to the notion that adult heart cells respond to pressure overload by calling upon the fetal hypertrophic growth program to provide compensatory growth and increased contractility. Indeed, many contractile protein genes originally expressed during myofibrillogenesis in fetal cardiomyocytes are reexpressed during adaptive hypertrophic growth in the adult heart.[1,2] The similarities in gene expression profiles between fetal and adult hypertrophic myofibrillogenic programs suggest that many of the transcription factors, signal transduction proteins, and other regulatory molecules responsible for fetal hypertrophy may also be active in adult cardiac hypertrophy. In this review, we will discuss the recent progress made in the study of the transcription and signal transduction mechanisms responsible for expression of contractile proteins and maintenance of the hypertrophic state in cardiac muscle cells.

HYPERTROPHIC GROWTH OF CARDIAC CELLS DURING DEVELOPMENT AND IN THE ADULT

During fetal development, the heart undergoes rapid growth and increased pumping capacity to meet the needs of the expanding cardiovascular system. To increase muscle mass and contractility, cardiac muscle cells first undergo cell division to generate increased numbers of cells that populate and expand the myocardium. This phase of cardiac cell proliferation or hyperplasia ends some time after birth, and all subsequent myocardial growth is via the enlargement or hypertrophic growth of individual muscle cells.[3] Given the different demands placed upon the cardiac muscle cell by these growth requirements, it is likely that two different genetic programs are operating to bring about hyperplasia and hypertrophic growth. In an adult with a normally functioning heart, neither one of these developmental programs is expressed in cardiac muscle cells after the heart is fully formed. However, in certain individuals, overburdening the heart with an excessive workload triggers an adaptive response that involves reactivation of the fetal program of hypertrophic growth and myofibrillogenesis. Under these conditions, heart muscle cells undergo an increase in mass and reexpress proteins that form the contractile apparatus. Increased cardiac mass and contractile protein expression do not lead to an increase in normal cardiac function, however, and despite this adaptive growth response, the heart fails as a result of its decreased ability to adequately pump blood.

From the preceding observations, it is apparent that the adaptive response of cardiac muscle cells to pressure overload is directed toward increasing contractility through reexpression of the developmental programs of cardiac growth and myofibrillogenesis. The reexpression of a developmental growth program suggests that all the regulatory molecules responsible for coordinated and controlled gene expression within the program are also reexpressed during adaptive growth. For example, the appropriate assembly of a functioning sarcomere in adult cardiac muscle cells is likely to require the same coordinated expression of contractile protein and enzyme genes regulating sarcomere formation and function during development. Thus, recapitulation of the myofibrillogenic and hypertrophic growth programs is likely to involve reexpression of the transcription factors and regulatory molecules that control fetal expression of contractile protein genes as well as the reactivation of signal transduction pathways involved in the growth of cardiac myocytes. In this review, we consider these two aspects of cardiac hypertrophy in the context of both developmental growth and adaptive hypertrophic growth. Because the number of contractile proteins and signal transduction pathways involved is too large and complex to be adequately described in this review, we will focus on two representative examples of each, expression of the gene encoding the cardiac contractile protein myosin light chain 2 and the Jak/Stat signal transduction pathway involved in angiotensin-dependent cardiac hypertrophy. The study of these specific examples may provide more general insights into the molecular mechanisms underlying the adaptive response of cardiac muscle cells to hypertrophic stimuli.

DEVELOPMENTAL GROWTH OF THE HEART AND INITIATION OF MYOFIBRILLOGENESIS

The heart is the first functional organ to develop in vertebrate embryos and arises from a complex series of cellular and morphogenetic interactions. Much of heart embryology has been studied using the avian embryo as a model system. In the avian embryo, heart development begins with the onset of gastrulation at stage 4 when the committed cells of the splanchnic mesoderm migrate through the primitive streak and come to lie on either side of Hensen's node, the avian equivalent of Spemann's organizer in *Xenopus* embryos.[4,5,6] The mesodermal cells in the two bilateral heart-forming regions undergo proliferation and differentiation and form two independent heart primordia, which fuse at about stage 9 to form a single beating heart tube at stage 10.[7] In the mouse embryo, the formation of the heart tube occurs through essentially identical steps between 7 and 8.5 days postcoitum.[8]

The onset of a beating heart tube implies that the myofibrillogenic program of gene expression leading to production of contractile proteins and assembly of functional sarcomeres has already commenced by this stage of development. In chicken embryos, contraction of cardiomyocytes can be observed after 36 hours in ovo at stage 9 to 10. At this point, cardiac myocytes are positive for a number of contractile proteins, such as desmin, α-actinin, titin, myosin, α-actin, troponin, and myomesin.[9] Most myofibrillar proteins are in their proper place in the sarcomere as soon as the first contractions appear in the developing heart. Soon thereafter, the myofibrils acquire their definitive orientation along the cell axis, probably by interaction with the intermediate filament protein, desmin, and with vinculin in the costameres, which constitute important components of the cardiac cytoskeleton responsible for transmission of force generated by the contractile apparatus.[10,11] At around 48 hours of development, the flow of blood within the nascent circulatory system is controlled by the rhythmically contracting meshwork of myofibrils in the cardiomyocytes.[10]

Shortly after birth, the neonatal heart grows by hypertrophy of the cardiomyocytes, due in part to the pressure overload engendered by the expanding circulatory system. A striking feature of the ventricular hypertrophic response at this stage is the expression of genes not normally expressed in the adult heart (e.g., skeletal α-actin, atrial myosin light chain 1, β-myosin heavy chain, and atrial natriuretic factor) and the upregulation of the constitutively expressed ventricular myosin light chain 2 (vMLC-2) protein.[12] The expression of vMLC-2 is particularly interesting because it appears to play a role in the adaptive response of both developing and adult cardiac muscle cells to hypertrophic growth.

GENETIC MECHANISMS UNDERLYING THE ADAPTIVE RESPONSE TO HYPERTROPHIC STIMULI

Fetal cardiomyocytes respond to the pressure overload of the expanding circulatory system of the fetus by undergoing hypertrophic growth and expressing an array of contractile protein genes. This same response to pressure overload can be seen in cardiac muscle cells of an adult heart. As with developmental hypertrophic growth, the response of adult cardiac muscle cells to hypertrophic stimuli has been well studied and delineated in terms of its underlying genetic program.[3,13,14] Within the first 30 minutes of sustained pressure overload, an immediate–early set of genes encoding proto-oncogene transcription factors such as *egr*-1, *c-fos*, *c-jun*, and *c-myc*, among others, is rapidly expressed. After this immediate–early gene-expression phase, the next 12 hours of the hypertrophic response are characterized by the reexpression of "fetal" contractile protein genes. These genes, which include those encoding β-myosin heavy chain, skeletal α-actin, and β-tropomyosin, were first expressed during heart development to impart contractility to embryonic and fetal cardiomyocytes. In addition to these genes, continued exposure to hypertophic agents elicits expression of constitutively expressed contractile protein genes such as myosin light chain 2 and cardiac α-actin. Thus, the genetic program underlying the adaptive response of cardiac muscle cells to hypertrophic stimuli such as pressure overload is biphasic, with immediate–early proto-oncogenes expressed first, followed by the reexpression of a fetal program that recapitulates the developmental growth and myofibrillogenesis that cardiomyocytes undergo to acquire contractile properties.

The activation of these genetic programs occurs when hypertrophic stimuli impinge upon cardiac muscle cells and initiate a signaling cascade that leads from the hypertrophic signal at the cell membrane to the activation of genes in the cell nucleus. The most widely accepted model of the genetic and subcellular events underlying hypertrophy of cardiac muscle cells suggests that hypertrophic stimuli, either mechanical or hormonal, activate a variety of cell surface receptors that are linked to signal transducers.[3,14] These transducers can directly activate and cross-activate multiple signal transduction pathways that lead to the nucleus, where they activate nuclear genes. In addition to the proto-oncogene and contractile protein genes mentioned earlier, growth factors and signaling molecules such as angiotensin II are also expressed by hypertrophic cardiac muscle cells. These molecules appear to sustain the adaptive response of hypertrophic cardiac muscle cells by acting in an autocrine fashion to maintain the hypertrophic state.[15] Although the signal transduction pathways responsible for transmitting the hypertrophic signal from membrane to nucleus have been the subject of intense study, how these pathways lead to activation of the genes involved in implementing and maintaining hypertrophic growth is only now being addressed. Our laboratory has focused on this aspect of cardiac hypertrophy by examining the genetic mechanisms controlling expression of two genes involved in this process, cardiac myosin light chain 2 and angiotensinogen. We have taken an approach that first seeks to identify the *cis*-acting promoter elements that control the expression of these genes and then characterizing the *trans*-acting factors that interact with these elements to mediate the signal-to-gene interaction.

VMLC-2 AND MYOFIBRILLOGENESIS IN STRIATED MUSCLE

Until recently, the physiological role of MLC-2 in striated muscle had not been as clearly established as its role in smooth muscle, where phosphorylation of MLC-2 by MLC-2 kinase serves as a switch for turning on the actin-activated myosin ATPase activity and controlling muscle contraction.[16] Later studies, however, have clearly shown that in striated muscles MLC-2 undergoes phosphorylation and that both its phosphorylated and nonphosphorylated forms play important developmental as well as functional roles, first in the organization of the cytoskeleton and myofibrillogenesis and then in the mechanics of filament contraction of fully differentiated muscle cells.[17] These observations have been confirmed and extended by genetic studies showing MLC-2 to be a critical component of myofibrillogenesis and cardiac cell contractility. In genetically altered mice lacking the vMLC-2 gene, the atrial form of MLC-2 was found to be upregulated in the ventricles of the embryonic heart.[18] Despite this increased ventricular expression, atrial MLC-2 was not able to functionally compensate for the missing ventricular form of MLC-2: ultrastructural analysis revealed defects in sarcomere assembly and an embryonic form of dilated cardiomyopathy. It was concluded that there was a unique requirement for vMLC-2 in myofibrillogenesis as well as in the functional maturation of the ventricular chamber. More recently, Aoki and coworkers have clearly demonstrated that during development of cardiac hypertrophy at least one signaling pathway for assemby of new sarcomeres is through phosphorylation of vMLC-2.[19] These observations are supported by a study in which transgenic mice expressing a transgene encoding a nonphosphorylatable form of vMLC-2 exhibited sarcomeric disorganization and developed cardiomyopathy.[20] Together, these observations suggest that vMLC-2 is a key contributor to the myofibrillogenic program during development as well as to the adaptive response of hypertrophic adult cardiac muscle cells to hemodynamic stress overload.

CONTRACTILE PROTEIN GENE EXPRESSION: MYOSIN LIGHT CHAIN 2 GENE

The overriding question concerning cMLC-2 and all other contractile proteins reenlisted into the service of hypertrophic cardiac muscle cells is how their genes are reactivated by hypertrophic stimuli. One possibility is that the transcriptional apparatus that once activated these genes during development becomes reactivated in hypertrophic cardiac muscle cells. Addressing this possibility requires knowledge of the gene promoters driving cardiac gene expression and the transcription factors and associated proteins interacting with promoter elements to activate their transcription. In the case of the cMLC-2 gene, our laboratory as well as others have delineated both the *cis*- and *trans*-acting elements responsible for cMLC-2 gene expression and have conducted studies of cMLC-2 gene expression in heart tissue during development and under hypertrophic conditions.[21-24]

Expression of the chicken cMLC-2 gene is controlled by both positive and negative regulatory elements within the promoter.[22,25,26] Among the *cis*-acting promoter elements that positively regulate the cMLC-2 gene are the MEF-2 binding site element and the CArG box.[23] The CArG box and MEF-2 binding-site elements appear to be involved in the developmental regulation of the MLC-2 gene. Our laboratory has shown that during early stages of chicken development the MEF-2 binding site is bound by a nuclear protein called BBF-1.[23] BBF-1 appears to be immunologically distinct from the major MEF-2 protein isoforms and to bind the MEF-2 binding site prior to binding by known MEF-2 proteins. BBF-1 binding occurs as early as stage 5 of chicken development, a stage that coincides with the onset of cardiac MLC-2 mRNA expression (as determined by RT-PCR analysis of early-stage chicken embryo RNA[23]) and specification of the precardiac mesoderm as cardiogenic and well before the first overt signs of cardiac cell differentiation. Thus, BBF-1 may be an important regulator of chicken cMLC-2 gene expression during development of the cardiovascular system.

The upregulation of the cMLC-2 gene and reexpression of fetal contractile protein genes during hypertrophy raises two possibilities as to the *trans*-activating factors responsible for their expression. The first possibility is that one of the proto-oncogene transcription factors expressed early in the progression of hypertrophy acts to turn on the cMLC-2 gene. There are presently no experimental data that either favor or preclude this possibility. The second possibility is that the transcription factors that regulate cMLC-2 expression during development are upregulated and/or reexpressed in hypertrophic cardiac cells. To investigate this possibility for the cMLC-2 gene, our laboratory compared the binding activity of nuclear proteins to both the MEF-2 binding site and the CArG box of the cMLC-2 promoter in normal (WKY) and genetically hypertensive (SHR) strains of rats.[24] Nuclear extracts from SHR rat hearts exhibited significantly higher binding activity than those of WKY rats for both the MEF-2 binding site and the CArG box. For the MEF-2 binding site, this increase resulted from increased binding by BBF-1, BBF-3 (another binding protein of the MEF-2 binding site), and MEF-2 proteins. Interestingly, binding by BBF-1 and MEF-2 appeared to coincide with the development of hypertrophy in the SHR rats. The onset of BBF-1 binding activity for the MEF-2 binding site has been shown to occur as early as the formation of the cardiogenic mesoderm in chicken development.[23] This finding, together with the observed increase in BBF-1 binding activity in the myocardium of hypertrophic adult rats, suggests that the upregulation of cMLC-2 gene expression during hypertrophy is dependent upon the reassembly of the same transcriptional apparatus responsible for cMLC-2 gene expression during development. Although the gel-mobility shift assays used in these studies indicate an increased binding activity for BBF-1 and MEF-2, it is presently unclear whether this increase indicates a similar upregulation (i.e., increased transcription) of the genes encoding these transcription factors. Addressing this question requires the cloning and characterization of the gene encoding the BBF-1 activity.

Our laboratory has attempted to clone the cDNA encoding the BBF-1 binding protein.[27] An oligonucleotide representing the MEF-2 binding site to which BBF-1 binds was used to screen a cDNA expression library made from stage 6 chicken heart-forming region mRNA. This expression-cloning approach yielded a single cDNA, called CLP-1, that encodes a novel protein. Antibodies made to the recombinant CLP-1 protein were able to disrupt and supershift the MEF-2 binding site probe in gel-mobility shift assays. In addition, RT-PCR analysis of the early expression pattern of the CLP-1 mRNA showed that CLP-1 mRNA is expressed as early as stage 2 to 3, prior to stage 4 when precardiac mesodermal cells are assigned a cardiac fate. Immunohistochemistry of both isolated cardiomyocytes and chicken embryos shows CLP-1 to be a nuclear protein expressed in the precardiac mesoderm. These data suggest that CLP-1 is a candidate cDNA encoding the BBF-1 binding activity and may be a transcription factor involved in the expression of cardiac genes. Because the MEF-2 binding site to which BBF-1 binds is in the cMLC-2 gene promoter, this also suggests that CLP-1 is a candidate transcription factor for the developmental activation of the cMLC-2 gene. If this same developmental program of myofibrillogenesis is reactivated in hypertrophic cardiac muscle cells, then an important part of the hypertrophic adaptive response must be the reexpression of cardiogenic transcription factors such as CLP-1. It will be interesting to determine whether CLP-1 reexpression coincides with myofibrillogenesis during cardiac hypertrophy and whether CLP-1 is responsible for the reexpression of the cMLC-2 gene.

RAS, ANGIOTENSIN II GENE EXPRESSION, AND THE JAK/STAT SIGNAL TRANSDUCTION PATHWAY

From both clinical and basic science studies, it is becoming increasingly evident that aberrations in the renin-angiotensin system, or RAS, are significant contributors to the etiology of hypertension and cardiac hypertrophy.[28-33] The RAS controls blood pressure through complex biochemical and signal transduction pathways that together regulate the production of angiotensinogen and its conversion to the signaling peptide angiotensin II. It is angiotensin II that acts as a first messenger by binding to its receptor, the AT1 receptor, on the surface of target cells and increasing intracellular Ca++ levels and activating second messengers such as protein kinase C as well as other signaling pathways.[34] Although it has been known for some time that angiotensinogen produced by the liver acts to control blood pressure and may be involved in hypertension, more recent evidence from our laboratory and others have shown that angiotensinogen (and presumably an active RAS) can be found in heart tissue as well as other tissues.[35-39] These findings suggest that regulation of angiotensinogen expression is tissue-specific and that local sites of synthesis, such as the heart, may play biological roles independent of the circulating angiotensinogen from liver. Different biological roles suggest that different cardiopathies may result from aberrancies in heart (or local) RAS function, and this has been found to be the case. When cardiac cells are treated with angiotensin II, not only is angiotensinogen gene expression found to be stimulated, but the cardiac cells become hypertrophic. At the molecular level, these findings suggest a number of differences between local RAS and renal RAS. First, they suggest that angiotensin II receptors reside on heart cells and, unlike the kidney, form the basis for an autocrine type of RAS for the heart.[35,40] Second, it appears that angiotensin II stimulation of heart cells leads primarily to expression of the angiotensinogen gene, whereas in noncardiac cells such as renal cells, angiotensin II stimulates a protein kinase C signaling pathway that promotes Na+ uptake. This suggests that while heart and renal cells express the same type of cell surface angiotensin II receptors, these receptors are linked to different signal transduction pathways that mediate different effector functions.

Among the pressor-growth hormones known to be involved in hypertension, our laboratory has focused on angiotensin II and has provided some insight into how angiotensin II activates a normal cellular signaling pathway to bring about activation of the angiotensinogen gene.[35] Angiotensin II promotes myocardial hypertrophy via activation of the RAS.[30] When angiotensin II binds to its cell surface receptors, it activates G proteins that trigger a cascade of multiple second messenger systems.[41] This second messenger cascade leads to the nucleus, where it ultimately causes the activation of the angiotensinogen gene, which leads presumably to more production of angiotensin II and activation of RAS (angiotensinogen is the prohormone precursor to angiotensin II). Our laboratory and others have shown that angiotensin II activates the angiotensinogen gene through the Janus kinase (Jak)/STAT pathway of signal transduction and gene activation.[35,41]

The Jak/STAT signal transduction pathway leading from the cell membrane to the cell nucleus has a very interesting means of regulation.[42,43] In the absence of an activating signal, STAT proteins reside in the cytoplasm. Upon activation by a ligand, cell surface receptors activate Jak kinases, which then phosphorylate the cytoplasmic STAT proteins. Upon phosphorylation, the STAT proteins migrate to the nucleus, where they activate the transcription of target genes. The specificity of this signal transduction pathway resides in promoter elements of the target genes that specifically bind the STAT proteins. In the case of angiotensin II, it its believed that this hypertrophic agent promotes myocardial hypertrophy by activating the Jak/STAT signal transduction pathway to specifically induce expression of the angiotensinogen gene. Exactly how a general signal transduction pathway such as the Jak/STAT pathway specifically activates the angiotensinogen gene in cardiomyocytes has been a focus of our laboratory's research.[35] Sequence analysis of the angiotensinogen gene promoter has revealed the presence of a conserved sequence element (called the GAS domain for gamma (interferon) activated sequence) originally found in the promoter of the gamma interferon gene and required for its activation by STAT proteins. By using the STAT binding domain (St-domain) of this element as a probe in gel-electromobility shift assays (GMSA), this promoter element was shown to bind STAT proteins in cardiomyocytes treated with angiotensin II; comparison with untreated cardiomyocytes indicated that STAT3 and STAT6 are selectively and functionally activated in response to angiotensin II. Further linking angiotensin II with activation of STAT3 and STAT6 is the finding that upstream components of the Jak/STAT pathway, specifically the Jak kinases required for STAT phosphorylation and activation, were also activated in cardiomyocytes treated with angiotensin II (as indicated by phosphorylation of their tyrosine residues).

These studies support the idea that normal cell signal transduction pathways, such as the Jak/STAT pathway, are used by hypertrophic agents to transmit their signals to the cell nucleus, where they activate nuclear genes. Our studies show that the peptide hormone angiotensin II turns on the angiotensinogen gene through activation of the Jak/STAT pathway. Activation of the angiotensinogen gene by angiotensin II to produce more of the angiotensinogen prohormone for processing could lead to the formation of an autocrine loop that positively reinforces production of angiotensin II in cardiomyocytes. This feedback loop may lead to an "overactive" RAS, which in turn maintains the hypertrophic state. A schematic of the events involved in this process is depicted in Figure 9.1 (see Color Plate 3).

Although these studies provide support for focusing on the angiotensinogen gene as a key factor in the etiology of hypertension and cardiac hypertrophy, they have yet to provide any indication as to the molecular mechanisms involved. What can be postulated is based on what is presently known or confirmed by these and other studies. Essential hypertension appears to result from a genetic predisposition to have an elevated or overactive RAS. Although this could be due to elevated levels or activities of any of the enzymatic or substrate components of RAS, genetic studies suggest that the "molecular lesion" may reside within or around the angiotensinogen gene.[28,31] This is supported by animal studies in which overexpression of an angiotensinogen transgene in transgenic mice leads to hypertension.[44] Together, these observations suggest that angiotensinogen-dependent hypertension may result from alterations in the genetic mechanisms controlling angiotensinogen gene expression and that these alterations contribute to the abnormal overexpression of the angiotensinogen gene. Our laboratory has focused on the expression of the angiotensinogen gene in the etiology of cardiac hypertrophy. Our findings point to the involvement of the STAT3 and STAT6 transcription factors in linking angiotensin II stimulation of cardiac cells with upregulation of the angiotensinogen gene. This signal transduction pathway provides a novel avenue leading to cardiac hypertrophy and may provide insights into aberrant angiotensinogen expression associated with hypertension as well. A more complete description of the mechanisms controlling angiotensinogen gene expression will provide potentially useful new genetic markers (with defined function) for linking aberrant angiotensinogen gene expression with elevated blood pressure in populations at risk for hypertension and cardiac hypertrophy.

REFERENCES

1. Simpson, P.C., Long, C.S., Waspe, L.E., Henrich, C.J., and Ordahl, C.P. 1989. Transcription of early developmental isogenes in cardiac myocytes exhibiting hypertrophy. *Mol. Cell. Cardiol.* 21:79–89.
2. Parker, T.G., Parker, S.E., and Schneider, M.D. 1990. Peptide growth factors can provoke "fetal" contractile protein gene expression in rat cardiac myocytes. *J. Clin. Invest.* 85: 507–514.
3. Opie, L.H. 1998. Overload hypertrophy and its molecular biology. In: L.H. Opie, ed. *The Heart: Physiology, from Cell to Circulation*, 3rd edition, Lippincott-Raven, New York.
4. Hamburger, V. and Hamilton, H.L. 1951. A series of normal stages in the development of the chick embryo. *J. Morphol.* 88:49–92.
5. Rawles, M.E. 1943. The heart forming regions of the early chick blastoderm. *Physiol. Zool.* 16:22–42.
6. Spemann, H. 1938. *Embryonic Development and Induction*, Yale University Press, New Haven.
7. Rosenquist, G.C. and De Haan, R.L. 1966. Migration of precardiac cells in the chick blastoderm: a radioautographic study. *Carnegie Inst. Wash. Contrib. Embryol.* 3:111–121.

8. Kaufman, M.H. and Navaratnam, V. 1981. Early differentiation of the heart in mouse embryos. *J. Anat.* 133:235–246.

9. Ehler, E., Rothen, B.M., Hammerle, S.P., Komiyama, M., and Perriard, J.C. 1999. Myofibrillogenesis in the developing chicken heart: assembly of Z-disk, M-line and the thick filaments. *J. Cell. Sci.* 112:1529–1539.

10. Tokuyasu, K.T. and Maher, P.A. 1987. Immunocytochemical studies of cardiac myofibrillogenesis in early chick embryos. I. Presence of immunofluorescent titin spots in premyofibril stages. *J. Cell. Biol.* 105:2781–2793.

11. Shiraishi, L., Takamatsu, T., Price, R.L., and Fugita, S. 1997. Temporal and spatial patterns of phosphotyrosine immunolocalization during cardiac myofibrillogenesis of the chicken embryo. *Anat. Embryol. Berlin* 196:81–89.

12. Johnatty, S.E., Dyck, J.R., Michael, L.H., Olson, E.N., and Abdellatif, M. 2000. Identification of genes regulated during mechanical load-induced cardiac hypertrophy. *J. Mol. Cell. Cardiol.* 32:805–815.

13. Glennon, P.E., Sugden, P.H., and Poole-Wilson, P.A. 1995. Cellular mechanisms of cardiac hypertrophy. *Br. Heart J.* 73:496–499.

14. Schaub, M.C., Hefti, M.A., Harder, B.A., and Eppenberger, H.M. 1997. Various hypertrophic stimuli induce distinct phenotypes in cardiomyocytes. *J. Mol. Med.* 75:901–920.

15. Sadoshima, J., Xu, Y., Slayter, H.S., and Izumo, S. 1993. Autocrine release of angiotensin II mediates stretch-induced hypertrophy of cardiac myocytes in vitro. *Cell.* 75:977–984.

16. Gorecka, A., Aksoy, M.O., and Hartshorne, D.J. 1976. The effect of phosphorylation of gizzard myosin on actin activation. *Biochem. Biophys. Res. Commun.* 71:325–331.

17. Morano, I. 1999. Tuning the human heart molecular motors by myosin light chains. *J. Mol. Med.* 77:544–555.

18. Chen, J., Kubalak, S.W., Minamisawa, S., Price, R.L., Becker, D.K., Hickey, R., Ross, J., and Chien, K.R. 1998. Selective requirement of myosin light chain 2V in embryonic heart formation. *J. Biol. Chem.* 273:1252–1256.

19. Aoki, H., Sadoshima, J., and Izumo, S. 2000. Myosin light chain kinase mediates sarcomere organization during cardiac hypertrophy in vitro. *Nat. Med.* 6:183–188.

20. Sanbe, A., Fewell, J.G., Gulick, J., Osinska, H., Lorenz, J., Hall, D.G., Murray, L.A., Kimball, T.R., Witt, S.A., and Robbins, J. 1999. Abnormal cardiac structure and function in mice expressing non-phosphorylatable cardiac regulatory myosin light chain 2. *J. Biol. Chem.* 274:21085–21094.

21. Qasba, P., Lin, E., Zhou, M.D., Kumar, A., and Siddiqui, M.A.Q. 1992. A single transcription factor binds two divergent sequence elements with a common function in myosin light chain-2 promoter. *Mol. Cell. Biol.* 12:1107–1116.

22. Zhou, M.D., Goswami, S.K., Martin, M.E., and Siddiqui, M.A.Q. 1993. A new serum responsive, cardiac tissue-specific transcription factor recognizes the MEF-2 site in the myosin light chain-2 promoter. *Mol. Cell. Biol.* 13:1222–1231.

23. Goswami, S., Qasba, P., Ghatpande, S., Deshpande, A.K., Baig, M., and Siddiqui, M.A.Q. 1994. Differential expression of myocyte enhancer factor 2 family of transcription factors in development: the cardiac factor BBF-1 is an early marker for cardiogenesis. *Mol. Cell. Biol.* 14:5130–5138.

24. Doud, S.K., Pan, L.-X., Carleton, S., Marmorstein, S., and Siddiqui, M.A.Q. 1995. Adaptational response in transcription factors during development of myocardial hypertrophy. *J. Mol. Cell. Cardiol.* 27:2359–2372.

25. Shen, R., Goswami, S.K., Mascareno, E., Kumar, A., and Siddiqui, M.A.Q. 1991. Tissue-specific transcription of the cardiac myosin light chain 2 gene is regulated by an upstream repressor element. *Mol. Cell. Biol.* 11:1676–1685.

26. Dhar, M., Mascareno, E.M., and Siddiqui, M.A.Q. 1997. Two distinct factor-binding DNA elements in cardiac myosin light chain 2 gene are essential for repression of its expression in skeletal muscle. Isolation of a cDNA clone for repressor protein Nished. *J. Biol. Chem.* 272:18490–18497.

27. Ghatpande, S., Goswami, S., Mathew, W., Rong, G., Cai, L., Shafiq, S., and Siddiqui, M.A.Q. 1999. Identification of a novel cardiac lineage-associated protein (cCLP-1): A candidate regulator of cardiogenesis. *Dev. Biol.* 208:210–221.

28. Jeunemaitre, X., Soubrier, F., Kotelevtsev, Y.V., Lifton, R.P., Williams, C.S., Charru, A., Hunt, S.C., Hopkins, P.N., Williams, R.R., Lalouel, J., and Corvol, R. 1992. Molecular basis of human hypertension: role of angiotensinogen. *Cell.* 71:169–180.

29. Corvol, P., Jeunemaitre, X., Charu, A., Kotelevtsev, Y., and Sourbrier, F. 1995. Role of the renin-angiotensin system in blood pressure regulation and in human hypertension: New insights from molecular genetics. *Recent Prog. Hormone Res.* 50:287–308.

30. Raizada, M., Phillips, M., and Summers, C. 1993. *Cellular and Molecular Biology of the Renin-Angiotensin System*, CRC Press, Boca Raton, FL.

31. Caulfield, M., Lavender, P., Farrall, M., Munroe, P., Lawson, M., Turner, P., and Clark, A.J.L. 1994. Linkage of the angiotensinogen gene to essential hypertension. *New Engl. J. Med.* 330:1629–1633.

32. Sadoshima, J. and Izumo, S. 1993. Molecular characterization of angiotensin II-induced hypertrophy of cardiac myocytes and hyperplasia of cardiac fibroblasts. *Circ. Res.* 73: 413–423.

33. Lijnen, P. and Petrov, V. 1999. Renin-angiotensin system, hypertrophy and gene expression in cardiac myocytes. *J. Mol. Cell. Cardiol.* 31:949–970.

34. Braunwald, E. 1992. In: E. Braunwald, ed. *Heart Disease—A Textbook of Cardiovascular Medicine*, Vol. 1, 4th Edition, W.B. Saunders Company, Philadelphia.

35. Mascareno, E., Dhar, M., and Siddiqui, M.A.Q. 1998. Signal transduction and activator of transcription (STAT) protein-dependent activation of angiotensinogen promoter: A cellular signal for hypertrophy in cardiac muscle. *Proc. Natl. Acad. Sci. USA* 95:5590–5594.

36. Clauser, E., Gaillard, I., Wei, L., and Corvol, P. 1989. Regulation of angiotensinogen gene. *Am. J. Hypertens.* 2(5 Pt. 1):403–410.

37. Lynch, K.R. and Peach, M.J. 1991. Molecular biology of angiotensinogen. *Hypertension* 17(3):263–269.

38. Campbell, D.J. and Habener, J.F. 1986. Angiotensinogen gene is expressed and differentially regulated in multiple tissues of the rat. *J. Clin. Invest.* 78(1):31–39.

39. Lindpaintner, K., Takahashi, S., and Ganten, D. 1990. Structural alterations of the renin gene in stroke-prone spontaneously hypertensive rats: examination of genotype–phenotype correlations. *J. Hypertens.* 8(8):763–773.

40. Lee, A.A., Dillmann, W.H., McCulloch, A.D., and Villarreal, F.J. 1995. Angiotensin II stimulates the autocrine production of transforming growth factor-beta1 in adult rat cardiac fibroblasts. *J. Mol. Cardiol.* 27:2347–2357.

41. Marrero, M.B., Schieffer, B., Paxton, W.G., Duff, J.L., Berk, B.C., and Bernstein, K.E. 1995. The role of tyrosine phosphorylation in angiotensin II-mediated intracellular signalling. *Cardiovasc. Res.* 30:530–536.

42. Ihle, J.N. 1996. STATs: Signal transducers and activators of transcription. *Cell.* 84:331–334.

43. Horvath, C.M. and Darnell, J.E. 1997. The state of the STATs: recent developments in the study of signal transduction to the nucleus. *Curr. Opin. Cell. Biol.* 9:233–239.

44. Kimura, S., Mullins, J.J., Bunnemann, B., et al. 1992. High blood pressure in transgenic mice carrying the rat angiotensinogen gene. *EMBO J.* 11:821–827.

CHAPTER 10

Cytoskeletal Gene Expression in the Developing Cardiac Conduction System

Robert E. Welikson and Takashi Mikawa

INTRODUCTION

The rhythmic contraction of the mature heart is initiated, perpetuated, and co-ordinated by the cardiac excitation–conduction system (Figure 10.1; Tawara, 1906; Goldenberg and Rothberger, 1936; Bozler, 1943). This impulse-conducting and pacemaking system is composed of a specialized subset of cardiac cells that form a network of fibers. The pacemaking action potential of the heart is generated in the sinoatrial (SA) node located in the wall of the right atrium (Keith and Flack, 1907; Brooks and Lu, 1972). This action potential is then spread through the myocardium of both atria, resulting in contraction of atrial myocytes (Wenckebach, 1906; Thörel, 1909; Robb and Petri, 1961; Brooks and Lu, 1972). The activating impulse in the atria is not allowed, however, to spread directly to the ventricles. Instead, the impulse is propagated to the atrioventricular (AV) node located at the junction of the atria and ventricles (Tawara, 1906). The AV node delays the activating impulse before propa-gating the signal to the highly coupled and fast conducting cells of the AV bundle (His, 1893) and bundle branches (Tawara, 1906). This delay results in the separate and sequential contraction of the atria and ventricles. After the impulse is conducted through the above central conduction system, it is finally spread though the ventric-ular muscle via a peripheral conduction system composed of a ramified network of subendocardial and intramural conductive fibers called Purkinje fibers (Purkinje, 1845; Kölliker, 1902; Tawara, 1906). Activation of the ventricular myocardium through the Purkinje system produces a synchronized contraction of the ventricular chambers of the heart.

Conduction cells can be identified in the heart as morphologically distinct tissue or by unique immunological and molecular markers (Lamers et al., 1991; Schiaffino, 1997; Alyonycheva et al., 1997; Moorman et al., 1998). They also contain a unique set of cytoskeletal components: poorly developed myofibrils (Oliphant and Loewen, 1976), abundant intermediate filaments (Thornell et al., 1978), and neurofilaments (Gorza et al., 1988; Vitadello et al., 1990). Microtubules, leptemeres, and intercalated disks are rarely found (Oliphant and Loewen, 1976). Evidence now indicates that conduction cells gain these unique characteristics during embryogenesis through a terminal diversification process from parental contractile myocytes (Mikawa and Fischman, 1996; Mikawa, 1998; Gourdie et al., 1999). In this chapter, we discuss the

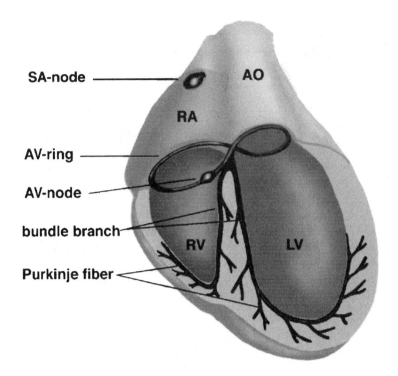

FIGURE 10.1. Diagram of the conduction system of the chicken heart. The central conduction network contains both atrial and ventricular elements, which include the sinoatrial (SA) node, atrioventricular (AV) ring, node, and bundle branches. The peripheral conduction system is composed of subendocardial and periarterial Purkinje fiber networks.

ontogeny of the cardiac conduction system, the expression and localization of cytoskeletal proteins in conduction cells, and the potential mechanisms that regulate conduction cell differentiation and gene expression.

SUBCOMPONENTS OF THE CONDUCTION SYSTEM

Sinoatrial Node

The SA node is the pacemaker of the heart and is located at the junction of the superior vena cava and the right atrium (Figure 10.1). In mammals, the SA node can be distinguished from the working myocardium as a distinct cluster of cells with smaller diameter (Truex and Smythe, 1965). The cells generally contain irregular and sparse amounts of contractile material surrounded by fine intermediate-type filaments. Glycogen is very prominent in many of the cells. Although the lack of a T tubule system is striking, these fibers do, however, form peripheral couplings. Despite these distinctions, the difference of the node from the ordinary myocardium is not that distinct. Instead, there is a gradual morphological transition from the nodal tissue to the myocardium (Sommer and Johnson, 1979).

Atrioventricular Node

The AV node is located at the base of the interatrial septum, near the origin of the posterior leaflet of the tricuspid valve (Figure 10.1). It is positioned relatively deep in the tissue, near the endocardium. The electophysiological function of this tissue is to slow the excitatory impulse from the SA node to the ventricles. Cells of the AV node are similar to those found in the SA node. They are interspersed with connective tissue and extensive vascularization. Included in this are venous sinusoids, arterioles, and an extended capillary network. In mammals, the AV node is separated from the endocardium by a thin wedge of atrial myocardium. The inferior region of the AV node is compact and sits on the annulus. The compact portion in some species is composed of two distinct layers, with nodal organization in both longitudinal and transverse directions. The cells in the periphery of this region are flat and spindle-like, whereas more irregularly shaped fibers are located more deeply (Thaemert, 1973). The superior and right margins are composed of more loosely connected fibers and tend to mix phenotypically with the muscle fibers of the right atrium and interatrial septum. The inferior region of the node becomes more regularly aligned as it becomes continuous with the atrioventricular bundle. The peripheral cells in the medial portion of the node are globular-like and contain hardly any myofibrils. The layering of tissue in the AV node may be responsible for the delay in conduction through the AV node.

Atrioventricular Bundle (Bundle of His)

Impulses leaving the AV node are conducted into the AV bundle. The transition from AV node to AV bundle is gradual and not easily distinguished. The proximal cells of the AV bundle are the same size as those in the AV node. The cells at the lower and anterior portion of the AV node become narrowed and aligned to form the AV bundle. The bundle is rooted in a loose mass of fibrous vascular tissue along the upper part of the interventricular septum, below the noncoronary cusp of the aorta (Figure 10.1). In humans, the AV bundle gives rise to two left bundle branches, a superior and inferior limb, and one right bundle branch. The bundle branches undergo extensive branching within the interventricular septum before moving in all directions in the ventricular walls to form the Purkinje system. As the fibers become more distal, they become larger and contain a greater amount of cytoplasm. Typically, the left and right bundle branches lie near the subendocardium (Figure 10.1). The bundle branch fibers proximal to the AV bundle are the same size as the bundle. As they move toward the apex, these fibers rapidly increase in size to become larger than the ordinary myocardial fibers of the ventricles.

Purkinje Fibers

The left and right bundle branches terminate to become the Purkinje fibers. Purkinje fibers are distributed widely throughout the subendocardium of the left and right ventricles (Figure 10.1). The Purkinje fibers can be distinguished from ordinary cardiac myocytes physiologically and morphologically. Although different species display a wide morphological variation in Purkinje fibers, they are classified into three broad categories (Truex and Smythe, 1965). In the first category, the Purkinje fibers have large diameters and are typically well differentiated. Birds, egg-laying mammals (monotremes), and hoofed mammals (ungulates) belong to this first group. In the avian heart, Purkinje fibers follow coronary arteries as they descend near the epicardial surface of the ventricles, while others run under the endocardium of the septum and ventricular walls. In the second group, which includes humans, monkeys, squirrels, cats, and dogs, Purkinje fibers have intermediate or small diameters and are difficult to distinguish from cardiac fiber of the working ventricular myocardium. In the final category, the Purkinje fibers show little cellular differentiation. These fibers are again difficult to distinguish from the ventricular myocardium and are usually placed close to the endocardium. Species belonging to this group are bats, rats, guinea pigs, and rabbits.

CONDUCTION CELL MYOFIBRILS AND CYTOSKELETON

Several studies have examined the myofibrils of the Purkinje fibers in the cow heart (Oliphant and Loewen, 1976; Thornell et al., 1978; Eriksson and Thornell, 1979). Similar to the ventricular myocardial fibers, Purkinje fibers contain myofibrils. The myofibrils in the Purkinje cells are generally sparse and are not as well organized as those in the working myocardium (Figure 10.2). Likewise, the mitochondria content in Purkinje fibers is lower than in ordinary cardiac fibers. The myofibrils in

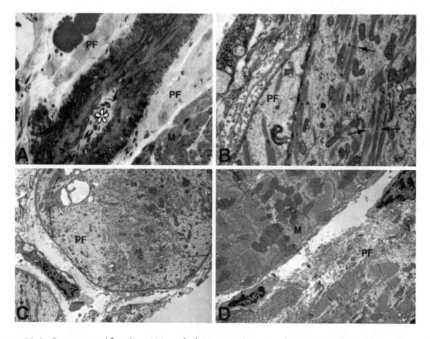

FIGURE 10.2. Low magnification (A) and electron micrographs (B–D) of Purkinje fibers from an adult chicken heart. (A) Purkinje fibers (PF) bordering an artery within the ventricular septum. The asterisk indicates the lumen of the artery. (B) Survey micrograph of Purkinje fiber (PF) juxtaposed to an endothelial cell (E). The Purkinje fiber myofibrils are typically sparse, loosely packed, and are located in the periphery of the cell. Many myofibrils in this cell have thickened, incomplete, or absent Z-bands (arrow). (C) Cross-section of a Purkinje fiber bordering an endothelial cell. The Purkinje cell shown contains an enormous amount of intermediate filaments that occupy most of the central portion of the cytoplasm. (D) A longitudinal section of a Purkinje fiber cell neighboring an ordinary working cardiac myocyte (M). The working cardiac myocyte is typified by dense packing of myofibrils, and mitochondria. In contrast, the Purkinje fiber has fewer mitochondria, sparse distribution of myofibrils, and a greater amount of cytoplasm.

Purkinje fibers are typically near the cell border, interspersed in a pale cytoplasm full of glycogen (Thornell and Sjostrom, 1975) and intermediate filaments (Thornell et al., 1978). These intermediate filaments, measuring 10 nm in diameter, are abundant and occupy most of the central cytoplasm. They intermingle with the myofibrils, mitochondria, and other cell organelles. Near the myofibrils, the intermediate filaments are arranged in bundles parallel to the myofibrils (Eriksson and Thornell, 1979).

Although some of these myofibrils appear ultrastructurally identical to those in the working ventricle, others appear aberrant. The myofibrils branch and interconnect with each other and often appear loosely packed. Unlike those of the ventricular myocardium, the myofilaments in Purkinje fibers are often misaligned. Intercalated disks, again typical of the working myocardium, are rarely observed. Other structures not often observed are leptemeres and microtubules. In addition, myofibrils in one cell generally lack an opposing myofibril in the neighboring cell (Oliphant and Loewen, 1976). Several Z-band alterations have been observed in Purkinje myofibers, including either thickened or incomplete structures.

GENE EXPRESSION IN THE CARDIAC CONDUCTION SYSTEM

Analysis of the proteins in the conduction system and the ordinary working myocardium suggests that the conductive tissue has a gene expression pattern distinct from the rest of the heart. Prior work on cow Purkinje fiber proteins by Thornell et al. (1978) has identified structural proteins similar to ordinary myocytes, with marked quantitative differences. Purkinje fibers contain a prominent protein of 55 kd, which comprises approximately 50% of the stained structural proteins. This protein corresponds to the cytoplasmic intermediate filaments found throughout the cytoplasm. In contrast, troponin, tropomyosin, myosin light chain 1 (MyLC-1), and myosin light chain 2 (MyLC-2) are present in less proportions in Purkinje than in ordinary myocytes. The myosin content in conduction system also is much less than in myocytes (Saito et al., 1981). Myosin extraction from conductive tissue (AV node, AV bundle, and AV branches) of the cow yielded only 20% of the myosin per wet weight than was obtained from the left ventricular myocardium.

Since then, several unique genes, including myofibrillar, structural, conductive, and transcription factor proteins, have been identified as markers for the developing and adult conduction tissues. Several studies have shown that conduction tissue shares many myofibrillar proteins with the ordinary myocytes (Thornell et al., 1978; Sartore et al., 1978; Saito et al., 1981). Some of these proteins are only transiently expressed in the myocardium but persist in conduction fibers. In contrast, cardiac MyBP-C, which is expressed in ordinary myocytes, is downregulated in Purkinje cells. Finally, markers that are unique to skeletal or neuronal lineages are expressed in the cardiac conduction system. There is also species variation of myofibrillar protein expression within the conduction system and ordinary myocardium, and the expression of some genes may differ by either alternative splicing or posttranslational modification. Table 10.1 lists mostly myofibrillar proteins expressed in cells of the adult and embryonic cardiac conduction systems in several different species.

TABLE 10.1. Proteins expressed in cells of adult and embryonic cardiac conduction systems in various species.

	Adult				
	Human	Bovine	Rat	Chicken	Rabbit
myosin heavy chain					
α	SAN, AVN, AVB, BB, P (Kuro et al., 1986)	A, V, SAN, AVN (Gorza et al., 1986) V, P (Komuro et al., 1986; Sartore et al., 1981)	A, V, AVB, BB, P (Dechesne et al., 1987)		A, V, P (Sartore et al., 1981)
β	SAN, AVN, AVB, BB, P (Kuro et al., 1986)	A, V, SAN, AVN (Gorza et al., 1986; Komuro et al., 1986)	A, V, AVB, BB, P (Dechesne et al., 1987)		
fetal		A, V, SAN, AVN (Gorza et al., 1986)			
atrial				A, P (Sartore et al, 1978)	
ventricular slow tonic				V, P (Sartore et al., 1978) P (Sartore et al., 1978; Gonzalez and Bader, 1985)	
MLC-1'		AVN, AVB, BB (Saito et al., 1981)			
MyBP-H				P (Alyonycheva et al., 1997)	
troponin I (cardiac)			A, V, AVN (Gorza et al., 1993)		
troponin I (slow)			AVN, (Gorza et al., 1993)		
α-smooth muscle actin			AVN, AVB, BB (Ya et al., 1997)		
neurofilament M					SAN, AVN, AVB, P (Gorza et al., 1988; Vitadello et al., 1990; Vitadello et al., 1996)
neurofilament L					SAN, AVN, AVB, P (Vitadello et al., 1990)

TABLE 10.1. *Continued.*

	Human	Bovine	Embryonic		Rabbit
			Rat	Chicken	
connexin 42					AVB, BB, P (Gourdie et al., 1993)
loss of polysialyated NCAM	AVB, BB, P (Watanabe et al., 1992)				
myosin heavy chain					
α					A, V (Sartore et al., 1981)
slow tonic MyBP-H				(Gonzalez and Bader, 1985) A, V, P (UR)	
troponin I (cardiac)			A, V, AVN (Gorza et al., 1993)		
troponin I (slow)			A, V, AVN (Gorza et al., 1993)		
Loss of cardiac MyBP-C				BB, P (Gourdie et al., 1998)	
α-smooth muscle actin			A, V, SAN, AVN, AVB, BB (Woodcock-Mitchell et al., 1988; Ya et al., 1997)		
neurofilament M				SA, P (Gorza et al., 1988; Vitadello et al., 1990)	
connexin 42				P (Gourdie et al., 1993)	
polysialyated NCAM	AVB, BB, P (Chuck et al., 1997; Watanabe et al., 1992)				

A—atrial myocardium.
V—ventricular myocardium.
SAN—sinoatrial node.
AVN—atrioventricular node.
AVB—atrioventricular (His) bundle.
BB—branch bundle.
P—Purkinje.
UR—unpublished results.

GENE EXPRESSION UNIQUE TO
THE CONDUCTION SYSTEM

Myosin

One of the best-characterized markers of the cardiac tissue phenotype is myosin, the major chemomechanical molecule responsible for muscle contraction and cell motility (Harrington and Rodgers, 1984). Myosin is a hexameric molecule consisting of two heavy chains and two nonidentical pairs of light chains (Warrick and Spudich, 1987). Several isoforms of myosin have been identified from cardiac and skeletal muscles, making up a gene family whose members are related both structurally and functionally. The myosin isoforms in striated muscles are highly conserved and appear to be functionally diverse rather than just redundant (reviewed in Weiss and Leinwand, 1996).

Immunocytochemical studies have convincingly demonstrated the heterogeneity of myosin heavy chain (MyHC) isoforms expressed in the avian heart. These different isoforms are variably distributed within the atrial, ventricular, and conductive myocardia. One of the first examinations of myosin isoforms in working and conductive myocardia was by Sartore and coworkers (Sartore et al., 1978). Antibodies against skeletal fast, skeletal slow, and ventricular myosin have demonstrated that all three epitopes are present in the adult chicken heart. The ventricular myocytes react with the antibody specific for ventricular myosin but not fast or slow skeletal myosin antibodies. In contrast, atrial myocytes react only with the antibody against fast skeletal myosin. Later studies have identified that the epitope identified by the fast skeletal antibody is present in the atrial-specific myosin heavy chain, AMHC1 (Yutzey et al., 1994). Purkinje fibers are reactive with all three antibodies, but most intensely with anti-slow skeletal myosin antibody. Careful analysis in consecutive sister sections has demonstrated that a single Purkinje cell expresses at least two myosin isotypes: slow and fast skeletal myosin or slow skeletal and ventricular myosin. Thus, in the adult chicken heart, conductive cells express myosins found in atria and ventricles as well as a slow skeletal myosin. The slow skeletal myosin epitope is not detectable in either the atria or ventricles until E17 (Gonzalez and Bader, 1985). The levels of slow skeletal myosin gradually reach to adult levels in posthatched chickens. These studies demonstrate that MyHC present in adult Purkinje fibers is not present in early embryonic hearts at a time when cellular precursors of the Purkinje fibers are present.

In the mammalian heart, two MyHC genes are expressed, α-MyHC and β-MyHC. The pattern of expression of these two genes is regulated developmentally, hormonally, and hemodynamically (Mahdavi et al., 1982). Owing to the dimerization of the α- and β-MyHC, three isoforms of myosins can be generated in the mammalian heart—an $\alpha\alpha$-MyHC homodimer (V1), an $\alpha\beta$-MyHC heterodimer (V2), and a $\beta\beta$ homodimer (V3) (Hoh et al., 1978). α-MyHC is the major isoform expressed in the adult ventricles and atria of small mammals, such as mice and rats (Lompre et al., 1984), and, with the exception of extraocular and masseter muscle, is cardiac-specific. In the bovine and human hearts, both α-MyHC and β-MyHC are present in the atrial myocardium. Gorza and coworkers have demonstrated that a fetal MyHC is additionally present in the nodal tissue but not in ordinary myocytes of the bovine heart (Gorza et al., 1986), confirming the heterogeneous nature of MyHC expression in the heart. The central portion of the SA node contains cells expressing mainly α- and fetal MyHC and a variable amount of β-MyHC. The fibers on the periphery of the

SA node do not express fetal MyHC at all, just α- and β-MyHC. The AV node also shows a regional difference in MyHC expression. The central and lower portions of the node expressed α- and fetal MyHC but not β-MyHC. At the region where the AV node is similar to the AV bundle, fetal MyHC can be found in the smaller nodal-like cells and not in the larger bundle branch cells, which are positive for β-MyHC. The remaining components of the ventricular conduction system, the AV bundle branches and Purkinje fibers, are negative for the fetal MyHC epitope. Thus, an MyHC isoform, antigenically related to fetal MyHC, is specifically expressed in the nodal conduction tissue of the bovine heart.

The heavy chains are not the only components of myosin to demonstrate heterogeneous expression within the cardiac conduction system. SDS PAGE analysis of myosin extracts has demonstrated a difference in MyLC composition between myocytes and cells of cardiac conduction tissue, including the AV node, AV bundle, and bundle branches (Saito et al., 1981). Two myosin light chains, MLC-1 of 25 kd and MLC-2 of 18 kd, are found in both ordinary myocytes and conduction cells. A third light-chain subunit, MLC-1' of 22.5 kd, is associated uniquely with myosin from the conduction system. The molar ratios of MLC-1, MLC-1', and MLC-2 to myosin are 1.5, 1.1, and 0.9, respectively. Thus, approximately one-third of the total MyHC of the bovine conduction tissue myosin contains MLC-1' as a subunit. Although myosin heterogeneity, with respect to both the heavy and light chains, is apparent in the bovine heart, no significant differences have been found in K^+-EDTA or Ca^{2+}-activated myofibrillar ATPase from myocytes and conduction cells (Saito et al., 1981).

Neuronal Proteins and Epitopes

Cardiac conduction cells exhibit some neuronal properties such as specific acetylcholinerase activity and inward Na- and K-current. They also express neurofilaments, the major intermediate filaments (IF) in neurons (for a review, see Xu et al., 1994). They are comprised of three subunits: neurofilament light (NF-L), neurofilament medium (NF-M), and neurofilament heavy (NF-H). Typical of IF proteins, neurofilament proteins contain a 310 amino acid central domain, which participates in the self-assembly of the protein into 10-nm filaments. When the adult rabbit heart is reacted with antibodies specific for NF-M, positive cells can be found within the SA node, AV node and AV bundle, bundle branches, and Purkinje fibers (Gorza et al., 1988; Vitadello et al., 1990). In contrast, the ordinary myocytes of the atria and ventricles are not reactive with the NF-M antibody. In situ hybridization of adult rat heart with an NF-M riboprobe has confirmed that NF-M transcripts, like the protein, are confined to the conduction system (Vitadello et al., 1996).

During the development of the rabbit heart, NF-M mRNA is first detected at embryonic day (ED) 9.5 in a few ventricular cells near the right-hand side of the AV junction. At ED 10, after the heart loops, NF-M transcripts are found in a small population of the AV junction. By ED 11, the number of cells transcribing NF-M has increased and localizes to the subendocardium in the AV junction and within the developing interventricular septum. NF-M protein is first detected in just a few cells in the AV junction and in the atrial wall at ED 11. By ED 15, NF-M is found in the conduction tissue in a pattern similar to that observed in the adult. Immunoelectron microscopy of the rabbit AV node with anti-NF-M antibodies showed that most of the NF-M organizes into filamentous bundles localized in the cytoplasmic spaces separating myofibrils and mitochondria. Similar to NF-M, NM-L can be detected with specific antibodies throughout all segments of the rabbit cardiac conduction system (Vitadello et al., 1990). Neither conduction nor ordinary cardiac myocytes demonstrated reactivity with NF-H or vimentin antibodies.

Another marker of the neuronal cell type is HNK-1. HNK-1 is a carbohydrate epitope, originally discovered on human natural killer cells (Abo and Balch, 1981) but also expressed on neural cells (Schachner and Martini, 1995), including neural crest (Tucker et al., 1984). Antibodies against HNK-1 have been used to discover several neural cell adhesion molecules, including N-CAM, myelin associated glyco-protein, L1, contactin, and P0 (Schachner and Martini, 1995). Speculation that pre-cursors of the cardiac conduction system were neural in origin was heightened when cells adjacent to the atrial wall of the rabbit heart at ED 11 were positive for the HNK-1 epitope as well as NF-M (Gorza et al., 1988). Although these cells also express sarcomeric myosin, the expression of HNK-1 in the heart is transient and can no longer be detected by ED 15.

Neural cell adhesion molecule (NCAM) is a cell surface glycoprotein and a member of the immunoglobulin cell adhesion molecule family. It has been shown to play a role in cell growth and migration in the brain and is expressed on a variety of cell types, including neurons, glial cells, and skeletal muscle (Walsh and Doherty, 1997). NCAM is posttranslationally modified with varying chain lengths of the car-bohydrate polysialic acid (PSA) and is regulated in a developmental manner (Kiss and Rougon, 1997). The function of PSA on NCAM is thought to be that of an attenu-ator of cell-to-cell adhesion forces, thereby influencing cell morphogenesis and migration. PSA-NCAM can be detected in the developing ventricular conduction system, as well as other regions, in the embryonic chick heart. At the tubular heart, Hamburger-Hamilton (H-H) stage 11, NCAM and PSA can be detected in the myocardium by antibody staining (Watanabe et al., 1992). By H-H stage 20, NCAM is still present throughout the myocardium, but the PSA moiety becomes restricted to cells within the AV junction, endocardial surface of the AV cushions, and at the edge of the atrial septum. After the heart septates, PSA-positive cells are organized into bundles within the myocardium of the septum. This PSA-positive bundle gives rise to two branches, which become connected to positive regions in the subendo-cardial myocardium. This staining pattern is consistent, temporally and spatially, with the developing conduction system. At H-H stage 44, PSA is still present in cells of the subendocardium. Immunostaining of alternate sections shows that these cells also express slow muscle MyHC, a marker for Purkinje fiber. In the adult chicken heart, however, NCAM is still ubiquitous, but PSA is no longer associated with cells expressing slow muscle MyHC. A possible role of PSA-NCAM in the conduction system is to attenuate their ability to make contacts with the ordinary myocardium while they are differentiating. Contractile cardiac myocytes, not expressing high levels of PSA, would then be allowed to make gap junctions or cadherin contacts among themselves, thus separating them from presumptive conductive cells (Watanabe et al., 1992).

EAP-300 is another neuronal marker of the cardiac conduction tissue. EAP-300 is a developmentally regulated protein expressed by the radial glia in the central nervous system (McCabe et al., 1992) as well as tissues of the peripheral nervous system and cardiac muscle (Kelly et al., 1995). Although EAP-300 is found in both contractile and specialized conductive tissues of the developing chicken heart, its expression in these tissues is differentially regulated, spatially and temporally. By immunostaining, EAP-300 can be detected as early as ED 2 (H-H stage 13) in the chick tubular heart and is readily detected in both atrial and ventricular myocardia from ED 5 and ED 6 (H-H stage 28) (McCabe et al., 1995). By ED 13, EAP-300 is preferentially expressed in the cardiac conduction tissue, including the AV bundle, bundle branches, and Purkinje fibers. However, just before hatching, an EAP-300 expression drops to the level of the ordinary working myocardium.

Gap Junctions

Gap junctions are the intracellular channels that mediate intercellular communication by the diffusion of small molecules and propagation of electrical impulses. They are found in practically all cell types and are responsible for the spread of excitation in smooth muscle and neurons as well as cardiac tissue. Gap junctions are composed of connexins (Cxs), a family of highly related proteins. In the heart, differential expression of connexin isoforms in the conduction system electrically separates the specialized conductive tissue from the ordinary working myocytes and thus promotes coordinated excitation and contraction. In the mammalian heart, four connexins are known to be expressed—Cx37, Cx40, Cx45, and Cx46 (reviewed in Moorman et al., 1998). In general, nodal tissues express connexin proteins at a significantly lower level. In the human SA and AV nodes, only faint expression of Cx40 and Cx45 can be found. This may account, in part, for slow conduction velocities in the SA and AV nodes.

Cx42, the chicken homolog of Cx40 (Beyer, 1990), has been immunolocalized to the conduction system in the embryonic and adult chicken hearts (Gourdie et al., 1993). Cx43, an isoform that is found ubiquitously in the myocardia of mammals, is restricted to smooth muscle coronary arteries in the chicken heart. The expression of Cx42 in the chicken heart is developmentally regulated. It is first detectable in peri-arterial cardiac myocytes and vascular endothelial cells at ED 9 to ED 10. Later, between ED 14 and ED 20, Cx42 is found only in the Purkinje fibers and the distal portions of the bundle branches. Just posthatching, Cx42 expression starts to spread proximally up the conduction system into the entire bundle branches and AV bundle. The levels of Cx42 at this time show a gradient pattern, with the more distal portions of the bundle branches expressing higher levels than the proximal regions. By 14 days posthatch, Cx42 protein expression becomes uniform along the bundle branches. In the adult, Cx42 expression is highest in the Purkinje fibers but also detectable in bundle branches. At no time is preferential expression of Cx42 found in working myocytes or other components of the conduction system.

The mammalian Cx42 homolog, Cx40, is the predominant isoform in the ventricular conduction systems of cows, pigs, dogs, and humans. In rats, Cx40 is also preferentially expressed in the conduction system, rather than Cx43 (Gourdie et al., 1992). Cx40 is probably not the only gap junction component functioning in the mammalian cardiac conduction system. Simon et al. created a Cx40 gene knockout in the mouse, and the mice were still viable (Simon et al., 1998). They did, however, display cardiac conduction abnormalities reminiscent of AV and bundle branch block.

PERSISTENCE OF EMBRYONIC CARDIAC GENE EXPRESSION IN THE CONDUCTION SYSTEM

Thick Filament Proteins

During the development of the heart, a variety of isoforms of contractile proteins are expressed in the avian and mammalian hearts. These genes either become restricted to different regions of the heart or are downregulated and give way to adult isoforms. Some myofibrillar protein isoforms present in the conduction system are expressed in myocytes only transiently during earlier development. This remnant gene expression does not necessarily mean that cardiac conductive tissue is embryonic in nature. Alternatively, this gene expression pattern could imply a functionally distinct and specialized cell type resulting from an alternative developmental pathway than contractile myocytes. For example, the expression of a thick filament associated protein, myosin protein binding H (MyBP-H), has been thought to be specifically expressed in skeletal muscle but not adult cardiac muscle. Immunohistological analysis of the adult chicken heart using a polyclonal antibody raised against chicken skeletal MyBP-H has demonstrated that MyBP-H is expressed in the heart preferentially in intramural Purkinje fibers (Alyonycheva et al., 1997). Biochemical analysis of the cardiac transcript and protein demonstrated that they were indistinguishable from their skeletal muscle counterparts. Our recent analysis of chick embryonic hearts has identified that MyBP-H is expressed throughout the ventricular myocardium earlier in development. However, later in development and after birth, it is confined to the Purkinje fibers of the peripheral conduction system (Figure 10.3).

Thin Filament Proteins

Components of the troponin complex are another example of structural proteins that undergo developmental regulation in the heart. The troponin complex is composed of three subunits, troponin T (TnT), troponin C (TnC), and troponin I (TnI). Along with tropomyosin (Tm), troponin regulates thin and thick filament interaction in a

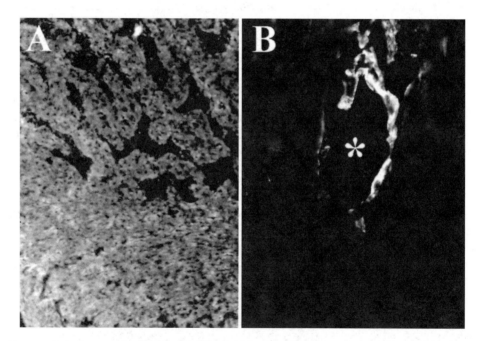

FIGURE 10.3. Immunolocalization of MyBP-H in the developing chicken heart. Frozen sections of an ED 9 (A) and ED 20 (B) embryonic chicken heart were immunostained with a polyclonal antibody specific for MyBP-H. At ED 9, MyBP-H is expressed ubiquitously throughout the ventricular myocardium. By ED 20, MyBP-H expression is restricted to Purkinje fibers, illustrated by perivascular localization of the antibody staining. Asterisk indicates the lumen of coronary artery.

calcium-dependent manner (reviewed in Farah and Reinach, 1995). TnI inhibits actin–myosin interaction. TnC binds calcium and releases the inhibitory effect of TnI. TnT binds the complex to Tm. Although only one gene encodes cardiac TnT, multiple isoforms, generated by alternative splicing, are expressed in the mammalian heart during development (Jin et al., 1992, 1996). In the hearts of rats and chickens, two isoforms of TnI have been detected, slow skeletal muscle (TnI_{slow}) and cardiac-specific ($TnI_{cardiac}$) (Sabry and Dhoot, 1989; Saggin et al., 1989). In the rat heart, TnI_{slow} mRNA is expressed during embryonic development, whereas $TnT_{cardiac}$ is most prominent in the adult. In situ hybridization demonstrated that both transcripts are present in the heart during embryogenesis but differ in their spatial and temporal expression patterns (Gorza et al., 1993). TnI_{slow} mRNA is the first of the two to appear, at the 10th day of development (13 somites). The TnI_{slow} signal persists throughout all of the heart through development and then progressively disappears with a different time course in the atrial and ventricular myocardia. $TnI_{cardiac}$ transcripts are not detected until later stages of development and then persist throughout the atria and ventricles in neonatal and adult rat hearts. The rise of $TnI_{cardiac}$ transcripts and the reciprocal fall of TnI_{slow} transcripts are coordinate in their spatial patterning. TnI_{slow} mRNA starts to disappear in the atria and then vacates rostrally from the ventricle to the outflow tract. During this transition, $TnI_{cardiac}$ appears in the same spatial pattern. The only regions of the heart where TnI_{slow} persists are the AV node and the other regions of the conduction system. In these regions, $TnI_{cardiac}$ is still detectable, but its signal is weaker.

α-smooth muscle actin (α-SMA) is also expressed in the myocytes of the developing rat heart (Ya et al., 1997). By immunostaining, its appearance is first detected in the heart at ED 9 and localizes mainly at the periphery of the cardiac myocytes. The protein levels of α-SMA then increase until ED 14. At this time, the protein can be found in all regions of the developing rat heart. The expression of α-SMA, however, is transient in cardiac myocytes. By ED 12, the α-SMA signal is already decreasing in cardiac myocytes in the interventricular septum and is then followed by decreases in the ventricle at ED 16. Between ED 16 and seven days after birth, α-SMA expression is lost in the atria, AV canal, and the ventricular trabeculae. The only region where α-SMA persists in the neonatal rat heart is in the ventricular conduction system, including the AV bundle and bundle branches. Similarly, desmin expression accumulates early in development and starts to decline by ED 14. Immunostaining demonstrates that the desmin signal, however, remained stronger in cells of the conduction system than in the working myocardium (Ya et al., 1997). Because the difference of the desmin signal between the working and conductive myocardia is a degree of intensity, desmin does not serve as an optimal marker of the cardiac conduction as do other proteins, which are restricted to the conduction myofibers.

DOWN REGULATION OF CONTRACTILE MYOCYTE-SPECIFIC PROTEIN EXPRESSION IN THE CONDUCTION SYSTEM

One example of loss of protein expression in the conduction system is the thick filament protein myosin binding protein C (MyBP-C). MyBP-C is a large protein of about 130 kd and represents approximately 2% of myofibril protein mass (Offer et al., 1973). At least three isoforms of MyBP-C have been identified in humans, mice, and chickens—fast skeletal, slow skeletal, and cardiac. In the adult chicken, the MyBPs become regionally distinct. By immunostaining, cardiac MyBP-C has been shown to be expressed broadly in the ordinary working myocardium of the heart (Yasuda et al., 1995). Cardiac MyBP-C is not expressed in the peripheral Purkinje fibers of the conduction system (Gourdie et al., 1998). At the ED 18 developing chick heart, immunostaining of MyBP-C was specifically downregulated in myocytes as they differentiated into Purkinje fibers. It is not known whether the other MyBP-C isoforms are expressed in the chicken Purkinje fibers. In contrast, MyBP-H expression in the heart is only found in the Purkinje fibers (Alyonycheva et al., 1997). Although a cardiac-specific isoform of MyBP-H has not been identified, the skeletal isoform is expressed in the Purkinje fibers.

PROTEINS COMMON TO CONDUCTION FIBERS AND MYOCYTES

In the ordinary myocytes of mammals, β-MyHC is the major cardiac MyHC isoform expressed in large adults such as humans and cows. In the bovine heart, very little β-MyHC is expressed in the atria. Likewise, little α-MyHC can be found in the ventricles. However, the Purkinje fibers in the bovine heart show a variable expression of α-MyHC when probed with an anti–α-MyHC antibody (Sartore et al., 1981). A quantitative study using competitive ELISA has shown that both α- and β-MyHC are expressed in the AV node and Purkinje fibers, but at different ratios than found in the ordinary myocardium of the atrium or ventricle (Komuro et al., 1986). More β-MyHC is expressed in the AV node than in the atrium. Furthermore, α-MyHC is much more abundant in the Purkinje fibers than in the ventricles.

The myosin content in the adult rat heart is almost purely α-MyHC. Immunohistological examinations have revealed that over 90% of the cells of 21- to 31-day rat ventricles are purely α-MyHC (Dechesne et al., 1987). The remaining fibers, expressing both α- and β-MyHC, are localized near vessels and the endocardial surface, especially in the intraventricular septum. The location and morphology of these fibers with both MyHC epitopes indicate that they are part of the AV bundle branches and Purkinje network. In addition to the presence of the homodimeric MyHC forms, V1 and V3, in the conduction system of the rat heart, immunoaffinity purification studies have shown that at least a portion of the myosin is in the heterodimeric form, V2 (Dechesne et al., 1987).

In humans, almost all the muscle fibers in the atria express α-MyHC, and many express β-MyHC. All the muscle fibers in the human ventricle express β-MyHC and very few express α-MyHC (Tsuchimochi et al., 1984). One of the regions in the ventricle where α-MyHC can be found is in the conduction system. Immunofluorescence of human heart sections with monoclonal antibodies against α-MyHC has revealed that all muscle fibers in the SA node and AV node express α-MyHC (Kuro et al., 1986). In addition to the SA and AV nodes, many fibers in the AV bundle and bundle branches also contained α-MyHC. Only a few fibers of the Purkinje system are positive for α-MyHC. On the other hand, β-MyHC is found in all fibers of the human cardiac system except for the SA node, which contained hardly any of the protein.

ORIGIN OF THE CONDUCTION SYSTEM

As early as the tubular stage, myocytes of the heart become electrically active (Kamino et al., 1981). The pacemaking impulses in the tubular heart are first detected in the posterior inflow tract, the presumptive sinus venous, and atrium and travel rostrally through the tubular heart. This caudal to rostral propagation of impulses maintains a unidirectional peristeric wave of contraction. Except for the pacemaker cells, the conduction system has not developed yet at this stage. The primitive tubular heart undergoes looping and partitions into four chambers. Once the four-chambered heart is established, the propagation pathway of pacemaking impulses in the ventricle needs to change to be transmitted to the ventricular apex, avoiding direct propagation to the basal part of the ventricle (Chuck et al., 1997). This topological shift of the impulse-transmission pathway in the ventricle depends on the differentiation and patterning of the cardiac conduction system (Mikawa, 1998).

The origin of cardiac conduction system cells has been under debate since the mid-1980s. Evidence for either or both neural crest or myocyte derivations has been proposed (Gorza et al., 1988; Mikawa and Fischman, 1996; Gourdie et al., 1995, 1999; Vitadello et al., 1996; Moorman, 1998). Evidence for the neural crest origin comes largely from the expression of neuronal genes in certain populations of conduction cells (Gorza et al., 1988; Vitadello et al., 1990). Concurrent to the neuronal gene expression, numerous sarcomeric proteins, which are muscle cell specific, are differentially expressed in the central and peripheral conduction systems (Schiaffino, 1997).

Retroviral cell marking methods have been used for fate analyses of individual cells to determine the origin of several cardiac cell lineages (reviewed in Mikawa and Fischman, 1996; Mikawa, 1998). Replication-defective retroviruses were originally developed to study cell fate in the central nervous system and the eye (Cepko, 1988; Sanes, 1989). Stable integration of the retroviral genome into an infected cell results in the genetic tagging of that cell and its progenitors. Manipulation of the proviral DNA with the subsequent insertion of reporter genes, such as β-galactosidase (β-gal), has produced replication-defective retroviruses that provide a reliable and reproducible tool to study cell lineages in higher vertebrates in vivo. To address the question of whether working myocytes and conduction cells share a common myogenic precursor, our group used the retroviral cell marking method with β-gal as a reporter in the developing chick heart.

In the chick, mesodermal cells are committed to the cardiac lineage by H-H stage 4 (ED 1) (Garcia-Martinez and Schoenwolf, 1993) and complete their differentiation by H-H stage 15 (ED 2) (Gonzalez and Bader, 1985). On the other hand, cardiac neural crest cells do not begin to migrate until between ED 2 and ED 3 and do not enter the tubular heart until ED 4. Therefore, if cardiac conduction cells were neurogenic in origin, their precursors would be absent in the embryonic heart before ED 3. Conversely, if the conduction cells were myogenic in origin, their precursors would be present at ED 3. We tagged individual contractile myocytes in the ED 3 tubular-stage heart with β-gal retrovirus. Analysis of the clonal tagged populations was performed at ED 14 and ED 18, time points within a period during which significant changes are occurring in the development of the peripheral conduction system (Gourdie et al., 1995; Cheng et al., 1999). This is evidenced by the expression of Cx42, a gap junction protein, which becomes restricted to both periarterial and subendocardial Purkinje fibers, signifying an electrophysiological distinction of Purkinje fiber and working myocardium. From the retroviral tagging experiments, we consistently observed a subset of clonally related myocytes that differentiated into conductive Purkinje fibers. In the same study, tagging of the neural crest at ED 2 resulted in labeling of parasympathetic postganglionic cardiac neurons but never any cardiac conduction tissues (Gourdie et al., 1995; Cheng et al., 1999). These data demonstrate, for the first time, that Purkinje fiber cells are recruited locally from already differ-

entiated and beating myocytes. This model is in contrast to the one previously proposed in which the conduction system is developed by the outgrowth of predestined
progenitors distinct from working myocytes (Moorman et al., 1998).

VESSEL-DERIVED SIGNALS AS INDUCERS OF THE CONDUCTION CELL FATE

In the avian heart, Purkinje fibers form a network in close spatial relationship to
the endocardium and coronary arterial vessels (Truex and Smythe, 1965; reviewed in
Lamers, 1991; Gourdie et al., 1993). The retroviral cell lineage experiments demonstrated that Purkinje fibers differentiate from a subset of working embryonic cardiac
myocytes juxtaposed to the endocardium and developing coronary arteries (Gourdie
et al., 1996; Cheng et al., 1999). This led to the hypothesis that vessel-derived
paracrine signals induce adjacent contractile myocytes to differentiate toward the
conductive Purkinje fibers (Mikawa and Fischman, 1996; Gourdie et al., 1998). To
test whether the selection of the conductive phenotype may be mediated by signals
from the developing arteries, we have both inhibited and promoted coronary artery
growth in the embryonic chick heart (Hyer et al., 1999). If an induction of the
Purkinje phenotype is dependent on instructive cues from vascular tissue, then the
density and location of Purkinje fibers would be predicted to change in parallel with
changes in coronary artery growth. To suppress normal coronary arterial branching,
the cardiac neural crest was ablated at ED 2, prior to migration, by a laser. To hyper-
vascularize the embryonic heart, fibroblast growth factor (FGF), a cytokine
shown to induce blood vessel development (Mikawa, 1995), was ectopically expressed
in the embryonic heart using a recombinant retrovirus. Neural ablation resulted
in a reduction in coronary arteries within the ventricular myocardial wall. Similarly,
using immunological markers unique to Purkinje fiber cells, Purkinje fibers
could only be identified in areas near the remaining histologically normal arteries.
Infection of the proepicardial organ and hence coronary artery precursors at H-H
stage 17 to 18 produced regional overexpression of FGF and an increase in the
branching and density of coronary arteries. Immunohistochemical analysis revealed
that ectopic Purkinje fibers were found adjacent to the FGF-induced coronary
arteries. These results demonstrate that the developmental growth pattern of the
coronary arteries in the avian heart provides not only an inductive signal of Purkinje
fiber differentiation but also a blueprint for the network's organization (Hyer et al.,
1999).

To determine which paracrine signals could induce working myocytes into
Purkinje fiber phenotype, several known blood vessel factors were tested on isolated
embryonic chick cardiac myocytes in vitro (Gourdie et al., 1998). Of the factors
tested, which included endothelin (ET), its precursor big ET, FGF, and platelet-
derived growth factor, only ET converted the beating myocytes to the Purkinje fiber
phenotype. ET is a shear-stress–induced cytokine abundant in the arterial system of
the myocardium (reviewed in Masaki et al., 1991). Within a few days of addition of
recombinant ET to the embryonic cardiac myocytes, Cx42, a Purkinje-fiber–specific
marker, was induced accompanied by the eventual loss of expression of a contractile
myocyte marker, MyBP-C. ET antagonists neutralized the ability of ET to induce
this response. Thus, embryonic myocytes are competent to respond to a specific
paracrine factor prominently secreted from coronary arterial tissue and can be
induced to exhibit a Purkinje fiber phenotype. This leads to the model illustrated in

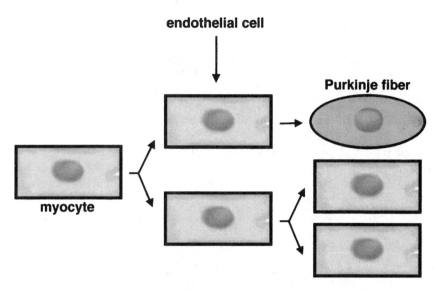

FIGURE 10.4. A model for Purkinje fiber differentiation within the myocyte lineage in the developing chick heart. A subpopulation of clonally related contractile myocytes adjacent to developing arteries and subendocardium receive an inductive cue to differentiate into Purkinje fiber cells. Contractile myocytes continue to undergo cell division while Purkinje fiber cells withdraw from the proliferate state.

Figure 10.4 in which Purkinje fibers are recruited from contractile myocytes along the developing coronary bed.

REGULATION OF GENE EXPRESSION IN THE CONDUCTION SYSTEM

Trans-activators

Thus far, the molecular mechanism responsible for conduction cell differentiation is largely unknown. RNA from Msx-2, a homeobox domain gene homologous to the *Drosophila* muscle segment homeobox gene (*msh*), has been found to be expressed transiently in conduction cell progenitors in developing chick hearts (Chan-Thomas et al., 1993). Homeobox genes encode a large class of transcription factors that play an important role in pattern formation in early development. In situ hybridizations show that during development (H-H 23) Msx-2 becomes restricted to a distinct myocardial cell population, which forms an incomplete AV ring early in heart development. During later development (H-H stages 34 and 37), cells expressing Msx-2 could be morphologically identified as cardiac conduction tissue. The spatial and temporal expression of Msx-2 indicates that in addition to serving as a convenient marker for the identification of developing conduction cells, it may also play a role in the formation and patterning of the conduction system by activating precursors to assume this specialized cell phenotype. The formation of an Msx-2 expression ring is consistent with the concept of a ring-like structure at the AV junction as the initiation site of conduction system formation (Wessels et al., 1992). However, Cheng and coworkers showed that in 196 retrovirally infected hearts, no common progenitors were shared between the central and peripheral conduction cells (Cheng et al., 1999).

Moreover, infected cells could incorporate within any part of the conduction system. This suggests that conduction cells share a closer lineage to neighboring working myocyte cells than to a more distal portion of the conduction network.

In addition to Msx-2, another homeobox transcription factor, Nkx2.5, has been linked to the conduction system. Nkx2.5, the mammalian homolog of the *tinman* gene in *Drosophila*, is required for the differentiation of mesoderm into heart muscle (Bodmer, 1993). Gene knockouts in mice of the Nkx2.5 gene resulted in embryonic lethality before the tubular heart began to loop (Lyons et al., 1995). The mouse homolog has been shown to be expressed in cardiac myocardial cells from 8.5 days postcoitum through adulthood (Komuro and Izumo, 1993). Schott and coworkers have reported that mutations in the human Nkx2.5 have been linked to an autosomal dominant atrial septal defect (ASD), an inherited disease associated with atrioventricular conduction delays (Schott et al., 1998). Several families in which most of the family members have congenital heart disease were examined. Most of the individuals were affected by ASD. Three different mutations in the Nkx2.5 were identified in the families studied. Many of the ASD family members exhibited progressive conduction disease, with prolonged conduction through the AV node, while other electrophysiological properties appeared normal. These AV conduction abnormalities persisted in affected individuals with either spontaneous or surgical correction of their ASD and even in family members with normal heart structure. These observations indicate an additional role for Nkx2.5 in the adult conduction system for the maintenance of normal physiological AV node function.

Cis-elements

Purkinje fiber cells originate from working myocytes that differentiate into specialized conduction cells with a unique pattern of gene expressions. As discussed, many of these genes are typical of neuronal or skeletal muscle linkages. Cardiac muscle is structurally, functionally, and developmentally distinct from skeletal muscle. The basic/helix-loop-helix (bHLH) family of muscle determination transcription factors regulates the expression of many of the muscle-specific proteins (reviewed in Buckingham, 1994). These factors, which are solely expressed in skeletal muscle cell and their precursors, bind to the *cis*-element, E box, in the upstream region of muscle structural proteins. These factors include MyoD, Myf-5, myogenin, and MRF-4. Another group of myogenic bHLH transcription factors, belonging to the myogenic enhancer factor 2 (MEF-2) family, are believed to activate muscle genes by binding to their own specific DNA recognition sites but by acting as a cofactor in other bHLHs' muscle-specific gene activation. Cardiac muscle normally does not express any of the members of the MyoD family of myogenic transcription factors.

The ability to initiate and activate a skeletal muscle program with the cardiac tissue in the embryonic heart has been tested in a mouse transgenic model. MyoD has been ectopically expressed in the developing mouse under the direction of the muscle creatine kinase enhancer/promoter (Miner et al., 1992). The exogenous MyoD expression in the heart gives rise to lethality in utero in several founder lines, with the transgenic pups displaying grossly misshapen hearts, which do beat but are unable to pump blood. On the molecular level, ectopic MyoD fails to activate Myf-5 or MRF-4. The only myogenic counterpart induced in the transgenic mouse by MyoD is myogenin. Transcripts from skeletal α-actin, a gene that is expressed in early heart development (Sassoon et al., 1988), and cardiac hypertrophy (Schwartz et al., 1986) are found in the late-stage embryonic heart of the MyoD transgenic mouse. In addition, immunostaining and immunoblots have demonstrated the presence of two skeletal-muscle–specific MyHCs, embryonic and perinatal, in the hearts of these mice. These data demonstrate that MyoD can activate skeletal-muscle–specific genes in the developing cardiac myocardium.

In another transgenic mouse model, a muscle-specific enhancer/promoter was able to direct specific expression in the peripheral conduction system. Desmin is a muscle-specific cytoskeletal protein belonging to the same family of intermediate filaments as NF-M and is characterized by its ability to aggregate into 10-nm-diameter structures (Lazarides, 1982). Desmin is expressed in cardiac, skeletal, and smooth muscle and has been shown to be one of the first muscle proteins expressed in both the heart and somites (Hill et al., 1986; Furst et al., 1989; Babai et al., 1990). Investigation of the proximal 5′ region of the desmin gene has identified several *cis*-acting elements responsible for inducible expression of the desmin gene in myogenic and nonmyogenic cell lines (Li and Paulin, 1991). Several elements that conferred muscle cell specificity and regulated level of expression have been mapped within one kilobase (kb) of the transcriptional start site of the desmin gene (Figure 10.5). Included in this DNA sequence is a 280-bp enhancer that promotes high expression and contains several binding sites of myogenic bHLH transcription factors, including MEF-2 and MyoD. MEF-2 expression is found in the developing skeletal and cardiac muscles and nervous systems of both invertebrates and vertebrates (for a review, see Black and Olson, 1998). MyoD expression, on the other hand, has yet to be detected in the heart. Other elements within the promoter include CArG and SP1. The CArG element (CC(A/T)6GG) had been first described as the core sequence of the serum response element (SRE) within early-response genes, such as c-fos (Treisman, 1992). A CArG element is also present in the 5′-flanking region of the cardiac α-actin, skeletal α-actin, β-actin, α-myosin heavy chain, cardiac myosin light chain 2, and troponin T genes (Muscat et al., 1988; Mably et al., 1993; Papadopoulos and Crow, 1993; Molkentin et al., 1994; Mably, 1996). In addition to these striated muscle genes, it has been shown that multiple *cis* CArG elements within the smooth muscle α-actin promoter are necessary for smooth muscle cell specific expression (Mack and Owens,

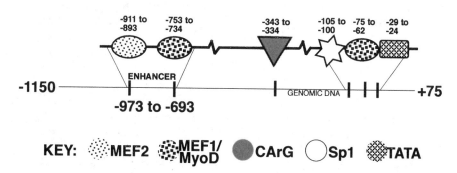

FIGURE 10.5. Schematic of the 5′ upstream region of the human desmin gene. High-level expression of the gene depends on a 280-bp muscle-specific enhancer located between −973 and −693 base pairs upstream of the transcriptional start site. The promoter sequence is characterized by several consensus-binding sequences recognized by muscle-specific transcription factors, including MEF-1, MEF-2, myogenin, and MyoD. Transgenic mice bearing a β-gal gene driven by the 5′ upstream region of the human desmin gene illustrated here express β-galactosidase in the conduction system of the heart.

1999). Sp1 (specificity protein 1) is a general transcriptional factor and member of the Sp transcription factors family, a group of proteins that are structurally and evolutionarily related (for a review, see Suske, 1999). Sp1 binds to GC boxes, *cis*-elements that are required for expression in many ubiquitous, tissue-specific, and viral genes (Suske, 1999).

To determine whether the same or different *cis*-elements directed desmin expression in cardiac, skeletal, and smooth muscles, transgenic mice were created expressing a reporter gene, β-gal, under the direction of the 1 kb DNA 5′ regulatory sequence of the desmin gene (DES1) (Li et al., 1993). In the DES1 mice, β-gal expression coincided with the endogenous desmin gene expression. No transgene gene expression was detected in smooth muscle or the working myocardium of the heart. The DES1 promoter did direct β-gal expression in the conduction system of the heart. β-gal is first observed in a few scattered cells of the heart in an 8-day embryo. These cells were identified phenotypically as being part of the developing conduction system. The transgenic expression continued in the peripheral conduction system through embryonic development and after birth. These experiments demonstrate that the 1 kb DNA regulatory sequences of the desmin gene are capable of directing skeletal-muscle–specific expression in vivo, and other *cis*-elements within the desmin 5′ regulatory region are responsible for cardiac and smooth muscles. This DNA region, which contains binding sites for multiple muscle determination transcription factors, is competent to direct expression in the mouse cardiac conduction system. One of these binding sites binds to MyoD, a transcription factor that is skeletal-muscle–specific. This suggests that a skeletal-muscle–specific program could be active in working cardiac myocytes that differentiate into specialized conduction cells.

CONCLUDING REMARKS

As immunological and molecular probes become more available, the cytoskeletal and noncytoskeletal gene expression profiles in the cardiac conduction system are becoming better characterized. Myofibril proteins typical of cardiac and skeletal muscles can be found throughout the cardiac conduction system. Neuronal proteins, such as neurofilament intermediate proteins, can also readily be identified in different components of the conductive network. The function of these several proteins expressed in three distinct cellular phenotypes in conduction cells is still unknown. Retroviral tagging experiments have conclusively shown that contractile myocytes, when induced from instructive cues from the developing coronary arterial system, differentiate into conduction cells of both central and peripheral conduction systems. Hence, we have evaluated here the gene expression pattern of the conduction system relative to the working cardiac myocyte gene expression pattern instead of patterns typical of neuronal or skeletal muscle phenotypes. With the exception of unique gap junction proteins, the functional consequences of the activation or loss of gene expression discussed earlier are still unclear. Therefore, the proteins only serve as immunological and molecular markers of the conduction cell phenotype. As we learn more about the transcriptional control of conduction cell gene expression, the function of these diverse sets of gene products will undoubtedly become more evident. The question arises whether unique conduction cell transcription factors exist and if so how they are activated. The identification of additional cardiac conduction-cell–specific genes and a characterization of their *cis*-regulatory regions should add to our understanding of the development and function of the conduction cell phenotype.

ACKNOWLEDGMENT

Supported by the National Institutes of Health. The excellent technical assistance of Ms. Lydia Miroff and Leona Cohen-Gould is acknowledged with gratitude. T.M. is an Irma T. Hirschl Scholar.

REFERENCES

Abo, T. and Balch, C.M. 1981. A differentiation antigen of human NK and K cells identified by a monoclonal antibody (HNK-1). *J. Immunol.* 127:1024–1029.

Alyonycheva, T., Cohen-Gould, L., Siewert, C., Fischman, D.A., and Mikawa, T. 1997. Skeletal muscle-specific myosin binding protein-H is expressed in Purkinje fibers of the cardiac conduction system. *Circ. Res.* 80:665–672.

Babai, F., Musevi-Aghdam, J., Schurch, W., Royal, A., and Gabbiani, G. 1990. Coexpression of alpha-sarcomeric actin, alpha-smooth muscle actin and desmin during myogenesis in rat and mouse embryos I. Skeletal muscle. *Differentiation* 44:132–142.

Beyer, E.C. 1990. Molecular cloning and developmental expression of two chick embryo gap junction proteins. *J. Biol. Chem.* 265:14439–14443.

Black, B.L. and Olson, E.N. 1998. Transcriptional control of muscle development by myocyte enhancer factor-2 (MEF2) proteins. *Annu. Rev. Cell Dev. Biol.* 14:167–196.

Bodmer, R. 1993. The gene tinman is required for specification of the heart and visceral muscles in *Drosophila* [published erratum appears in *Development* 1994 Nov;119(3):969]. *Development* 118:719–729.

Botzler, E. 1942. The initiation of impulses in cardiac muscle. *Am. J. Physiol.* 138:273–282.

Brooks, C. McC. and Lu, H.-H. 1972. *The Sinoatrial Pacemaker of the Heart*, Charles C. Thomas, Springfield, MA.

Buckingham, M.E. 1994. Muscle: the regulation of myogenesis. *Curr. Opin. Genet. Dev.* 4:745–751.

Cepko, C. 1988. Retrovirus vectors and their application in neurobiology. *Neuron* 1:345–353.

Chan-Thomas, P.S., Thompson, R.P., Robert, B., Yacoub, M.H., and Barton, P.J.R. 1993. Expression of homeobox genes Msx-1 (Hox 7) and Msx-2 (Hox 8) during cardiac development in the chick. *Dev. Dyn.* 197:203–216.

Cheng, G., Litchenberg, W.H., Cole, G.J., Mikawa, T., Thompson, R.P., and Gourdie, R.G. 1999. Development of the cardiac conduction system involves recruitment within a multipotent cardiomyogenic lineage. *Development* 126:5041–5049.

Chuck, E.T., Watanabe, M., Timm, M., and Fallah-Najmabadi, H. 1997. Differential expression of PSA-NCAM and HNK-1 epitopes in the developing cardiac conduction system of the chick. Cardiac expression of polysialylated NCAM in the chicken embryo: correlation with the ventricular conduction system. *Dev. Dyn.* 209:182–195.

Dechesne, C.A., Leger, J.O., and Leger, J.J. 1987. Distribution of α- and β-myosin heavy chains in the ventricular fibers of the postnatal developing rat. *Dev. Biol.* 123:169–178.

Eriksson, A. and Thornell, L.E. 1979. Intermediate (skeleton) filaments in heart Purkinje fibers. A correlative morphological and biochemical identification with evidence of a cytoskeletal function. *J. Cell Biol.* 80:231–247.

Farah, C.S. and Reinach, F.C. 1995. The troponin complex and regulation of muscle contraction. *FASEB J.* 9:755–767.

Furst, D.O., Osborn, M., and Weber, K. 1989. Myogenesis in the mouse embryo: differential onset of expression of myogenic proteins and the involvement of titin in myofibril assembly. *J. Cell Biol.* 109:517–527.

Garcia-Martinez, V. and Schoenwolf, G.C. 1993. Primitive-streak origin of the cardiovascular system in avian embryo. *Dev. Biol.* 159:706–719.

Goldenberg, M. and Rothberger, C.J. 1936. Über des Elektrogramm der spezifischen Herzmuskulatur. *Pflügers Arch.* 237:295–306.

Gonzalez, S. and Bader, D. 1985. Characterization of a myosin heavy chain in the conductive system of the adult and developing chicken heart. *J. Cell Biol.* 100:270–275.

Gorza, L., Ausoni, S., Merciai, N., Hastings, K.E.M., and Schiaffino, S. 1993. Regional differences in troponin I isoform switching during rat heart development. *Dev. Biol.* 156: 253–264.

Gorza, L., Sartore, S., Thornell, L.E., and Schiaffino, S. 1986. Myosin types and fiber types in cardiac muscle, III: nodal conduction tissue. *J. Cell Biol.* 102:1758–1766.

Gorza, L., Schiaffino, S., and Vitadello, M. 1988. Heart conduction system: a neural crest derivative? *Brain Res.* 457:360–366.

Gourdie, R.G., Green, C.R., Severs, N.J., Anderson, R.H., and Thompson, R.P. 1993. Evidence for a distinct gap-junctional phenotype in ventricular conduction tissues of the developing and mature avian heart. *Circ. Res.* 72:278–289.

Gourdie, R.G., Green, C.R., Severs, N.J., and Thompson, R.P. 1992. Immunolabelling patterns of gap junction connexins in the developing and mature rat heart. *Anat. Embryol. Berling* 185:363–378.

Gourdie, R.G., Kubalak, S., and Mikawa, T. 1999. Conducting the embryonic heart: Orchestrating development of the conduction system. *Trends in Card. Med.* 9:18–26.

Gourdie, R.G., Mima, T., Thompson, R.P., and Mikawa, T. 1995. Terminal diversification of the myocyte lineage generates Purkinje fibers of the cardiac conduction system. *Development* 121:1423–1431.

Gourdie, R.G., Wei, Y., Kim, D., Klatt, S.C., and Mikawa, T. 1998. Endothelin-induced conversion of embryonic heart muscle cells into impulse-conducting Purkinje fibers. *Proc. Natl. Acad. Sci. USA* 95:6815–6818.

Harrington, W.F. and Rodgers, M.E. 1984. Myosin. *Annu. Rev. Biochem.* 53:35–73.

Hill, C.S., Duran, S., Lin, Z.X., Weber, K., and Holtzer, H. 1986. Titin and myosin, but not desmin, are linked during myofibrillogenesis in postmitotic mononucleated myoblasts. *J. Cell. Biol.* 103:1–96.

His, Wm., Jr. 1893. Die Tätigkeit des embryonalen Herzens und deren Bedeutung für die Lehre von der Herzbewegung beim Erwachsenen. *Arb. Med. Klin. Leipzig* 14.

Hoh, J.F., McGrath, P.A., and Hale, P.T. 1978. Electrophoretic analysis of multiple forms of rat cardiac myosin: effects of hypophysectomy and thyroxine replacement. *J. Mol. Cell. Cardiol.* 10:1053–1076.

Hyer, J., Johansen, M., Prasad, A., Wessels, A., Kirby, M.L., Gourdie, R.G., and Mikawa, T. 1999. Induction of Purkinje fiber differentiation by coronary arterialization. *Proc. Natl. Acad. Sci. USA* 96:13214–13218.

Jin, J.P., Huang, Q.Q., Yeh, H.I., and Lin, J.J. 1992. Complete nucleotide sequence and structural organization of rat cardiac troponin T gene. A single gene generates embryonic and adult isoforms via developmentally regulated alternative splicing. *J. Mol. Biol.* 227: 1269–1276.

Jin, J.P., Wang, J., and Zhang, J. 1996. Expression of cDNAs encoding mouse cardiac troponin T isoforms: characterization of a large sample of independent clones. *Gene* 168:217–221.

Kamino, K., Hirota, A., and Fujii, S. 1981. Localization of pacemaking activity in early embryonic heart monitored using voltage-sensitive dye. *Nature* 290:595–597.

Keith, A. and Flack, M. 1907. The form and nature of the muscular connections between the primary divisions of the vertebrate heart. *J. Anat. Physiol.* 41:172–189.

Kelly, M.M., Phanhthourath, C., Brees, D.K., McCabe, C.F., and Cole, G.J. 1995. Molecular characterization of EAP-300: a high molecular weight, embryonic polypeptide containing an amino acid repeat comprised of multiple leucine-zipper motifs. *Dev. Brain Res.* 85: 31–47.

Kiss, J.Z. and Rougon, G. 1997. Cell biology of polysialic acid. *Curr. Opin. Neurobiol.* 7: 640–646.

Kölliker, A. 1902. Gewebeslehre. 6 Aufl. Lpz.

Komuro, I. and Izumo, S. 1993. Csx: a murine homeobox-containing gene specifically expressed in the developing heart. *Proc. Natl. Acad. Sci. USA* 90:8145–8149.

Komuro, I., Tsuchimochi, H., Ueda, S., Kurabayashi, M., Seko, Y., Takaku, F., and Yazaki, Y. 1986. Isolation and characterization of two isozymes of myosin heavy chain from canine atrium. *J. Biol. Chem.* 261:4504–4509.

Kuro-o, M., Tsuchimochi, H., Ueda, S., Takaku, F., and Yazaki, Y. 1986. Distribution of cardiac myosin isozymes in human conduction system. Immunohistochemical study using monoclonal antibodies. *J. Clin. Invest.* 77:340–347.

Lamers, W.H., De Jong, F., De Groot, I.J., and Moorman, A.F. 1991. The development of the avian conduction system, a review. *Eur. J. Morphol.* 29:233–253.

Lazarides, E. 1982. Intermediate filaments: A chemically heterogeneous, developmentally regulated class of proteins. *Ann. Rev. Biochem.* 51:219.

Li, Z., Marchand, P., Humbert, J., Babinet, C., and Paulin, D. 1993. Desmin sequence elements regulating muscle-specific expression in transgenic mice. *Development* 117:947–959.

Li, Z.L. and Paulin, D. 1991. High level desmin expression depends on a muscle-specific enhancer. *J. Biol. Chem.* 266:6562–6570.

Lompre, A.M., Nadal, G., and Mahdavi, V. 1984. Expression of the cardiac ventricular alpha- and beta-myosin heavy chain genes is developmentally and hormonally regulated. *J. Biol. Chem.* 259:6437–6446.

Lyons, I., Parsons, L.M., Hartley, L., Li, R., Andrews, J.E., Robb, L., and Harvey, R.P. 1995. Myogenic and morphogenetic defects in the heart tubes of murine embryos lacking the homeobox gene Nkx2–5. *Genes Dev.* 9:1654–1666.

Mahdavi, V., Periasamy, M., and Nadal-Ginard, B. 1982. Molecular characterization of two myosin heavy chain genes expressed in the adult heart. *Nature* 297:659–664.

Mably, J.D., Sole, M.J., and Liew, C.C. 1993. Characterization of the GArC motif. A novel cis-acting element of the human cardiac myosin heavy chain genes. *J. Biol. Chem.* 268: 476–482.

Mack, C.P. and Owens, G.K. 1999. Regulation of smooth muscle alpha-actin expression in vivo is dependent on CArG elements within the 5′ and first intron promoter regions. *Circ. Res.* 84:852–861.

Masaki, T., Kimura, S., Yanagisawa, M., and Goto, K. 1991. Molecular and cellular mechanism of endothelin regulation: Implications for vascular function. *Circulation* 84:1457–1468.

McCabe, C.F., Gourdie, R.G., Thompson, R.P., and Cole, G.J. 1995. Developmentally regulated neural protein EAP-300 is expressed by myocardium and cardiac neural crest during chick embryogenesis. *Dev. Dyn.* 203:51–60.

McCabe, C.F., Thompson, R.P., and Cole, G.J. 1992. Distribution of the novel developmentally-regulated protein EAP-300 in the embryonic chick nervous system. *Dev. Brain Res.* 66:11–23.

Mikawa, T. 1995. Retroviral targeting of FGF and FGFR in cardiomyocytes and coronary vascular cells during heart development. *Ann. N.Y. Acd. Sci.* 752:506–516.

Mikawa, T. and Fischman, D.A. 1996. The polyclonal origin of myocyte lineages. *Annu. Rev. Physiol.* 58:509–521.

Mikawa, T. 1998. Cardiac lineages. In: *Heart Development*, Eds. R.P. Harvey and N. Rosenthal. Academic Press, pp. 19–33.

Miner, J.H., Miller, J.B., and Wold, B.J. 1992. Skeletal muscle phenotypes initiated by ectopic MyoD in transgenic mouse heart. *Development* 114:853–860.

Molkentin, J.D., Kalvakolanu, D.V., and Markham, B.E. 1994. Transcription factor GATA-4 regulates cardiac muscle-specific expression of the alpha-myosin heavy-chain gene. *Mol. Cell. Biol.* 14:4947–4957.

Moorman, A.F., de Jong, F., Denyn, M.M., and Lamers, W.H. 1998. Development of the cardiac conduction system. *Circ. Res.* 82:629–644.

Muscat, G.E., Gustafson, T.A., and Kedes, L. 1988. A common factor regulates skeletal and cardiac alpha-actin gene transcription in muscle. *Mol. Cell. Biol.* 8:4120–4133.

Offer, G., Moos, C., and Starr, R. 1973. A new protein of the thick filaments of vertebrate skeletal myofibrils. Extraction, purification, and characterization. *J. Mol. Biol.* 74:653–676.

Oliphant, L.W. and Loewen, R.D. 1976. Filament systems in Purkinje cells of the sheep heart: possible alterations of myofibrillogenesis. *J. Mol. Cell. Cardiol.* 8:679–688.

Papadopoulos, N. and Crow, M.T. 1993. Transcriptional control of the chicken cardiac myosin light-chain gene is mediated by two AT-rich *cis*-acting DNA elements and binding of serum response factor. *Mol. Cell. Biol.* 13:6907–6918.

Purkinje, J. 1845. Mikroskopisch-neurologische Beobachtungen. *Arch. Anat. Physiol. Wiss. Med.* 12:281–295.

Robb, J.S. and Petri, R. 1961. Expansions of the atrio-ventricular system in the atria. In: *The Specialized Tissue of the Heart*, Eds. A. Paes de Carvalho et al., pp. 1–18.

Sabry, M. and Dhoot, G. 1989. Identification and pattern of expression of a developmental isoform of troponin I in chicken and rat cardiac muscle. *J. Muscle Res. Cell Motil.* 10: 85–91.

Saito, K., Tamura, Y., Saito, M., Matsumura, K., Niki, T., and Mori, H. 1981. Comparison of the subunit compositions and ATPase activities of myosin in the myocardium and conduction system. *J. Mol. Cell. Cardiol.* 13:311–322.

Sanes, J.R. 1989. Analysing cell lineage with a recombinant retrovirus. *TINS* 12:21–28.

Sartore, S., Gorza, L., Pierobon, S., Bormioli, L., Dalla, L., and Schiaffino, S. 1981. Myosin types and fiber types in cardiac muscle. I. Ventricular myocardium. *J. Cell. Biol.* 88: 226–233.

Sartore, S., Pierobon, B., and Schiaffino, S. 1978. Immunohistochemical evidence for myosin polymorphism in the chicken heart. *Nature* 274:82–83.

Sassoon, D.A., Garner, I., and Buckingham, M. 1988. Transcripts of alpha-cardiac and alpha-skeletal actins are early markers for myogenesis in the mouse embryo. *Development* 104:155–164.

Schachner, M. and Martini, R. 1995. Glycans and the modulation of neural-recognition molecule function. *Trends Neurosci.* 18:183–191.

Schiaffino. S. 1997. Protean patterns of gene expression in the heart conduction system. *Circ. Res.* 80:749–750.

Schwartz, K., de la, B., Bouveret, P., Oliviero, P., Alonso, S., and Buckingham, M. 1986. Alpha-skeletal muscle actin mRNA's accumulate in hypertrophied adult rat hearts. *Circ. Res.* 59: 551–555.

Simon, A.M., Goodenough, D.A., and Paul, D.L. 1998. Mice lacking connexin40 have cardiac conduction abnormalities characteristic of atrioventricular block and bundle branch block. *Curr. Biol.* 8:295–298.

Sommer, J.R. and Johnson, E.A. 1979. Ultrastructure of cardiac muscle. In: *The Cardiovascular System*, Ed. R.M. Berne, American Physiological Society, Bethesda, MD, pp. 113–186.

Suske, G. 1999. The Sp-family of transcription factors. *Gene* 238:291–300.

Tawara, S. 1906. *Das reizleitungssystem des Sägetierherzens*, Gustav Fischer, Jena.

Thaemert, J.C. 1973. Fine structure of the atrioventricular node as viewed in serial sections. *Am. J. Anat.* 136:43–66.

Thörel, C. 1909. Vorläufige Mitteilung über eine besondere Muskelverbindung zwischen der Cava superior und dem His'schen Bündlel. *Münch. Med. Wochenschr.* 56:2159.

Thornell, L.E., Eriksson, A., Stigbrand, T., and Sjostrom, M. 1978. Structural proteins in cow Purkinje and ordinary ventricular fibres—a marked difference. *J. Mol. Cell. Cardiol.* 10: 605–616.

Thornell, L.E. and Sjostrom, M. 1975. Purkinje fibre glycogen. A morphologic and biochemical study of glycogen particles isolated from the cow conducting system. *Basic Res. Cardiol.* 70:661–670.

Treisman, R. 1992. The serum response element. *Trends Biochem. Sci.* 17:423–426.

Truex, R.C. and Smythe, M.Q. 1965. Comparative morphology of the cardiac conduction tissue in animals. *Ann. NY. Acad. Sci.* 127:19–33.

Tsuchimochi, H., Sugi, M., Kuro-o, M., Ueda, S., Takaku, F., Furuta, S., Shirai, T., and Yazaki, Y. 1984. Isozymic changes in myosin of human atrial myocardium induced by overload. Immunohistochemical study using monoclonal antibodies. *J. Clin. Invest.* 74:662–665.

Tucker, G.C., Aoyama, H., Lipinski, M., Tursz, T., and Thiery, J.P. 1984. Identical reactivity of monoclonal antibodies HNK-1 and NC-1: conservation in vertebrates on cells derived from the neural primordium and on some leukocytes. *Cell. Differ.* 14:223–230.

Vitadello, M., Matteoli, M., and Gorza, L. 1990. Neurofilament proteins are coexpressed with desmin in heart conduction system myocytes. *J. Cell. Sci.* 97:11–21.

Vitadello, M., Vettore, S., Lamar, E., Chien, K.R., and Gorza, L. 1996. Neurofilament M mRNA is expressed in conduction system myocytes of the developing and adult rabbit heart. *J. Mol. Cell. Cardiol.* 28:1833–1844.

Walsh, F.S. and Doherty, P. 1997. Neural cell adhesion molecules of the immunoglobulin super-family: role in axon growth and guidance. *Annu. Rev. Cell. Dev. Biol.* 13:425–456.

Warrick, H.M. and Spudich, J.A. 1987. Myosin structure and function in cell motility. *Annu. Rev. Cell. Biol.* 3:379–421.

Watanabe, M., Timm, M., and Fallah-Najmabadi, H. 1992. Cardiac expression of polysialy-lated NCAM in the chicken embryo: correlation with the ventricular conduction system. *Dev. Dyn.* 194:128–141.

Weiss, A. and Leinwand, L.A. 1996. The mammalian myosin heavy chain gene family. *Annu. Rev. Cell. Dev. Biol.* 12:417–439.

Wenckebach, K.F. 1906. Beiträge zur Kenntnis der menschlichen Hetztätigkeit. *Arch. Anat. Physiol.* 1–2:297–354.

Wessels, A., Vermeulen, J.L.M., Verbeek, F.J., Virgh, S., Klmn, F., Lamers, W.H., and Moorman, A.F.M. 1992. Spatial distribution of "tissue-specific" antigens in the developing human heart and skeletal muscle. III. An immunohistochemical analysis of the distribution of the neural tissue antigen G1N2 in the embryonic heart; Implications for the development of the atrioventricular conduction system. *Anat. Rec.* 232:97–111.

Xu, Z., Dong, D.L., and Cleveland, D.W. 1994. Neuronal intermediate filaments: new progress on an old subject. *Curr. Opin. Neurobiol.* 4:655–661.

Ya, J., Markman, M.W., Wagenaar, G.T., Blommaart, P.J., Moorman, A.F., and Lamers, W.H. 1997. Expression of the smooth-muscle proteins alpha-smooth-muscle actin and calponin, and of the intermediate filament protein desmin are parameters of cardiomyocyte maturation in the prenatal rat heart. *Anat. Rec.* 249:495–505.

Yasuda, M., Koshida, S., Sato, N., and Obinata, T. 1995. Complete primary structure of chicken cardiac C-protein (MyBP-C) and its expression in developing striated muscles. *J. Mol. Cell. Cardiol.* 27:2275–2286.

Yutzey, K.E., Rhee, J.T., and Bader, D. 1994. Expression of the atrial-specific myosin heavy chain AMHC1 and the establishment of anteroposterior polarity in the developing chicken heart. *Development* 120:871–883.

PART IV

Heart Development and Myofibrillar Organization

CHAPTER 11

Onset of a Cardiac Phenotype in the Early Embryo

Leonard M. Eisenberg and Carol A. Eisenberg

INTRODUCTION

Soon after fertilization, vertebrate embryos grow very rapidly. Thus, very early in gestation a sizeable yet underdeveloped organism requires circulating blood. This need dictates the early appearance of a contractile heart, which is the first functional organ in both the bird and mammalian embryos. Incipient heart tissue makes its arrival within the mesoderm layer during the onset of gastrulation. The process whereby nondifferentiated cells of primary mesoderm give rise to contractile cardiomyocytes is a subject that has greatly intrigued developmental biologists throughout the twentieth century. Since the early 1990s, a number of regulatory molecules have been identified that are important players in these events. Yet, how these molecular parts fit into the total story is still far from understood. In this chapter, we will discuss what is known about the morphological events that underlie the formation of the primitive heart, relate that information to the identification of candidate regulators of cardiogenesis in the early embryo (and description of their presumptive roles), and finally bring this information on cardiac development in context with the overall diversification of the primary mesoderm. Like many topics in biology, the study of cardiac development has profited both from tissue culture and in vivo experimentation. Because the onset of cardiogenesis occurs so early during embryogenesis, avian embryos have proven to be the most practical model system for studying these events in higher vertebrates, especially with regard to examining the behavior of precardiac tissue in isolation from the embryo. Thus, much of the discussion will be dominated by avian development. It has only been during recent years with the advent of transgenic and gene-targeted mice that mammalian models have made major contributions to our understanding of the primary events in cardiogenesis. Additionally, studies using frog and zebrafish embryos have contributed to this field. Surprisingly, an animal model that has yielded much information on early cardiogenesis is the fruit fly, *Drosophila melanogaster*. Despite the significant morphological differences between vertebrate and invertebrate hearts, there appears to be at least some homology of the molecular events that mold their respective cardiac tissue. Among vertebrate species, the molecular biology of early cardiogenesis seems to be totally conserved. This has allowed a fuller picture of early cardiogenesis to be compiled with information gathered from these various animal models, a cross-reference necessitated by the various strengths and weaknesses of each of the experimental systems.

THE BEGINNINGS OF THE HEART

The modern period of the study of heart formation begins with an observation of Francis Sabin (1920) that the first contractions of the heart occur during the 10 somite stage of the chick embryo—later defined as H-H stage 10 (Hamburger and Hamilton, 1951). Another landmark in the field of early cardiogenesis was Patten and Kramer's description of the progressive fusion and coincident contractility of the primitive cardiac tube (Patten and Kramer, 1933). The study of precardiac tissue commenced in the 1930s, a period when several laboratories set out to examine the organ-forming capabilities of early embryonic tissue (summarized in Rawles, 1943). It was through the efforts of Mary Rawles (1943) that the heart-forming areas were initially described for precontractile chick embryos. This map was delineated by culturing tissue fragments isolated from gastrulating chick embryos that displayed the head process—now referred to as H-H stage 5—and examining for subsequent development of contractile tissue. She was able to show that heart-forming potential was possessed by two large embryonic regions, lateral to Hensen's node (Figure 11.1A). In reporting these findings, Rawles speculated that the actual precardiac fields were probably narrower than the map she defined for cardiac potency of embryonic tissue. This was borne out by the investigations of Stalsberg and DeHaan (1969), who radiolabeled small tissue fragments consisting of the mesoderm and endoderm layers (mesendoderm) from H-H stage 5 embryos, which were subsequently transplanted into a second identically staged embryo, at the same position. These embryos were then allowed to develop further, sectioned, and analyzed for the radioactive label. Thus, by examining whether a particular region of the early embryo gave rise to heart tissue, they formulated a map for presumptive precardiac tissue (Figure 11.1B). Although these data are 30 years old, Stalsberg and DeHaan's study is to date the most comprehensive and thorough study of the precardiac mesoderm yet reported.

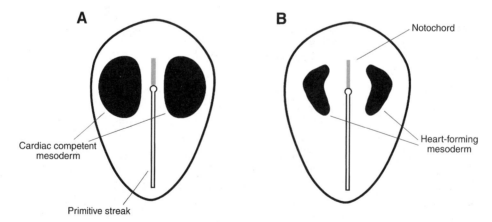

FIGURE 11.1. Cardiac fate maps of H-H stage 5 chick embryos according to the (A) explant culture studies of Mary Rawles (1943), or (B) in vivo labeling experimentation of Stalsberg and DeHaan (1969). Dark gray areas denote the location of mesoderm regions that possess cardiac potential.

EXPRESSION OF A CARDIAC PHENOTYPE BY MESODERM CELLS

The map of Stalsberg and DeHaan clearly shows that cells fated to become heart tissue reside within the anterior half of the bird embryo in mesodermal fields that are bilaterally distributed in reference to the primitive streak. Observations with the scanning electron microscope have indicated that precardiac cells comprise morphologically distinct cellular fields as early as H-H stage 5 (Drake and Jacobson, 1988). An H-H stage 5 embryo is a relatively simple structure consisting of the three primary germ layers. At H-H stage 6, lateral mesoderm begins to split into two distinct layers: the somatic and splanchnic mesoderms (DeHaan, 1965). It is within the splanchnic layer that the precardiac cells continue to reside, as shown by the time-lapse photographic studies of DeHaan (1963). The mesodermal origin of the heart has also been demonstrated with tissue explants from precardiac-stage embryos (H-H stages 5 to 6), as only the mesoderm layer will by itself give rise to cardiac tissue (Sugi and Lough, 1995). About H-H stage 8 to 9, the paired cardiac primordia of the splanchnic layer begin to fuse at the embryonic midline, with beating of the primitive heart commencing by H-H stage 10 (DeHaan, 1965).

This brief account describes the temporal sequence of the development of precardiac mesoderm to cardiac tissue. A major question that has interested cardiac developmental biologists is when these cells actually become specified to the cardiac lineage. In other words, when do they show evidence of expressing a cardiac phenotype? Striated muscle first appears in the chick heart by H-H stage 9 to 10 (10 somite stage), an observation first reported in 1919 (Lewis, 1919) and consistent with the first display of beating by the primitive heart (Patten and Kramer, 1933; DeHaan, 1965). However, some contractile proteins are expressed within the heart-forming fields much earlier. Detection of an embryonic isoform of sarcomeric myosin heavy chain (sMyHC), at both the RNA and protein levels, indicates that mesodermal cells can exhibit a cardiac phenotype as early as H-H stage 7 (Bisaha and Bader, 1991; Han et al., 1992). These cells display sMyHC in a prestriated pattern, with the organization of sarcomeric proteins into myofibrils beginning by H-H stage 9 to 10 (Han et al., 1992). By H-H stage 8 to 9, α-actin, titin, α-actinin, α-MyHC, and β-MyHC make their appearance in chick myocardial tissue (Tokuyasu and Maher, 1987a, 1987b; Ruzicka and Schwartz, 1988; de Jong et al., 1990). The pattern of α-actin expression is particularly interesting because the smooth muscle and cardiac isoforms are the first to be present within myocardial tissue, with the former exhibited at very high levels. In sequence, cardiac α-actin is upregulated, skeletal α-actin becomes expressed, and the smooth muscle isoform disappears from the developing heart (Ruzicka and Schwartz, 1988). The latter event occurs by H-H stage 12, at the same time when α-MyHC and β-MyHC expression becomes widespread by the myocardium (de Jong et al., 1987).

The preceding narrative of cardiac development as outlined in the bird appears very similar in the embryonic mouse, although early embryogenesis is relatively stretched out in the mammal. For example, fusion of paired mesodermal cardiac primordia and expression of the first sarcomeric proteins occur between embryonic days (ED) 7.5 and 8 (de Ruiter et al., 1992) in the mouse. In the bird, these events transpire within the first two days of development. Another organism that will enter the discussion is the frog. Despite differences in the structure of the mature amphibian heart as compared to birds and mammals (e.g., the ventricle is not septated), the sequence of events in the formation of the heart is nearly identical (Sater and Jacobson, 1989, 1990).

The temporal pattern of contractile proteins within the developing embryo indicates that a definitive cardiac phenotype is exhibited very early in embryogenesis. The question that always arises in developmental biology is what came before. In regard to the onset of cardiogenesis, what molecules control the transcription of contractile protein genes? Furthermore, is the expression of these regulatory genes the ultimate indicator that a cardiac phenotype has been specified? With the precedent of the MyoD family of transcription factors that specify the skeletal muscle phenotype (Molkentin and Olson, 1996)—the first example of lineage-specific master regulatory genes—the search for the cardiac counterparts was begun. The gene that has been touted as the master regulator of the cardiac phenotype is Nkx2.5. The evidence put forward for its prominence as the head of a hierarchy of cardiac transcription factors has been gathered from many species. Because great weight has been given to data obtained from *Drosophila*, we will briefly digress from our discussion of vertebrate cardiogenesis to describe the formation of the *Drosophila* heart.

MOLECULAR EVENTS OF EARLY CARDIOGENESIS: *CLUES FROM DROSOPHILA*

The mature *Drosophila* heart consists of a simple tube that lies at the dorsal midline (Bodmer, 1993; Zaffran et al., 1995; Curtis et al., 1999). Hence, it is often referred to as the dorsal vessel. This tube is comprised of an inner layer of contractile cardial cells and an outer pericardial layer (Figure 11.2). Like its vertebrate counterpart, the *Drosophila* heart is derived from the mesodermal layer (Azpiazu and Frasch, 1993; Bodmer, 1993). However, there are several features that distinguish the *Drosophila* and vertebrate hearts. The most obvious is that the dorsal vessel is derived from dorsal mesoderm, unlike the ventral mesoderm origin of the vertebrate heart. Also, the mature *Drosophila* heart is not a segmented organ. But perhaps most significant is that there are not separate myocardial and endocardial layers in the dorsal vessel. Rather, the inner layer is composed of myoendothelial cells, which exhibit both muscle and endothelial properties. Specifically, cardial cells express sarcomeric pro-

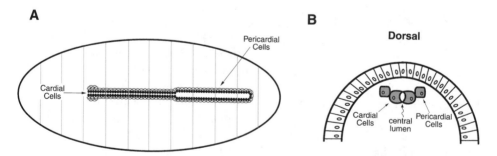

FIGURE 11.2. Structure of the *Drosophila* heart (dorsal vessel). (A) Dorsal view of a *Drosophila* embryo, following formation of the dorsal vessel. The *Drosophila* heart primarily consists of two cell types. The inner layer is comprised of contractile cardial cells, which are abutted by the outside layer of pericardial cells. (B) Transverse section of a similarly staged *Drosophila* embryo showing the outer pericardial and inner cardial cell layers of the dorsal vessel. Opposing cardial cells form a central lumen to allow blood flow, thereby functioning as both a myocardium and endocardium.

teins and contract, as well as forming lumens through which blood cells (hemocytes) pass. Despite these differences that suggest the analogy between the *Drosophila* and vertebrate hearts may have its limits, the study of dorsal vessel formation in *Drosophila* embryos has provided useful information for investigating cardiac development in vertebrates.

The *Drosophila* gene that has elicited the greatest interest from cardiac biologists is the homeobox gene *tinman* (Azpiazu and Frasch, 1993; Bodmer, 1993; Bodmer and Venkatesh, 1998). Like most genes, *tinman* was named for the embryonic phenotype that resulted from its mutation. Hence, the *tinman* mutant has no heart. The *tinman* gene is initially expressed throughout the mesoderm but becomes restricted to dorsal tissue by the end of gastrulation. Dorsal mesoderm subsequently partitions into somatic, visceral, and cardiac mesoderms, an event that is apparently dependent on *tinman* expression. In the absence of a functional *tinman* gene, neither cardiac, visceral, nor dorsal skeletal muscle will form (Bodmer, 1993; Bodmer and Venkatesh, 1998). As development proceeds, *tinman* disappears from the somatic mesoderm. It will persist in visceral tissue, although transiently. *Tinman* expression will, however, remain a permanent feature of the heart. Evidence for a functional role of *tinman* in cardiac myogenesis has been provided by studies of the *Drosophila* myocyte enhancer factor 2 (Dmef-2) gene (Gajewski et al., 1997, 1998). Dmef-2 is a MADS domain transcription factor that is required for the differentiation of all muscle cells. Transcription of the *dmef2* gene within cardial cells is dependent on *tinman*, a regulatory event that has been shown to be due to the interaction of *tinman* in a cardiac-specific enhancer complex. Thus, the evidence indicates that *tinman* plays a preeminent role in the formation of the heart, albeit its function at this point may be a broader one of subdividing the dorsal mesoderm into various tissue primordia. Also, its continued expression in the dorsal vessel and specific activation of *dmef2* indicate that it is substantially involved in promoting cardiomyogenic differentiation.

If *tinman* expression is in fact the primary event in *Drosophila* cardiogenesis, then what molecular events are responsible for its restricted expression in the dorsal mesoderm? Secreted signaling proteins that may be in large part responsible for eliciting *tinman* expression in precardiac mesoderm are decapentaplegic (Dpp) and wingless (Wg). Dpp belongs to the highly diverse TGFβ superfamily of protein signaling factors, with a sequence and structure that are most similar to the vertebrate members bone morphogenetic protein (BMP)-2 and -4. Dpp is produced by the dorsal ectoderm and is necessary for the maintenance of *tinman* expression within the dorsal mesoderm (Frasch, 1995; Yin and Frasch, 1998). Thus, *dpp* and *tinman* mutant embryos share phenotype, as cardiac and visceral mesodermal tissue are absent. The ectoderm also secretes a second BMP-2/BMP-4 related molecule, screw (Scw), which appears to act in consort with Dpp in regulating *tinman* expression. *Scw* mutants display reduced, although not totally absent, dorsal *tinman* expression. As with *dpp* mutations, embryos that lack functional *scw* do not form hearts (Yin and Frasch, 1998).

Wg is a member of the highly conserved WNT family of secreted signaling proteins, which have been shown to play crucial roles in patterning emerging embryonic tissues in both *Drosophila* and vertebrates (Parr and McMahon, 1994; Cadigan and Nusse, 1997; Moon et al., 1997). Wg is best known for its role in determining the anterior-to-posterior polarity of each segment within the *Drosophila* trunk (Klingensmith and Nusse, 1994). Within the ectoderm, Wg is expressed as a single broad stripe of cells within each segment (Baker, 1988). It is also exhibited within the mesoderm, although this latter expression disappears by the end of gastrulation

(Lawrence et al., 1994). As is the case with *tinman* mutations, embryos that lack functional Wg protein do not form hearts (Park et al., 1996; Bodmer and Venkatesh, 1998). However, *wg* mutant embryos will express *tinman* within the dorsal mesoderm prior to stages when heart tissue normally forms. Because heart tissue will not form in the absence of a Wg signal, *tinman* will therefore not persist within a subset of the mesoderm. Whether this indicates that Wg promotes heart development by either actively sustaining *tinman* expression or promoting *tinman*-independent signals has not been definitively resolved (Figure 11.3).

WNT functional activity has been ascribed in large part to a novel signal transduction pathway that has been conserved among invertebrate and vertebrate species (Cadigan and Nusse, 1997) (Figure 11.4). WNT binding to its cell membrane recep-

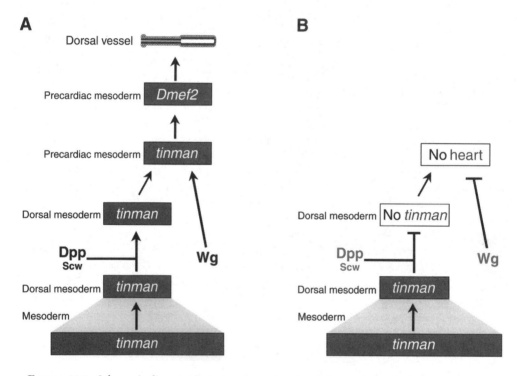

FIGURE 11.3. Schematic diagram depicting the molecular signaling pathway that elicits cardiac tissue formation in *Drosophila*. (A) The initial broad mesodermal expression of the homeobox gene *tinman* becomes constricted to the dorsal mesoderm. *Tinman* expression in the dorsal mesoderm is dependent on the extracellular signaling protein Dpp, and to a lesser extent on the Dpp-related protein Scw. The WNT protein Wg provides another extracellular signal crucial for heart formation. Although *tinman* expression in the dorsal mesoderm does not require a Wg signal, its continued expression in the precardiac areas of this mesodermal region is Wg-dependent. Precardiac mesodermal expression of *tinman* stimulates the expression of Dmef2, a MADS box transcription factor necessary for contractile gene expression. (B) In embryos lacking a functional *tinman* gene, no cardiac tissue will develop. In Dpp mutant embryos, *tinman* is not expressed in the dorsal mesoderm. Thus, in the absence of *tinman*, no heart will form. The cardiac phenotype of Scw mutants is similar to Dpp, although low levels of *tinman* will be displayed in the dorsal mesoderm. In the absence of a Wg signal, the heart will not form, although *tinman* will be expressed in the dorsal mesoderm. However, without the formation of cardiac tissue in Wg mutant embryos, there will be no continued expression of *tinman* (Gajewski et al., 1997; Park et al., 1998; Yin and Frasch., 1998).

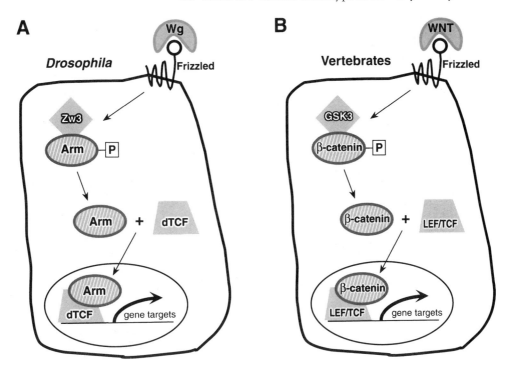

FIGURE 11.4. Schematic diagram of the WNT signal transduction pathway. (A) in *Drosophila*, the WNT protein Wg binds to the transmembrane protein Frizzled, which causes the inhibition of the kinase zw3. This in turn prevents the phosphorylation of the regulatory protein Armadillo (Arm), which allows its association with the transcriptional protein dTCF. Together, the Arm/dTCF complex enhances the transcription of specific gene targets. (B) The specific molecular events of the WNT signal transduction pathway have been conserved in vertebrates, although the names given for some of these molecules are different. The kinase GSK3 is the vertebrate homolog of zw3. β-catenin is homologous to the *Drosophila* protein Arm. dTCF is related to the LEF/TCF group of vertebrate transcription factors. Furthermore, gene duplication during evolution has increased the number of isoforms for several of the component proteins of this signaling pathway. There are at least 18 distinct WNT genes in the vertebrate genome. Also, both the vertebrate Frizzled and LEF/TCF protein families contain multiple members. For detailed information on WNT signal transduction, see the WNT Gene Homepage at www.stanford.edu/~rnusse/wntwindow.html.

tor (Frizzled; Dfz) inhibits the activity of the zeste-white3 encoded kinase (Zw3). In turn, this allows the multifunctional regulatory protein armadillo (Arm) to bind to the transcription factor dTCF (also known as pangolin), forming a complex which then translocates to the nucleus and exerts transcriptional enhancer activity. The importance of these various molecular components of Wg signaling for cardiogenesis has been demonstrated (Park et al., 1998). As expected, both *arm* and *dTCF* mutations exhibited a similar cardiac phenotype as shown for *wg*. Because Wg is thought to act by inhibiting Zw3, it was anticipated that enhancing or reducing activity of Zw3 would inhibit or increase cardiogenesis, respectively. Yet, the results produced by altering *zw3* expression were more complex. Conditional expression of an ectopically expressed *zw3* gene under control of a heat-shock promoter resulted in a mixture of phenotypes. Although most of the embryos analyzed did not exhibit any discernable changes in cardiac cell phenotype, many displayed either excess or reduced numbers of cardiac progenitors. A further surprise was that removal of *zw3* function by germ line mutation reduced the number of cardiogenic cells. In these latter embryos, dorsal mesoderm expression of *tinman* was greatly diminished. These results seem to imply that *zw3* may play two distinct roles in the formation of the *Drosophila* heart. First, functional *zw3* is necessary for the development of the dorsal mesoderm, a necessary first step in the specification of cardiac mesoderm. Subsequently, *zw3* activity may be required to be inhibited—presumably via a Wg signal— in order for the dorsal mesoderm to give rise to the dorsal vessel. In vertebrates, the putative involvement of WNT signaling pathways in cardiogenesis may be equally complex, as multiple WNT genes are expressed in the early heart. Included among these cardiac WNTs are those that exhibit activities antagonistic to classical WNT signaling (Torres et al., 1996; Moon et al., 1997). Thus, the twofold and contrasting roles that zw3 exhibits in *Drosophila* cardiogenesis may have relevance for vertebrate cardiogenesis.

TINMAN-RELATED GENES IN THE VERTEBRATE HEART

The regulatory molecule that has received the most interest in vertebrate cardiogenesis is Nkx2.5 (Harvey, 1996; Patterson et al., 1998). This molecule is the earliest and most persistently expressed representative of a subgroup of homeobox genes that are vertebrate homologs of *tinman*. Nkx2.5 expression is exhibited in the vertebrate heart throughout embryogenesis and adulthood (Komuro and Izumo, 1993; Lints et al., 1993; Tonissen et al., 1994; Schultheiss et al., 1995). Although the heart is the only organ that shows continuous expression of this molecule, its expression is not cardiac-limited. Actually, it appears in the early stages of organogenesis for other tissues, such as the pharynx, thyroid, tongue, and spleen (Lints et al., 1993; Tonissen et al., 1994; Kasahara et al., 1998). Additionally, five other *tinman*-related genes have been identified in vertebrates: Nkx2.3, Nkx2.6, Nkx2.7, Nkx2.8, and Nkx2.9. All these *tinman*-related genes appear to be expressed, at least transiently in myocardial tissues during development (Evans et al., 1995; Buchberger et al., 1996; Lee et al., 1996; Brand et al., 1997; Reecy et al., 1997; Biben et al., 1998; Newman and Krieg, 1998). However, unlike Nkx2.5, whose pattern of expression is highly conserved among all vertebrates, these other *tinman*-related genes display significant species-specific variations in their cardiac expression patterns.

Once expressed in cardiogenic tissue, Nkx2.5 remains an enduring feature of the vertebrate heart. There is a question as to when it first becomes expressed in cardiogenic tissue. There is no doubt that Nkx2.5 is transcribed in the heart-forming fields by cells that still exhibit a precardiac phenotype (i.e., prior to sarcomeric protein expression). In the H-H stage 6 chick, Nkx2.5 transcription is clearly displayed throughout the precardiac mesoderm (Schultheiss et al., 1995; Ehrman and Yutzey, 1999). Although this gene is expressed as early as H-H stages 4 to 5, the pattern it exhibits is rather curious in regard to the cardiac primordia (Figure 11.5). At this timepoint in early gastrulation, Nkx2.5 is exhibited as a crescent that resides along the anterior border of the embryo, immediately adjacent to the extraembryonic yolk sac (Schultheiss et al., 1995). Because by H-H stages 6 and 7 the Nkx2.5 gene expression pattern coincides precisely with the heart-forming fields, the temptation is to declare the early Nkx2.5 domain as precardiac. Yet, the H-H stage 5 Nkx2.5 crescent is markedly anterior of what has been determined experimentally as cardiac-forming

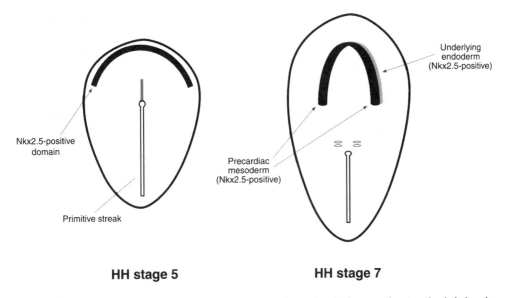

HH stage 5   **HH stage 7**

FIGURE 11.5. Illustration of Nkx2.5 expression in the early chick gastrula. On the left-hand side is a representation of an H-H stage 5 chick embryo. In dark gray is shown the Nkx2.5-positive domain that is exhibited as a crescent along the anterior border of the embryo. On the right-hand side is drawn an H-H stage 7 embryo, with the Nkx2.5 expression within the mesoderm displayed in dark gray. This Nkx2.5-positive domain at H-H stage 7 coincides exactly with the precardiac mesoderm. Underneath this region is depicted the endodermal expression domain of Nkx2.5 (light gray). This diagram is based on published in situ hybridization studies of Schultheiss et al. (1995).

mesoderm (Stalsberg and DeHaan, 1969; see Figure 11.1B). Moreover, the H-H stage 4 Nkx2.5 domain appears to be anterior to the expanse of the mesodermal layer. One possible explanation is based on the transient Nkx2.5 expression in other emerging tissues. This includes areas of the endoderm, which has been demonstrated to express Nkx2.5 at least as early as H-H stage 6 (Schultheiss et al., 1995; Figure 11.5). Therefore, it is possible that the H-H stage 5 Nkx2.5 crescent corresponds to the endoderm layer. Thereafter, Nkx2.5 begins to become expressed independently in the precardiac mesoderm by H-H stage 5+ or stage 6. Maybe now is the time to reapproach the cell fate mapping of the precardiac mesoderm and extend the analysis to H-H stage 4. Using modern techniques where mesoderm tissue could be labeled without removal from the embryo (i.e., retroviral labeling), and performed with the thoroughness and comprehensiveness that was the hallmark of the Staslberg and DeHaan study, should help resolve this issue about the early Nkx2.5 expression. This information would shed great light on the role of Nkx2.5 in cardiogenesis because expression in H-H stage 4 precardiac tissue would imply that it might play the primary role in specifying the cardiac fields. On the other hand, a delay of precardiac Nkx2.5 until H-H stage 6 might indicate that cardiac specification predates Nkx2.5 expression. Either way, it is clear that Nkx2.5 plays a major role in regulating the subsequent differentiation of precardiac cells to a definitive cardiomyocyte phenotype.

The elimination of functional Nkx2.5 in mice by targeted disruption of its gene produced an embryonic lethal phenotype due to severe heart defects (Lyons et al.,

1995). These homozygous Nkx2.5 mutant mice did, however, display normal cardiac development until the formation of the primary heart tube, including the differentiation of contractile cardiomyocytes. Subsequent remodeling of the heart during looping morphogenesis was highly disrupted. These Nkx2.5-minus hearts did not exhibit ventricular trabeculation, cardiac cushion formation, or septation. Although these data demonstrate the great importance of Nkx2.5 for cardiogenesis, the display of differentiated cardiac muscle in these mutant mice was a surprise—considering the complete absence of cardiogenesis in *Drosophila tinman* mutants (Bodmer, 1993). One possible explanation would be due to functional redundancy of multiple *tinman*-related genes. Though knockout mice with multiple Nkx2 genes targeted have not yet been produced, experiments with *Xenopus* embryos were performed where total Nkx2 activity was directly inhibited in cardiogenic tissue (Grow and Krieg, 1998). In these studies, the introduction of dominant-negative Nkx2 completely abolished mesodermal differentiation to cardiomyocytes. Moreover, this inhibition was rescued by ectopic Nkx2.5 expression, demonstrating that *tinman*-related genes are essential for vertebrate cardiogenesis.

OTHER TRANSCRIPTION FACTORS THAT PROMOTE A CARDIAC PHENOTYPE

Further demonstration of the functional importance of Nkx2.5 in cardiogenesis has come from studies of cardiac gene transcription. Several cardiac genes (e.g., cardiac α-actin and atrial natriuretic factor) have been shown to be directly upregulated by interaction of Nkx2.5 with a tissue-specific enhancer (Chen and Schwartz, 1995; Durocher et al., 1996). However, *tinman*-related transcription factors do not confer cardiac gene expression by themselves, as ectopic expression of Nkx2.5 in noncardiogenic tissue does not promote cardiogenesis (Cleaver et al., 1996). This is explained by the presence of several types of cardiac-associated transcriptional regulators that are required to promote cardiomyocyte differentiation. Prominent among these are members of the GATA and MADS box transcription factor families (Edmondson et al., 1994; Croissant et al., 1996; Jiang and Evans, 1996; Olson and Srivastava, 1996). The GATA group of zinc-finger–containing proteins consists of six members, of which three (GATA-4, -5, and -6) are highly expressed in cardiogenic tissue (Arceci et al., 1993; Heikinheimo et al., 1994; Laverriere et al., 1994; Jiang and Evans, 1996; Morrisey et al., 1997). MADS (MCM1, agamous, deficiens, serum response factor) box transcription factors expressed in the heart consist of the four vertebrate MEF-2 (A-D) proteins (Edmondson et al., 1994; Molkentin et al., 1996; Lin et al., 1997; Buchberger and Arnold, 1999) and serum response factor (SRF) (Croissant et al., 1996). Like Nkx2.5, the transcription factors GATA-4, GATA-5, MEF-2C, and SRF are all exhibited in precardiac mesoderm (Heikinheimo et al., 1994; Laverriere et al., 1994; Croissant et al., 1996; Lin et al., 1997; Jiang et al., 1998).

GATA proteins bind to *cis*-elements containing a WGATAR or nearly identical sequence (Arceci et al., 1993; Ko and Engel, 1993). All GATA proteins contain two zinc fingers within the DNA binding domain. These transcription factors appear to be lineage-specific, with GATA-1, -2, and -3 expressed primarily by blood lineage cells (Weiss and Orkin, 1995). GATA-4, -5, and -6 are often exhibited in endodermal tissues (Arceci et al., 1993; Laverriere et al., 1994), in addition to their cardiac expression—which for GATA-4 and GATA-5 includes both myocardium and endocardium (Heikinheimo et al., 1994; Morrisey et al., 1997). Like *tinman*-related genes, GATA factors are not exhibited in any muscle lineage but the heart. GATA factors have been shown to activate a number of cardiac promoters, including α-myosin heavy chain (MyHC), β-MyHC, cardiac troponin C, cardiac troponin I, atrial natriuretic factor (ANF), and B-type natriuretic peptide (BNP) (Molkentin et al., 1994; Thuerauf et al., 1994; Jiang and Evans, 1996; Durocher et al., 1997; Murphy et al., 1997; Charron et al., 1999). Mice homozygous for the GATA-4 null allele produced cardiomyocytes but were embryonic lethal, as the cardiac fields did not fuse to form the primary heart tube (Molkentin et al., 1997). Because endoderm development was also greatly disturbed, this outcome might have resulted from an overall disruption of ventral morphogenesis, which effected cellular movements (Narita et al., 1997). A similar phenotype was obtained with chick embryos when either GATA-4, GATA-5, or GATA-6 expression was specifically suppressed using antisense oligonucleotides (Jiang et al., 1998). Inhibition of GATA-4 expression in the mouse P19 embryonic stem-cell line or ectopic GATA gene expression (GATA-4, -5, or -6) in *Xenopus* embryos decreased or enhanced, respectively, the numbers of cardiomyocytes and transcription of cardiac-specific genes (Grepin et al., 1995; Jiang and Evans, 1996).

The expression of the MADS-domain–containing SRF and MEF-2 proteins are features of all myogenic lineages (Olson et al., 1995; Croissant et al., 1996). In this respect, they differ from *tinman*-related and GATA proteins. The MADS box is a highly conserved region of the protein that mediates both DNA binding and protein dimerization (Nurrish and Treisman, 1995). Additionally, MEF-2 factors possess a unique MEF-2 domain, which in combination with the adjacent MADS box confers a distinct DNA binding and protein dimerization specificity (Black and Olson, 1998). SRF recognizes variations on the sequence $CC(A/T)_6GG$, whereas MEF-2 factors bind to elements consisting of the $C/TTA(A/T)_4TAG/A$ consensus sequence. SRF is required for cardiac α-actin gene activation (Chen et al., 1996; Croissant et al., 1996), one of the earliest phenotypic genes to be exhibited in cardiogenic tissue (Ruzicka and Schwartz, 1988). MEF-2 factors have been implicated in the cardiac-specific expression of desmin, myosin light chain 2V, and α-MyHC (Navankasattusas et al., 1992; Kuisk et al., 1996; Black and Olson, 1998). Similar to Nkx2.5 and GATA-4, targeted disruption of the MEF-2C gene produced mice that were embryonic lethal due to a constraint on cardiac development (Lin et al., 1997). Again, cardiomyocytes arose—albeit with depressed levels of expression of some cardiac genes—but with major disturbances in cardiac morphogenesis occurring during heart looping stages.

The results generated from these investigations on the primary transcriptional regulators of vertebrate cardiogenesis were strikingly similar. Instead of cardiac development being completely prevented, it was arrested at an early stage. Possibly, if the activities of either all MEF-2 or all GATA genes were inhibited in the precardiac mesoderm, then—similar to the dominant inhibition studies of Nkx2 genes in *Xenopus*—cardiogenesis would be totally blocked. Similarities of phenotype may be indicative of cooperative regulation of early cardiogenesis by these groups of transcription factors. Studies on the transcriptional control of various individual cardiac genes indicate that these cardiac myogenic factors stimulate gene expression as components of multiprotein enhancer complexes (Chen et al., 1996; Chen and Schwartz, 1996; Durocher et al., 1997; Black and Olson, 1998; Lee et al., 1998; Sepulveda et al., 1998). For several cardiac genes, transcriptional enhancement requires the interaction among Nkx2 (*tinman*-related), GATA, and MADS box transcription factors. Thus, the cardiomyocyte lineage may be specified by the combination of these transcription factors (Figure 11.6). Additional levels of complexity may exist because many of

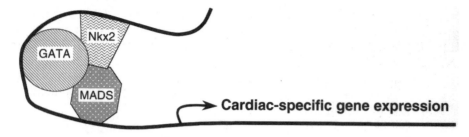

FIGURE 11.6. Transcriptional regulation of cardiac specification. Cardiac-specific gene expression in the early mesoderm appears to involve the combinatorial interactions of several types of transcriptional enhancer proteins. These include members of Nkx2 (*tinman*-related), GATA, and MADS families of transcriptional factors, which may interact to form multiprotein complexes that as a unit induce a cardiac phenotype.

these factors bind DNA as homo- or heterodimers, increasing the potential number of molecular players that contribute to transcriptional enhancement of cardiac genes. For example, GATA-4 and GATA-6 cooperatively regulate ANF and BNP genes, indicating combinatorial interactions among related transcription factors (Charron et al., 1999). These considerations on the potential complexity of cardiac regulation should be kept in mind in the study of secreted factors that define the heart-forming mesoderm. Although not as comprehensively studied as the transcriptional regulators, a large number of secreted signaling proteins have been identified as candidate regulators of the emerging cardiac fields.

SECRETED FACTORS THAT INITIATE CARDIOGENESIS IN THE VERTEBRATE EMBRYO

An important stimulus for heart formation appears to be the adjacent endodermal layer (Sugi and Lough, 1994; Schultheiss et al., 1995; Sugi and Lough, 1995). This functional cardiac-promoting activity is limited to the anterior endoderm; that is, only the endoderm underlying the precardiac regions can provide the appropriate signals for cardiac differentiation of mesodermal tissue. A number of secreted signaling proteins have been shown to be produced by the anterior endoderm during precardiac stages. These include fibroblast growth factor (FGF)-1, FGF-2, FGF-4, activin A, TGFβ1, TGFβ3, BMP-2, insulin-like growth factor II (IGF-II), and Cerberus (Dickson et al., 1993; Sanders et al., 1994; Sugi and Lough, 1995; Antin et al., 1996; Lough et al., 1996; Zhu et al., 1996; Schultheiss et al., 1997; Andrée et al., 1998; Schneider and Mercola, 1999). Additionally, WNT11, TGFβ2, and Cripto are expressed by anterior mesoderm and BMP-4 by the overlaying ectoderm (Dickson et al., 1993; Johnson et al., 1994; Eisenberg et al., 1997; Schultheiss et al., 1997). Although all of these proteins have been shown to play major roles in patterning developing embryonic tissue, a definitive demonstration of their importance for cardiac specification and/or stimulation of a cardiac phenotype has not been forthcoming for some of these molecules.

Interestingly, the cardiogenic stimulating activities of these factors in culture can differ dramatically depending on the stage and origin of the responding tissue. Activin, TGFβ1, or FGF-4, but neither BMP-2 nor BMP-4, will promote the differentiation of contractile tissue from pregastrula posterior chick epiblast (Ladd et al.,

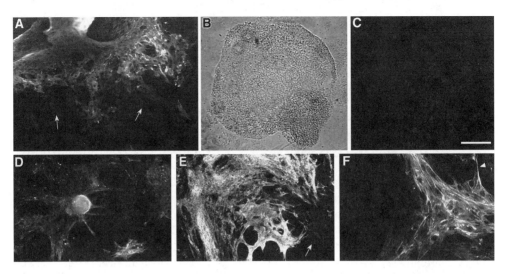

FIGURE 11.7. Cardiac differentiation of mesendoderm explants. Immunofluorescence (A,C–F) and phase (B) microscopic analysis of explanted mesoderm/endoderm (mesendoderm) tissue from H-H stage 5 quail embryos that were cultured 72 hours and immunolabeled for sarcomeric myosin heavy chain (sMyHC). Panel A shows tissue obtained from anterior precardiac areas that was beating within the first 30 hours of culture and displayed high levels of sMyHC staining. Panels B and C show a representative explant of noncardiogenic posterior mesendoderm that does not generate cardiac tissue, as shown by the absence of clustered sMyHC-positive cells. In contrast, treating noncardiogenic posterior mesendoderm with either (D) BMP-4 + FGF-2 or (E,F) WNT11-containing conditioned media (WNT11 CM) promoted the development of contractile tissue. These treated explants exhibited sMyHC-positive regions that corresponded to areas of beating tissue prior to fixation. Arrows in panels A and E indicate faintly sMyHC-positive cells within the lateral portions of the cardiac field, which seem to be fusing to the central group of definitive cardiomyocytes. The arrowhead in panel F shows the sarcomeric organization of sMyHC in the WNT11 CM treated posterior tissue. Scale bar: (A) 200 μm; (B,C) 400 μm; (D,E) 250 μm; (F) 55 μm.

1998). On the other hand, BMP-2 and BMP-4 are able to stimulate cardiac differentiation of gastrula-stage tissue (Figure 11.7). BMP-2 or BMP-4 can elicit a contractile phenotype from gastrula-stage noncardiogenic posterior mesendoderm tissue, provided that FGF-2 or FGF-4 is also present in these cultures (Lough et al., 1996; Ladd et al., 1998). Although activin is, like BMP-2 and BMP-4, a member of the TGFβ superfamily of signaling factors, it has not demonstrated a similar capability to promote cardiac differentiation of noncardiogenic gastrula-stage tissue (Ladd et al., 1998). Yet, in the absence of the endoderm layer, it can help support cardiac differentiation of H-H stage 6 precardiac mesoderm (Sugi and Lough, 1995).

As discussed earlier, the vertebrate proteins BMP-2 and BMP-4 are closely related to *Drosophila* Dpp and Scw—two molecules that appear to play major roles in heart formation of the fruit fly (Yin and Frasch, 1998). That BMPs are likewise specifically involved in promoting vertebrate cardiogenesis has been indicated by studies with noggin, a specific inhibitor of BMP signaling, which significantly decreased the prevalence of contractile tissue in H-H stage 5 precardiac mesendoderm cultures (Ladd et al., 1998). Additional support for the supposition that BMPs are necessary for heart formation has been obtained from whole embryo experimentation. Implantation of either BMP-2–coated heparin-acrylamide beads or BMP-2–secreting stably trans-fected cells into developing chick embryos ectopically stimulated cardiac-associated gene expression (Schultheiss et al., 1997; Andrée et al., 1998). In both cases, Nkx2.5 and GATA-4 were upregulated among cells adjacent to the implant. However, this Nkx2.5 response was limited to anterior portions of the embryo, whereas GATA-4 was able to be induced more caudally. Moreover, BMP-2 by itself was not sufficient to stimulate a definitive cardiac phenotype, as shown by the absence of sMyHC-expressing cells surrounding the implant (Andrée et al., 1998).

The secreted factor that appears to have the most dramatic effect on cardiomyocyte differentiation is Cripto-1, a member of the epidermal growth factor family. Cripto-1 is expressed in the primitive streak and is later restricted to the developing heart (Johnson et al., 1994). Mice lacking a functional *cripto-1* gene do not produce car-diomyocytes, one of the most severe cardiac phenotypes observed in any gene knock-out mouse strain (Xu et al., 1999). Embryoid bodies developed from homozygous *cripto-1*-negative mouse embryonic stem (ES) cells did not give rise to cardiomy-ocytes, although many other tissues appeared to develop normally (Xu et al., 1998). Introducing a *cripto-1* transgene into these mutant ES cells restored cardiac differen-tiation. Although the influence of Cripto on early avian cardiogenesis has not been examined, dramatic enhancement of cardiac differentiation was observed with aggregate cultures of chick blastoderm cells in response to TGFα (Eisenberg and Markwald, 1997), a molecule closely related to Cripto.

Many of the secreted factors discussed earlier also have been shown to be impor-tant for the cardiac differentiation of the mesoderm stem-cell line QCE6. The QCE6 cell line was derived from the anterior mesoderm of H-H stage 4 quail embryos and has been shown to be representative of early nondifferentiated mesoderm (Eisenberg and Bader, 1996; Eisenberg and Markwald, 1997). These cells will develop into fully differentiated contractile cardiomyocytes when incorporated into aggregate cultures derived from H-H stage 5 chick blastoderms (Eisenberg and Markwald, 1997). When cultured alone, these cells will exhibit a cardiac phenotype—as indicated by expres-sion of a variety of sarcomeric proteins, such as sMyHC, desmin, and cardiac tro-ponin I—in response to FGF-2, TGFβ2, TGFβ3, and retinoic acid (Eisenberg and Bader, 1995; Eisenberg and Bader, 1996). Further augmentation of this growth factor mixture with TGFβ1, IGF-II, and platelet-derived growth factor (PDGF) will begin to promote the organization of these sarcomeric proteins (Eisenberg and Bader, 1996). A molecule required for the cardiac differentiation of these cells is WNT11 (Eisenberg et al., 1997). QCE6 cells express both WNT11 mRNA and protein. Inhibition of WNT11 expression by these cells blocks their ability to undergo cardiac differentiation in response to the aforementioned secreted factors. We have shown that conditioned media from cells secreting high amounts of WNT11 protein can provoke cardiogenesis of noncardiogenic embryonic tissue (Eisenberg, and Eisenberg, 1999) (Figure 11.7C,D). Together with the QCE6 cell data, this result is the first indication that involvement of WNTs in *Drosophila* cardiogenesis has been conserved in vertebrates.

INVOLVEMENT OF WNTS IN VERTEBRATE CARDIOGENESIS

WNT11 is first exhibited in the early avian gastrula within Hensen's node. By H-H stage 4, the WNT11 expression has spread outward as a mesodermal crescent with Hensen's node as its apex. At H-H stage 5, this WNT11 domain is extended

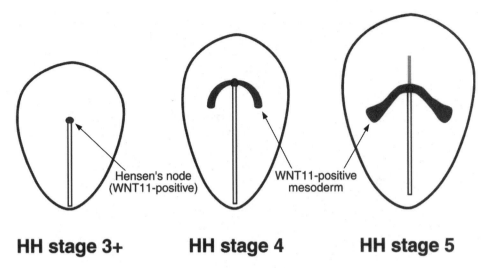

HH stage 3+   HH stage 4   HH stage 5

FIGURE 11.8. WNT11 expression in gastrula-stage avian embryos. WNT11 RNA is initially exhibited within Hensen's node (arrow) at H-H stage 3+. By H-H stage 4, WNT11 expression extends into the mesoderm as a crescent with Hensen's node at its apex. The WNT11 RNA domain continues to spread laterally within the embryo by H-H stage 5 of development. At this stage, the mesodermal region of WNT11 expression appears to overlap with the posterior portion of the precardiac mesodermal fields (see Fig. 11.1B). The WNT11 expression pattern is based on previously published in situ hybridization studies (Eisenberg et al., 1997).

more laterally (Eisenberg et al., 1997; Figure 11.8). Subsequently, WNT11 expression decreases and then reappears, first in axial mesoderm and then within cardiac tissue (personal observation). This expression pattern in the early avian gastrula identifies WNT11 as being the best candidate WNT for interacting with prospective heart cells in the mesoderm (Figure 11.8). To date, it is the only WNT described whose RNA expression overlaps with the precardiac mesoderm of the avian embryo. Comparison with Stalsberg and DeHaan's map of the developing avian heart (Stalsberg and DeHaan, 1969; see Figure 11.1) clearly indicates that WNT11 is expressed within the posterior portion of the stage 5 heart-forming fields. It is also worth keeping in mind that in situ hybridization for genes encoding secreted factors provides a potentially narrow view of the protein distribution. Thus, WNT11 protein may very well be dispersed among cells located more rostral than the domain of RNA expression. In mice, the WNT11 expression pattern in the early embryo appears to be totally conserved (Kispert et al., 1996). Likewise, the expression of WNT11 in the tubular mouse heart is also conserved (Christiansen et al., 1995). Interestingly, in the early mouse gastrula, the WNT2 gene transcription pattern (Monkley et al., 1996) appears to coincide exactly with WNT11. In contrast, WNT2 is not expressed in the early avian gastrula (Eisenberg et al., 1997), although it is exhibited at tubular heart stages (Figure 11.9A,B (see Color Plate 4)). During early gastrulation, at least two other WNTs are expressed: WNT8 in the primitive streak (Hume and Dodd, 1993) and WNT3a in the posterior portion of the embryo (Takada et al., 1994). In the tubular heart of the chick, five individual WNT genes have been detected: WNT2 (Figure 11.9A,B), WNT5a, WNT7a (Dealy et al., 1993), WNT11, and WNT14 (Figure 11.9C,D (see color insert)).

The role of these WNT molecules in cardiogenesis has not yet been defined. Although knockout mice have been derived for some of these molecules, their cardiac phenotype has not been analyzed in detail. Considering the identical expression of WNT2 and WNT11 in the early gastrula, it is very likely that both genes would need to be mutated in order to disrupt heart formation in mice. Because the expression of WNTs in other tissues (i.e., brain, limbs, kidney, reproductive organs) has been shown to be crucial for their development (Parr and McMahon, 1994; Cadigan and Nusse, 1997; Moon et al., 1997), it seems unlikely that their expression in the heart is incidental. Because WNTs have been shown to play significant roles in defining tissue borders and field formation, it seems reasonable to speculate that WNT11 is involved in establishing the posterior border of the emerging heart. This may be borne out by our explant studies that show that exposure of mesendoderm tissue to WNT11-containing media promotes the aggregation of precardiac cells (Eisenberg and Eisenberg, 1999).

One consequence of multiple WNTs being expressed in cardiogenic tissue is that antagonistic signaling pathways may come into play in patterning the emerging heart. Most investigations on the mechanisms of WNT signal transduction have focused on the pathway outlined in Figure 11.4. The details of this pathway in *Drosophila* and vertebrates appear to be identical, although the names for the molecular homologs differ. In vertebrates, WNT binding to its receptor (Frizzleds) inhibits the kinase GSK3 (zw3 homolog), which allows the multifunctional protein β-catenin (Arm homolog) to interact with LEF/TCF transcription factors (dTCF homolog). The subsequent translocation of this molecular complex allows for the transcriptional enhancement of downstream WNT target genes (Cadigan and Nusse, 1997). An important caveat to these findings is that WNTs can be divided into two broad classes—referred to as the WNT1 and WNT5a classes—based on distinct functional activities and signaling mechanisms (Moon et al., 1997; Slusarski et al., 1997; Sheldahl et al., 1999). In contrast to the WNT1 class proteins, the WNT5a class does not appear to signal via β-catenin. Apparently, this latter group of WNTs stimulates phosphatidylinositol (PI) signaling via a G-protein–linked mechanism. It is believed that in response to WNT5a-class proteins, the enzyme phospholipase C (PLC) hydrolyzes phosphatidylinositol 4,5-diphosphate (PIP_2) to generate the second messengers inositol-1,4,5-triphosphate (IP_3), and diacylglycerol (DAG). In turn, IP_3 causes the release of calcium ions (Ca^{2+}) from stores within the endoplasmic reticulum, and DAG activates protein kinase C (PKC) (Figure 11.10). WNT11 has been classified with WNT5a in its functional activity (Moon et al., 1997; Wang et al., 1997). Moreover, WNT11 expression by QCE6 cells appears to increase levels of intracellular Ca^{2+} (Figure 11.11). Although the functional implications of WNT-mediated PI signal transduction are not understood, an important observation was that both WNT5a and WNT11 can antagonize the activity of WNT1 class proteins (Torres et al., 1996; see Figure 11.10). Considering both the possible involvement of WNT11 in heart-field formation and the large number of WNTs expressed in the tubular heart, there may be parallels in vertebrate cardiogenesis of the multiple roles of zw3 signaling in the development of the *Drosophila* dorsal vessel (Park et al., 1998); that is, the function of primary signals in vertebrate cardiogenesis (e.g., WNT11) may be to allow the zw3 homolog GSK3 to remain active. Afterward, other WNTs expressed in the developing heart may act to inhibit GSK3, which would allow for further development of cardiogenic tissue.

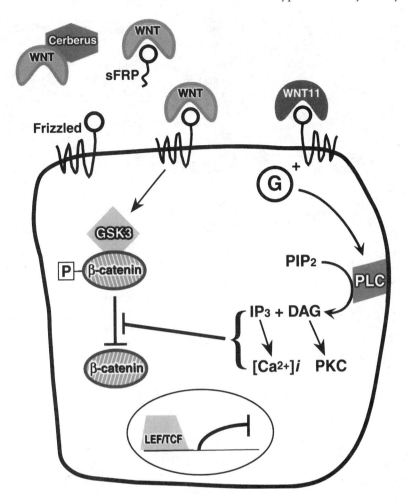

FIGURE 11.10. Antagonistic WNT signaling pathways. This diagram of a WNT11-mediated PI signaling is based on experimental studies with WNT5a (Moon et al., 1997; Slusarski et al., 1997; Sheldahl et al., 1999), a molecule that shares functional properties with WNT11 (Moon et al., 1997; Wang et al., 1997). Both of these molecules are able to antagonize the activity of WNT11 class proteins (Torres et al., 1996). Thus, it is proposed that the importance of WNT11 for early cardiogenesis may be to block β-catenin-mediated signal transduction. WNT11 interaction with Frizzled transmembrane proteins is hypothesized to activate PLC by a G-protein–linked mechanism. PLC hydrolyzes PIP_2 to generate the second messengers IP_3 and DAG, which in turn promotes Ca^{2+} fluxes and stimulates PKC activity. Although the importance of these signal transduction events for WNT5a-type (e.g., WNT11) functional activity has not yet been formally proven, there is evidence that PI signaling can prevent β-catenin-dependent embryonic patterning (Hedgepeth et al., 1997). Additionally, WNT-driven β-catenin signal transduction can be blocked by directly preventing WNT triggering of cellular responses. Several types of extracellular WNT inhibitors have been described, which include the secreted Frizzled-related proteins (sFRP; Rattner et al., 1997; Wang et al., 1997) and Cerberus (Piccolo et al., 1999). The WNT binding protein Cerberus is an antagonist of both WNT and TGFβ signaling. It is secreted by the anterior endoderm during precardiac stages, and studies with *Xenopus* embryos have indicated that it may be necessary for heart formation (Schneider and Mercola, 1999).

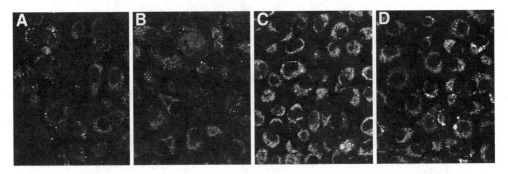

Figure 11.11. WNT11 enhances intracellular Ca^{2+} levels in QCE6 cells. QCE6 cells express WNT11 mRNA and protein. The effect of WNT11 on cellular responses was studied by stably inserting a WNT11 antisense or sense transgene, respectively, to either diminish or enhance WNT11 expression in QCE6 cells (Eisenberg et al., 1997). For these experiments, confluent cell cultures were briefly labeled with the Ca^{2+} indicator Oregon Green BAPTA (Molecular Probes, Inc.), prior to microscopic observation. Upon excitation with a confocal microscope laser, this reagent produces levels of fluorescent emissions that are enhanced as a function of increasing Ca^{2+} concentrations. Panels A and B show two independently derived WNT11 antisense QCE6 sublines that produce greatly reduced levels of WNT11 protein. In panels C and D are displayed, respectively, the parental QCE6 cells and a stably transfected variant subline that overexpresses WNT11 protein. Note that the two WNT11 antisense cell lines (A, B) exhibit much less Ca^{2+}-dependent fluorescence than cells that express much higher levels of WNT11 (C,D).

CARDIAC TISSUE FORMATION FROM EARLY MESODERM—OBSERVATIONS AND CONCLUSIONS

In investigating whether WNTs may influence cardiac tissue formation, we used explants from early gastrula-stage avian embryos as our experimental model for heart formation (Eisenberg and Eisenberg, 1999). As a necessary control for these studies, the cardiac potential of H-H stage 4 to 6 mesendoderm from anterior precardiac or posterior noncardiogenic embryonic regions was examined. Great care was taken when obtaining posterior mesendoderm so as to isolate only tissue that was considerably posterior to the presumptive heart-forming fields. Surprisingly, both anterior and posterior mesendoderm from H-H stage 4 embryos exhibited a similarly high occurrence of contractile tissue development (Figure 11.12A–C). While H-H stage 5- or stage 6-derived explants gave rise to cardiac tissue only if obtained from anterior precardiac regions (Figure 11.7A–C), cultures of posterior noncardiogenic mesendoderm were not totally devoid of cells that exhibited cardiac phenotypes. Rather, these latter explants contained cells scattered within the tissue that displayed low-level expression of sarcomeric proteins (Figure 11.12D,E). The phenotype of these faintly positive posterior cells seemed similar to cells observed in H-H stage 5 (Figure 11.7A) or H-H stage 6 anterior explants (Figure 11.12F), except that these faintly positive anterior cells were not randomly scattered. Instead, they were detected lateral to definitive heart fields and appeared to be in the process of fusing into the cardiac tissue. H-H stage 5 to 6 posterior mesendoderm exposed to WNT11 (either by plasmid transfection or with conditioned media) seemed to display a similar

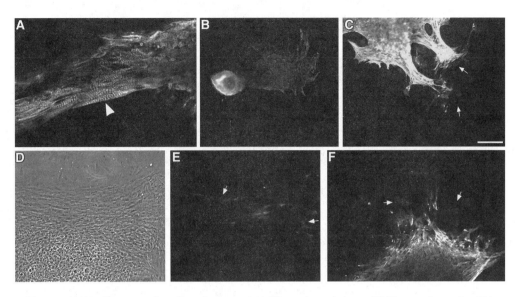

FIGURE 11.12. Patterns of cardiac phenotype within explanted tissue. H-H stage 4 anterior (A) or posterior (B,C) mesendoderm explants cultured for 72 hours were stained for sMyHC and visualized by immunofluorescence microscopy. Surprisingly, tissue from both anterior and posterior regions of the embryo produced contractile tissue and expressed sarcomeric myosin. Arrowhead in panel A shows the typical myofibrillar pattern of sMyHC staining. Arrows in panel C denote cells lateral to the contractile tissue that express low levels of sMyHC. Panels D and E show phase and fluorescent images of an H-H stage 6 posterior mesendoderm explant stained for sarcomeric desmin. Although cardiac fields were not exhibited, cells exhibiting low-level expression of sarcomeric proteins (E; arrows) were observed in a scattered distribution within the explant. In comparison, panel F displays a fluorescent image of sMyHC-stained H-H stage 6 anterior mesendoderm explant. Here is shown the border of a beating cardiac field exhibiting high-level sMyHC staining, surrounded by faintly sMyHC positive cells (arrows) that appear to be in the process of fusing to the contractile tissue. Scale bar: (A) 25 μm; (B) 250 μm; (C–F) 200 μm.

configuration, with faintly sarcomeric protein-positive cells surrounding definitive cardiac tissue (Figure 11.7E).

Three distinct observations can be distilled from these studies: (a) there is a narrowing of heart-forming potential among different mesodermal regions as the embryo ages from H-H stage 4 to H-H stage 5 to 6; (b) individual H-H stage 5 and stage 6 posterior mesoderm cells still possess potential to exhibit a cardiac phenotype but cannot organize into a cardiac tissue; and (c) factors capable of promoting cardiogenesis of H-H stage 5 to 6 posterior mesendoderm (e.g., WNT11) may act by organizing precardiac cells into a tissue. In *Drosophila*, *tinman* initially exhibits a broad mesodermal distribution, which narrows until it is displayed only in the heart. Thus, heart formation in the fruit fly may also involve a sequential restriction of potential. Hence, the formation of the heart may involve a two-step process involving cardiac field formation, whereby multiplicity of phenotypic potential becomes limited to a single option, and full differentiation, which follows or coincides with organogenesis. How the many transcriptional and secreted signaling proteins that have been implicated in heart formation fit into not only a linear pathway of cardiogenesis but also the larger morphological transitions of the early embryo should pose great challenges to developmental biologists.

ACKNOWLEDGMENTS

We thank Dr. Thomas Trusk for assistance with the Ca^{2+} studies. The authors' studies reported in this review were supported by American Heart Association Grant-in-Aid 9950638N and NIH Grant HL55923.

REFERENCES

Andrée, B., Duprez, D., Vorbusch, B., Arnold, H.H., and Brand, T. 1998. BMP-2 induces ectopic expression of cardiac lineage markers and interferes with somite formation in chicken embryos. *Mech. Dev.* 70:119–131.

Antin, P.B., Yatskievych, T., Dominguez, J.L., and Chieffi, P. 1996. Regulation of avian precardiac mesoderm development by insulin and insulin-like growth factors. *J. Cell. Physiol.* 168:42–50.

Arceci, R.J., King, A.A., Simon, M.C., Orkin, S.H., and Wilson, D.B. 1993. Mouse GATA-4: a retinoic acid-inducible GATA-binding transcription factor expressed in endodermally derived tissues and heart. *Mol. Cell. Biol.* 13:2235–2246.

Azpiazu, N. and Frasch, M. 1993. Tinman and bagpipe: two homeobox genes that determine cell fates in the dorsal mesoderm of *Drosophila*. *Genes Dev.* 7:1325–1340.

Baker, N.E. 1988. Localization of transcripts from the wingless gene in whole *Drosophila* embryos. *Development* 103:289–298.

Biben, C., Hatzistavrou, T., and Harvey, R.P. 1998. Expression of NK-2 class homeobox gene Nkx2-6 in foregut endoderm and heart. *Mech. Dev.* 73:125–127.

Bisaha, J.G. and Bader, D. 1991. Molecular analysis of myogenic differentiation: Isolation of a cardiac-specific myosin heavy chain. *Dev. Biol.* 148:335–364.

Black, B.L. and Olson, E.N. 1998. Transcriptional control of muscle development by myocyte enhancer factor-2 (MEF2) proteins. *Annu. Rev. Cell Dev. Biol.* 14:167–196.

Bodmer, R. 1993. The gene tinman is required for specification of the heart and visceral muscles in *Drosophila*. *Development* 118:719–729.

Bodmer, R. and Venkatesh, T.V. 1998. Heart development in *Drosophila* and vertebrates: conservation of molecular mechanisms. *Dev. Genetics* 22:181–186.

Brand, T., Andree, B., Schneider, A., Buchberger, A., and Arnold, H.H. 1997. Chicken Nkx2-8, a novel homeobox gene expressed during early heart and foregut development. *Mech. Dev.* 64:53–59.

Buchberger, A. and Arnold, H.H. 1999. The MADS domain containing transcription factor cMef2a is expressed in heart and skeletal muscle during embryonic chick development. *Dev. Genes Evol.* 209:376–381.

Buchberger, A., Pabst, O., Brand, T., Seidl, K., and Arnold, H.H. 1996. Chick NKx-2.3 represents a novel family member of vertebrate homologues to the *Drosophila* homeobox gene tinman: differential expression of cNKx-2.3 and cNKx-2.5 during heart and gut development. *Mech. Dev.* 56:151–163.

Cadigan, K.M. and Nusse, R. 1997. Wnt signaling: a common theme in animal development. *Genes Dev.* 11:3286–3305.

Charron, F., Paradis, P., Bronchain, O., Nemer, G., and Nemer, M. 1999. Cooperative interaction between GATA-4 and GATA-6 regulates myocardial gene expression. *Mol. Cell. Biol.* 19:4355–4365.

Chen, C.Y., Croissant, J., Majesky, M., Topouzis, S., McQuinn, T., Frankovsky, M.J., and Schwartz, R.J. 1996. Activation of the cardiac alpha-actin promoter depends upon serum response factor, Tinman homologue, Nkx-2.5, and intact serum response elements. *Dev. Genetics* 19:119–130.

Chen, C.Y. and Schwartz, R.J. 1995. Identification of novel DNA binding targets and regulatory domains of a murine tinman homeodomain factor, Nkx-2.5. *J. Biol. Chem.* 270:15628–15633.

Chen, C.Y. and Schwartz, R.J. 1996. Recruitment of the tinman homolog Nkx-2.5 by serum response factor activates cardiac alpha-actin gene transcription. *Mol. Cell. Biol.* 16:6372–6384.

Christiansen, J.H., Dennis, C.L., Wicking, C.A., Monkley, S.J., Wilkinson, D.G., and Wainwright, B.J. 1995. Murine wnt-11 and wnt-12 have temporally and spatially restricted expression patterns during embryonic development. *Mech. Dev.* 51:341–350.

Cleaver, O.B., Patterson, K.D., and Krieg, P.A. 1996. Overexpression of the tinman-related genes XNkx-2.5 and XNkx-2.3 in *Xenopus* embryos results in myocardial hyperplasia. *Development* 122:3549–3556.

Croissant, J.D., Kim, J.H., Eichele, G., Goering, L., Lough, J., Prywes, R., and Schwartz, R.J. 1996. Avian serum response factor expression restricted primarily to muscle cell lineages is required for alpha-actin gene transcription. *Dev. Biol.* 177:250–264.

Curtis, N.J., Ringo, J.M., and Dowse, H.B. 1999. Morphology of the pupal heart, adult heart, and associated tissues in the fruit fly, *Drosophila melanogaster. J. Morphol.* 240:225–235.

de Jong, F., Geerts, W.J., Lamers, W.H., Los, J.A., and Moorma, A.F. 1987. Isomyosin expression patterns in tubular stages of chicken heart development: a 3-D immunohistochemical analysis. *Anat. Embryol.* 177:81–90.

de Jong, F., Geerts, W.J., Lamers, W.H., Los, J.A., and Moorman, A.F. 1990. Isomyosin expression during pattern formation of the tubular heart: A three-dimensional immunohistochemical analysis. *Anat. Rec.* 226:213–227.

de Ruiter, M.C., Poelmann, R.E., van der Plas-de Vries, I., Mentink, M.M., and Gittenberger-de Groot, A.C. 1992. The development of the myocardium and endocardium in mouse embryos. Fusion of two heart tubes? *Anat. Embryol.* 185:461–473.

Dealy, C.N., Roth, A., Ferrari, D., Brown, A.M., and Kosher, R.A. 1993. *Wnt-5a* and *Wnt-7a* are expressed in the developing chick limb bud in a manner suggesting roles in pattern formation along the proximodistal and dorsoventral axes. *Mech. Dev.* 43:175–186.

DeHaan, R.L. 1963. Migration patterns of the precardiac mesoderm in the early chick embryo. *Exp. Cell. Res.* 29:544–560.

DeHaan, R.L. 1965. Morphogenesis of the vertebrate heart. In: *Organogenesis*, Eds. R.L. DeHaan and H. Ursprung, Holt, Rinehart and Winston, New York, pp. 377–419.

Dickson, M.C., Slager, H.G., Duffie, E., Mummery, C.L., and Akhurst, R.J. 1993. RNA and protein localisations of TGF beta 2 in the early mouse embryo suggest an involvement in cardiac development. *Development* 117:625–639.

Drake, C.J. and Jacobson, A.G. 1988. A survey by scanning electron microscopy of the extracellular matrix and endothelial components of the primordial chick heart. *Anat. Rec.* 222:391–400.

Durocher, D., Charron, F., Schwartz, R.J., Warren, R., and Nemer, M. 1997. The cardiac transcription factors Nkx2-5 and GATA-4 are mutual cofactors. *EMBO J.* 16:5687–5696.

Durocher, D., Chen, C.-Y., Ardati, A., Schwartz, R.J., and Nemer, M. 1996. The atrial natriuretic factor promoter is a downstream target for Nkx-2.5 in the myocardium. *Mol. Cell. Biol.* 16:4648–4655.

Edmondson, D.G., Lyons, G.E., Martin, J.F., and Olson, E.N. 1994. Mef2 gene expression marks the cardiac and skeletal muscle lineages during mouse embryogenesis. *Development* 120:1251–1263.

Ehrman, L.A. and Yutzey, K.E. 1999. Lack of regulation in the heart forming region of avian embryos. *Dev. Biol.* 207:163–175.

Eisenberg, C.A. and Bader, D.M. 1995. QCE-6: A clonal cell line with cardiac myogenic and endothelial cell potentials. *Dev. Biol.* 167:469–481.

Eisenberg, C.A. and Bader, D.M. 1996. The establishment of the mesodermal cell line QCE-6: A model system for cardiac cell differentiation. *Circ. Res.* 78:205–216.

Eisenberg, C.A. and Eisenberg, L.M. 1999. WNT11 promotes cardiac tissue formation of early mesoderm. *Dev. Dyn.* 216:45–58.

Eisenberg, C.A., Gourdie, R.G., and Eisenberg, L.M. 1997. *Wnt-11* is expressed in early avian mesoderm and required for the differentiation of the quail mesoderm cell line QCE-6. *Development* 124:525–536.

Eisenberg, C.A. and Markwald, R.R. 1997. Mixed cultures of avian blastoderm cells and the quail mesoderm cell line QCE-6 provide evidence for the pluripotentiality of early mesoderm. *Dev. Biol.* 191:167–181.

Evans, S.M., Yan, W., Murillo, M.P., Ponce, J., and Papalopulu, N. 1995. tinman, a *Drosophila* homeobox gene required for heart and visceral mesoderm specification, may be represented by a family of genes in vertebrates: XNkx-2.3, a second vertebrate homologue of tinman. *Development* 121:3889–3899.

Frasch, M. 1995. Induction of visceral and cardiac mesoderm by ectodermal DPP in the early *Drosophila* embryo. *Nature* 374:464–467.

Gajewski, K., Kim, Y., Choi, C.Y., and Schulz, R.A. 1998. Combinatorial control of *Drosophila* mef2 gene expression in cardiac and somatic muscle cell lineages. *Dev. Genes Evol.* 208:382–392.

Gajewski, K., Kim, Y., Lee, Y.M., Olson, E.N., and Schulz, R.A. 1997. D-mef2 is a target for tinman activation during *Drosophila* heart development. *EMBO J.* 16:515–522.

Grepin, C., Robitaille, L., Antakly, T., and Nemer, M. 1995. Inhibition of transcription factor GATA-4 expression blocks in vitro cardiac muscle differentiation. *Mol. Cell. Biol.* 15:4095–4102.

Grow, M.W. and Krieg, P.A. 1998. Tinman function is essential for vertebrate heart development: elimination of cardiac differentiation by dominant inhibitory mutants of the tinman-related genes, XNkx2-3 and XNkx2-5. *Dev. Biol.* 204:187–196.

Hamburger, V. and Hamilton, H.L. 1951. A series of normal stages in the development of the chick embryo. *J. Morphol.* 88:49–92.

Han, Y., Dennis, J.E., Cohen-Gould, L., and Bader, D.M., and Fischman, D.A. 1992. Expression of sarcomeric myosin in the presumptive myocardium of chicken embryos occurs within six hours of myocyte commitment. *Dev. Dyn.* 193:257–265.

Harvey, R.P. 1996. NK-2 homeobox genes and heart development. *Dev. Biol.* 178:203–216.

Hedgepeth, C.M., Conrad, L.J., Zhang, J., Huang, H.C., Lee, V.M., and Klein, P.S. 1997. Activation of the Wnt signaling pathway: a molecular mechanism for lithium action. *Dev. Biol.* 185:82–91.

Heikinheimo, M., Scandrett, J.M., and Wilson, D.B. 1994. Localization of transcription factor GATA-4 to regions of the mouse embryo involved in cardiac development. *Dev. Biol.* 164:361–373.

Hume, C.R. and Dodd, J. 1993. Cwnt-8C: A novel *Wnt* gene with a potential role in primitive streak formation and hindbrain organization. *Development* 119:1147–1160.

Jiang, Y. and Evans, T. 1996. The *Xenopus* GATA-4/5/6 genes are associated with cardiac specification and can regulate cardiac-specific transcription during embryogenesis. *Dev. Biol.* 174:258–270.

Jiang, Y., Tarzami, S., Burch, J.B., and Evans, T. 1998. Common role for each of the cGATA-4/5/6 genes in the regulation of cardiac morphogenesis. *Dev. Genetics* 22:263–277.

Johnson, S.E., Rothstein, J.L., and Knowles, B.B. 1994. Expression of epidermal growth factor family gene members in early mouse development. *Dev. Dyn.* 201:216–226.

Kasahara, H., Bartunkova, S., Schinke, M., Tanaka, M., and Izumo, S. 1998. Cardiac and extracardiac expression of Csx/Nkx2.5 homeodomain protein. *Circ. Res.* 82:936–946.

Kispert, A., Vainio, S., Shen, L., Rowitch, D.H., and McMahon, A.P. 1996. Proteoglycans are required for maintenance of Wnt-11 expression in the ureter tips. *Development* 122:3627–3637.

Klingensmith, J. and Nusse, R. 1994. Signaling by wingless in *Drosophila*. *Dev. Biol.* 166:396–414.

Ko, L.J. and Engel, J.D. 1993. DNA-binding specificities of the GATA transcription factor family. *Mol. Cell. Biol.* 13:4011–4022.

Komuro, I. and Izumo, S. 1993. Csx: A murine homeobox-containing gene specifically expressed in the developing heart. *Proc. Natl. Acad. Sci. USA* 90:8145–8149.

Kuisk, I.R., Li, H., Tran, D., and Capetanaki, Y. 1996. A single MEF2 site governs desmin transcription in both heart and skeletal muscle during mouse embryogenesis. *Dev. Biol.* 174:1–13.

Ladd, A.N., Yatskievych, T.A., and Antin, P.B. 1998. Regulation of avian cardiac myogenesis by activin/TGF beta and bone morphogenetic proteins. *Dev. Biol.* 204:407–419.

Laverriere, A.C., MacNeill, C., Mueller, C., Poelmann, R.E., Burch, J.B., and Evans, T. 1994. GATA-4/5/6, a subfamily of three transcription factors transcribed in developing heart and gut. *J. Biol. Chem.* 269:23177–23184.

Lawrence, P.A., Johnston, P., and Vincent, J.-P. 1994. Wingless can bring about a mesoderm-to-ectoderm induction in *Drosophila* embryos. *Development* 120:3355–3359.

Lee, K.H., Xu, Q., and Breitbart, R.E. 1996. A new tinman-related gene, nkx2.7, anticipates the expression of nkx2.5 and nkx2.3 in zebrafish heart and pharyngeal endoderm. *Dev. Biol.* 180:722–731.

Lee, Y., Shioi, T., Kasahara, H., Jobe, S.M., Wiese, R.J., Markham, B.E., and Izumo, S. 1998. The cardiac tissue-restricted homeobox protein Csx/Nkx2.5 physically associates with the zinc finger protein GATA4 and cooperatively activates atrial natriuretic factor gene expression. *Mol. Cell. Biol.* 18:3120–3129.

Lewis, M.R. 1919. The development of cross striation in the heart muscle of the chick embryo. *Johns Hopkins Hosp. Bull.* 30:176–181.

Lin, Q., Schwarz, J., Bucana, C., and Olson, E.N. 1997. Control of mouse cardiac morphogenesis and myogenesis by transcription factor MEF2C. *Science* 276:1404–1407.

Lints, T.J., Parsons, L.M., Hartley, L., Lyons, I., and Harvey, R.P. 1993. Nkx-2.5: a novel murine homeobox gene expressed in early heart progenitor cells and their myogenic descendants. *Development* 119:419–431.

Lough, J., Barron, M., Brogley, M., Sugi, Y., Bolender, D.L, and Zhu, X. 1996. Combined BMP-2 and FGF-4, but neither factor alone, induces cardiogenesis in non-precardiac embryonic mesoderm. *Dev. Biol.* 178:198–202.

Lyons, I., Parsons, L.M., Hartley, L., Li, R., Andrews, J.E., Robb, L., and Harvey, R.P. 1995. Myogenic and morphogenetic defects in the heart tubes of murine embryos lacking the homeobox gene Nkx2-5. *Genes Dev.* 9:1654–1666.

Molkentin, J.D., Firulli, A.B., Black, B.L., Martin, J.F., Hustad, C.M., Copeland, N., Jenkins, N., Lyons, G., and Olson, E.N. 1996. MEF2B is a potent transactivator expressed in early myogenic lineages. *Mol. Cell. Biol.* 16:3814–3824.

Molkentin, J.D., Kalvakolanu, D.V., and Markham, B.E. 1994. Transcription factor GATA-4 regulates cardiac muscle-specific expression of the alpha-myosin heavy-chain gene. *Mol. Cell. Biol.* 14:4947–4957.

Molkentin, J.D., Lin, Q., Duncan, S.A., and Olson, E.N. 1997. Requirement of the transcription factor GATA4 for heart tube formation and ventral morphogenesis. *Genes Dev.* 11:1061–1072.

Molkentin, J.D. and Olson, E.N. 1996. Defining the regulatory networks for muscle development. *Curr. Opin. Genet. Dev.* 6:445–453.

Monkley, S.J., Delaney, S.J., Pennisi, D.J., Christiansen, J.H., and Wainwright, B.J. 1996. Targeted disruption of the Wnt2 gene results in placentation defects. *Development* 122:3343–3353.

Moon, R.T., Brown, J.D., and Torres, M. 1997. WNTs modulate cell fate and behavior during vertebrate development. *Trends Genetics* 13:157–162.

Morrisey, E.E., Ip, H.S., Tang, Z., Lu, M.M., and Parmacek, M.S. 1997. GATA-5: a transcriptional activator expressed in a novel temporally and spatially-restricted pattern during embryonic development. *Dev. Biol.* 183:21–36.

Murphy, A.M., Thompson, W.R., Peng, L.F., and Jones, L. 1997. Regulation of the rat cardiac troponin I gene by the transcription factor GATA-4. *Biochem. J.* 322:393–401.

Narita, N. and Bielinska, M., and Wilson, D.B. 1997. Wild-type endoderm abrogates the ventral developmental defects associated with GATA-4 deficiency in the mouse. *Dev. Biol.* 189:270–274.

Navankasattusas, S., Zhu, H., Garcia, A.V., Evans, S.M., and Chien, K.R. 1992. A ubiquitous factor (HF-1a) and a distinct muscle factor (HF-1b/MEF-2) form an E-box-independent pathway for cardiac muscle gene expression. *Mol. Cell. Biol.* 12:1469–1479.

Newman, C.S. and Krieg, P.A. 1998. *Tinman*-related genes expressed during heart development in *Xenopus. Dev. Genetics* 22:230–238.

Nurrish, S.J. and Treisman, R. 1995. DNA binding specificity determinants in MADS-box transcription factors. *Mol. Cell. Biol.* 15:4076–4085.

Olson, E.N., Perry, M., and Schulz, R.A. 1995. Regulation of muscle differentiation by the MEF2 family of MADS box transcription factors. *Dev. Biol.* 172:2–14.

Olson, E.N. and Srivastava, D. 1996. Molecular pathways controlling heart development. *Science* 272:671–676.

Park, M., Venkatesh, T.V., and Bodmer, R. 1998. Dual role for the zeste-white3/shaggy-encoded kinase in mesoderm and heart development of *Drosophila. Dev. Genetics* 22:201–211.

Park, M., Wu, X., Golden, K., Axelrod, J.D., and Bodmer, R. 1996. The Wingless signaling pathway is directly involved in *Drosophila* heart development. *Dev. Biol.* 177:104–116.

Parr, B.A. and McMahon, J.P. 1994. *Wnt* genes and vertebrate development. *Curr. Opin. Genet. Dev.* 4:523–528.

Patten, B.M. and Kramer, T.C. 1933. The initiation of contraction in the embryonic chick heart. *Am. J. Anat.* 53:349–375.

Patterson, K.D., Cleaver, O., Gerber, W.V., Grow, M.W., Newman, C.S., and Krieg, P.A. 1998. Homeobox genes in cardiovascular development. *Curr. Topics Dev. Biol.* 40:1–44.

Piccolo, S., Agius, E., Leyns, L., Bhattacharyya, S., Grunz, H., Bouwmeester, T., and De Robertis, E.M. 1999. The head inducer Cerberus is a multifunctional antagonist of Nodal, BMP, and Wnt signals. *Nature* 397:707–710.

Rattner, A., Hsieh, J.-C., Smallwood, P.M., Gilbert, D.J., Copeland, N.G., Jenkins, N.A., and Nathans, J. 1997. A family of secreted proteins contains homology to the cysteine-rich ligand-binding domain of frizzled receptors. *Proc. Natl. Acad. Sci. USA* 94:2859–2863.

Rawles, M.E. 1943. The heart-forming areas of the early chick blastoderm. *Physiol. Zool.* 16:22–42.

Reecy, J.M., Yamada, M., Cummings, K., Sosic, D., Chen, C.Y., Eichele, G., Olson, E.N., and Schwartz, R.J. 1997. Chicken Nkx-2.8—a novel homeobox gene expressed in early heart progenitor cells and pharyngeal pouch-2 and -3 endoderm. *Dev. Biol.* 188:295–311.

Ruzicka, D.L. and Schwartz, R.J. 1988. Sequential activation of alpha-actin genes during avian cardiogenesis: vascular smooth muscle alpha-actin gene transcripts mark the onset of cardiomyocyte differentiation. *J. Cell Biol.* 107:2575–2586.

Sabin, F.R. 1920. Studies on the origin of blood-vessels and of red-blood corpuscles as seen in the living blastoderm of chicks during the second day of incubation. *Carnegie Cont. Embryol.* 9:215–262.

Sanders, E.J., Hu, N., and Wride, M.A. 1994. Expression of TGF beta 1/beta 3 during early chick embryo development. *Anat. Rec.* 238:397–406.

Sater, A.K. and Jacobson, A.G. 1989. The specification of the heart mesoderm occurs during gastrulation in *Xenopus laevis. Development* 105:821–830.

Sater, A.K. and Jacobson, A.G. 1990. The restriction of the heart morphogenetic field in *Xenopus laevis. Dev. Biol.* 140:328–336.

Schneider, V.A. and Mercola, M. 1999. Spatially distinct head and heart inducers within the *Xenopus* organizer region. *Curr. Biol.* 9:800–809.

Schultheiss, T.M., Burch, J.B., and Lassar, A.B. 1997. A role for bone morphogenetic proteins in the induction of cardiac myogenesis. *Genes Dev.* 11:451–462.

Schultheiss, T.M., Xydas, S., and Lassar, A.B. 1995. Induction of avian cardiac myogenesis by anterior endoderm. *Development* 121:4203–4214.

Sepulveda, J.L., Belaguli, N., Nigam, V., Chen, C.Y., Nemer, M., and Schwartz, R.J. 1998. GATA-4 and Nkx-2.5 coactivate Nkx-2 DNA binding targets: role for regulating early cardiac gene expression. *Mol. Cell. Biol.* 18:3405–3415.

Sheldahl, L.C., Park, M., Malbon, C.C., and Moon, R.T. 1999. Protein kinase C is differentially stimulated by Wnt and Frizzled homologs in a G-protein-dependent manner. *Curr. Biol.* 9:695–698.

Slusarski, D.C., Corces, V.G., and Moon, R.T. 1997. Interaction of Wnt and a Frizzled homologue triggers G-protein-linked phosphatidylinositol signaling. *Nature* 390:410–413.

Stalsberg, H. and DeHaan, R.L. 1969. The precardiac areas and formation of the tubular heart in the chick embryo. *Dev. Biol.* 19:128–159.

Sugi, Y. and Lough, J. 1994. Anterior endoderm is a specific effector of terminal cardiac myocyte differentiation in cells from the embryonic heart forming region. *Dev. Dyn.* 200:155–162.

Sugi, Y. and Lough, J. 1995. Activin-A and FGF-2 mimic the inductive effects of anterior endoderm on terminal cardiac myogenesis in vitro. *Dev. Biol.* 168:567–574.

Takada, S., Stark, K.L., Shea, M.J., Vassileva, G., McMahon, J.A., and McMahon, A.P. 1994. *Wnt-3a* regulates somite and tailbud formation in the mouse embryo. *Genes Dev.* 8:174–189.

Thuerauf, D.J., Hanford, D.S., and Glembotski, C.C. 1994. Regulation of rat brain natriuretic peptide transcription. A potential role for GATA-related transcription factors in myocardial cell gene expression. *J. Biol. Chem.* 269:17772–17775.

Tokuyasu, K.T. and Maher, P.A. 1987a. Immunocytochemical studies of cardiac myofibrillogenesis in early chick embryos. I. Presence of immunofluorescent titin spots in premyofibril stages. *J. Cell Biol.* 105:2781–2793.

Tokuyasu, K.T. and Maher, P.A. 1987b. Immunocytochemical studies of cardiac myofibrillogenesis in early chick embryos. II. Generation of alph-actinin dots within titin spots at the time of the first myofibril formation. *J. Cell Biol.* 105:2795–2801.

Tonissen, K.F., Drysdale, T.A., Lints, T.J., Harvey, R.P., and Krieg, P.A. 1994. XNkx-2.5, a *Xenopus* gene related to Nkx-2.5 and tinman: evidence for a conserved role in cardiac development. *Dev. Biol.* 162:325–328.

Torres, M.A., Yang-Snyder, J.A., Purcell, S.M., DeMarais, A.A., McGrew, L.L., and Moon, R.T. 1996. Activities of the *Wnt-1* class of secreted signaling factors are antagonized by the *Wnt-5A* class and by a dominant negative cadherin in early *Xenopus* embryos. *J. Cell. Biol.* 133:1123–1137.

Wang, S., Krinks, M., and Moos, Jr., M. 1997. Frzb-1, an antagonist of Wnt-1 and Wnt-8, does not block signaling by Wnts -3a, -5a, or -11. *Biochem. Biophys. Res. Commun.* 236:502–504.

Weiss, M.J. and Orkin, S.H. 1995. GATA transcription factors: key regulators of hematopoiesis. *Exp. Hematol.* 23:99–107.

Xu, C., Liguori, G., Adamson, E.D., and Persico, M.G. 1998. Specific arrest of cardiogenesis in cultured embryonic stem cells lacking Cripto-1. *Dev. Biol.* 196:237–247.

Xu, C., Liguori, G., Persico, M.G., and Adamson, E.D. 1999. Abrogation of the Cripto gene in mouse leads to failure of postgastrulation morphogenesis and lack of differentiation of cardiomyocytes. *Development* 126:483–494.

Yin, Z.Z. and Frasch, M. 1998. Regulation and function of *tinman* during dorsal mesoderm induction and heart specification in *Drosophila*. *Dev. Genetics* 22:187–200.

Zaffran, S., Astier, M., Gratecos, D., Guillen, A., and Semeriva, M. 1995. Cellular interactions during heart morphogenesis in the *Drosophila* embryo. *Biol. Cell.* 84:13–24.

Zhu, X., Sasse, J., McAllister, D., and Lough, J. 1996. Evidence that fibroblast growth factors 1 and 4 participate in regulation of cardiogenesis. *Dev. Dyn.* 207:429–438.

CHAPTER 12

Cellular, Molecular, and Developmental Studies on Heart Development in Normal and Cardiac Mutant Axolotls, *Ambystoma mexicanum*

Larry F. Lemanski, Xupei Huang, R.W. Zajdel,
Sharon L. Lemanski, Chi Zhang, Fanyin Meng,
Dalton Foster, Qing Li, and Dipak K. Dube

SUMMARY

The Mexican axolotl (*Ambystoma mexicanum*) provides an excellent model for studying heart development because it carries a simple recessive cardiac lethal mutation that results in a failure of the mutant embryonic myocardium to contract. In cardiac mutant axolotls, the hearts do not beat, apparently due to an absence of organized myofibrils. The mutant hearts can be rescued by coculturing them with normal anterior endoderm/mesoderm tissue, by a medium conditioned with normal anterior endoderm/mesoderm, or by an RNA isolated from the conditioned medium. We have previously isolated a single cDNA clone from a library prepared with total RNA from conditioned medium; this 166-nt-long in vitro synthesized RNA, directed by the unique cDNA clone (Clone #4), has the ability to correct the heart defect and promote myofibrillogenesis in mutant hearts. The criteria for rescue include contraction of mutant hearts throughout their lengths, an increase in sarcomeric tropomyosin arrays as shown by immunofluorescent confocal microscopy, and the ultrastructural appearance of organized sarcomeric myofibrils. More recently, we carried out RT-PCR with total RNAs extracted from normal or mutant axolotl embryos. A point mutation (G-T) was found in Clone #4 RNA derived from the mutant embryos at stages 20 and 30 as compared to normal axolotls at the same stages. Furthermore, we searched for the full-length Clone #4 gene in an axolotl genomic library. Primer extension yielded a product of ~500 bp at the 5′ end of the Clone #4 gene. The nucleotide sequence of the extended Clone #4 (Ext-Clone #4) was determined and was found to be unique because there was no significant homology with other known sequences available from the gene databases. Interestingly, T7 sense RNA from the Ext-Clone #4 (~500-nt RNA) showed a higher efficiency in rescuing mutant hearts than the T7 RNA from original Clone #4 (166-nt

RNA). Our working hypothesis is that the bioactive RNA is a regulatory RNA that may directly or indirectly upregulate tropomyosin production in the axolotl heart.

BACKGROUND

Dr. Rufus R. Humphrey (1968) of Indiana University reported his discovery of a mutant gene in the Mexican axolotl, *Ambystoma mexicanum*, which he designated as gene c, for "cardiac nonfunction." The gene is recessive and lethal. "Carrier" heterozygous (+/c) animals showed no ill effects. Mating between heterozygous parents (+/c × +/c) resulted in 25% of the offspring exhibiting hearts that form normally at first but then fail to initiate normal contractions (i.e., they do not beat throughout their lengths, nor do they establish vascular circulation).

In a series of experiments on this newly discovered natural mutation, Humphrey (1972) removed heart tissue from the cardiac regions of normal embryos and replaced it by the transplantation of mutant hearts into that area. The mutant heart tissue formed beating hearts when placed in this "normal" cardiac environment. Reciprocal transplants of normal hearts into mutant hosts did not beat. These results suggested that in the mutant embryos some factor necessary for normal differentiation is absent or there is some kind of inhibitory process present in mutants.

Jacobson and Duncan (1968) and Fullilove (1970) found in related species of urodele amphibians that the anterior endoderm is a potent heart inductor tissue, which promoted heart development. Modern studies in the chick (Sugi and Lough, 1994; Schultheisis et al., 1995) and *Xenopus* (Nascone and Mercola, 1995) also demonstrate that the anterior endoderm promotes cardiac differentiation in these species.

Lemanski et al. (1973, 1979, 1980) did studies using the hanging-drop organ culture technique to test the influence of culturing mutant hearts with a variety of tissues from normal embryos. When mutant hearts were placed in culture with tissues such as skeletal muscle, notochord, neural tissue, or posterior endoderm, no response was observed in the mutant hearts. They did not beat, nor could myofibrils be observed by electron microscopy. However, when mutant hearts were placed in hanging-drop cultures in combination with anterior endoderm from stage 26 to 27 normal embryos, mutant hearts could be seen to contract throughout their lengths in a large number of cases; without the presence of the anterior endoderm, the hearts failed to contract. The response appeared to be specific for the anterior endoderm. Was direct contact between the endoderm and mesoderm required to achieve the rescue? To answer this question, Dr. Lynn Davis (Davis and Lemanski, 1987) performed organ culture experiments of mutant hearts in Holtfreter's medium (amphibian saline solution) conditioned by normal anterior endoderm. When the mutant hearts were placed in the anterior endoderm conditioned medium, they beat rhythmically throughout their lengths, and studies with electron microscopy revealed the presence of organized sarcomeric myofibrils in the rescued mutant hearts. It appeared that the anterior endoderm had "conditioned" the medium with a diffusible substance that had the ability to rescue the mutant hearts by promoting the differentiation of functional contracting myofibrils.

Davis and Lemanski (1987) in further studies partially characterized the "active" material in the conditioned medium. In one set of experiments, they boiled the conditioned medium with no adverse affect on the rescuing activity, suggesting that the "factor" was probably not a complex protein. In addition, a series of enzymes were used to break down selected components of the conditioned medium (e.g., trypsin (protein) and ribonuclease (RNA)). None of the enzymes showed significant reductions in rescue activity except ribonuclease, which essentially eliminated the rescue phenomenon.

Thus, these results suggested that the rescuing factor was a ribonucleic acid (RNA). Corollary experiments were then performed in which RNA was extracted from the anterior endoderm of normal embryos and was used to treat the mutants in culture. As with the anterior endoderm itself, or anterior-endoderm conditioned medium, the RNA from the anterior endoderm also rescued the mutant hearts and resulted in rhythmic contractions. Electron microscopy confirmed these rescues by illustrating well-formed sarcomeric myofibrils of normal morphology in mutant heart cells.

Smith and Armstrong (1990) later corroborated the results of Davis and Lemanski (1987) when they reported that normal pharyngeal endoderm corrected the defect in mutant hearts in culture. In addition, they too found that RNA from normal embryonic endoderm had the ability to rescue mutant hearts (Smith and Armstrong, 1990). Although the numbers of mutant organs used in their experiments were too small to have statistical significance, their experimental results agreed completely with the earlier findings of Davis and Lemanski (1987) and further strengthened the conclusion that RNA promotes mutant heart rescue.

LaFrance et al. (1993) performed core-response experiments using RNA purified from normal endoderm/mesoderm conditioned medium. They found that the higher the RNA concentration in Steinberg's solution (amphibian saline), the higher the percentage of rescued mutant hearts in a given experiment. Confirmation of the results was accomplished by electron microscopy (to show organized myofibrils in rescued mutant heart cells) and by scanning confocal microcopy in combination with antitropomyosin staining of organized myofibrils in rescued mutant hearts (LaFrance and Lemanski, 1995).

For the past several years, we have continued research to elucidate the role of the bioactive RNA in rescuing mutant axolotl hearts, to understand the relationship between the active RNA and tropomyosin in myofibrillogenesis, and to further explore the molecular basis underlying the cardiac abnormality in Mexican axolotls. Significant progress has been made toward these aims.

RESCUE OF MUTANT HEARTS IN VITRO AND IN VIVO

By Ectopic Expression of Tropomyosin

The expression of tropomyosin protein, an essential component of the thin filament, has been found to be drastically reduced in cardiac mutant hearts of the Mexican axolotl. As a result, there is hardly any formation of sarcomeric myofibrils. Therefore, the naturally occurring cardiac mutation is an appropriate model for examining the effects of transfecting tropomyosin protein or tropomyosin cDNA into the deficient tissue. In fact, gene transfer is a potential method to correct biochemical defects by ectopic expression of a missing gene in the diseased tissues, such as the replacement of dystrophin in skeletal muscle affected by Duchenne muscular dystrophy (Trivedi and Dickson, 1995). The paucity of dystrophin protein in muscle cells is a useful model for gene transfer products in skeletal muscle. The mutant hearts of the Mexican axolotl lacking tropomyosin protein make this animal model uniquely suitable for studying the transfer of tropomyosin genes and the subsequent formation of cardiac myofibrils. In the transfection studies described here, we used cationic liposome transfection of heterologous tropomyosin cDNA or tropomyosin protein into intact mutant hearts in organ culture to determine whether increased tropomyosin expression would result in organized myofibrils.

First, we transfected purified tropomyosin protein from rabbits into whole embryonic mutant hearts at stages 36 and 39 by using tropomyosin-filled cationic liposomes. It is important to note that the embryonic hearts at these stages are very small and have only a single myocardial cell layer, which is not covered by epicardium (Fransen and Lemanski, 1988). Confocal microscopy was used to examine the mutant hearts transfected with the tropomyosin protein. We studied myofibril formation by indirect immunofluorescent procedures using a sarcomere-specific antitropomyosin antibody (CH1) and employing our standard protocol (Starr et al., 1989; LaFrance et al., 1993). An overall reduction of tropomyosin protein expression was confirmed in control mutant hearts. Transfection of tropomyosin protein derived from either rabbit or chicken skeletal muscles resulted in the formation of well-organized sarcomeric myofibrils where only amorphous collections or no tropomyosin would be normally visualized in mutant hearts. Interestingly, the more peripheral areas of the heart cells appeared to have formed more sarcomeric myofibrils than the deeper inner regions. Mutant hearts transfected with cardiac actin protein did not form organized myofibrils in the ventricle and were similar to the mutant hearts cultured in Steinberg's solution alone. Similarly, transfection of troponin or myosin light-chain proteins did not result in increased myofibril formation in the mutant hearts. The results of these protein transfection experiments reinforce the hypothesis that mutant embryonic hearts are indeed deficient in tropomyosin, and this is related to their failure to form organized myofibrils. Subsequent transfection of mouse α-tropomyosin cDNA under the control of a cardiac-specific promoter (Muthuchamy et al., 1995) into mutant hearts (stages 36 and 39) induced tropomyosin synthesis in the mutant hearts. This augmented tropomyosin synthesis significantly increased the formation of sarcomeric myofibrils in the mutant hearts. The control mutant hearts not transfected with the tropomyosin cDNA as well as the hearts transfected with eukaryotic expression vector containing the β-galactosidase reporter gene did not show increased transfected tropomyosin synthesis, nor were sarcomeric myofibrils found. Interestingly, mutant hearts, when transfected with tropomyosin cDNA, have more homogenous expression of tropomyosin throughout the individual hearts than hearts transfected with tropomyosin protein itself.

Rescue by RNA

Activity of RNA Purified from Normal Axolotl Endoderm-Conditioned Medium

We carried out experiments to further confirm the ability of RNA purified from normal endoderm/mesoderm conditioned medium to rescue mutant axolotl hearts in vitro. We found that total RNA purified from conditioned medium and treated with Proteinase K was specifically able to rescue mutant hearts in a dose-dependent manner (LaFrance et al., 1993); a concentration as low as 0.25 ng/μl showed bioactivity. Confocal microscopic analysis of c/c hearts immunostained after culture demonstrated that those treated with conditioned medium or RNA from conditioned medium contained significantly greater amounts of tropomyosin in sarcomeric arrays than untreated hearts, which contained virtually no sarcomeric myofibrils (LaFrance et al., 1993; Lemanski et al., 1995).

Subsequent tests of conditioned-medium RNA fractionated by sucrose density gradient centrifugation indicated that the bioactive RNA is a small stable molecule (Lemanski et al., 1995). We decided that the most efficient way to unequivocally identify the bioactive material as an RNA, but not resulting from contamination with a tightly bound protein in the molecules, would be to create a cDNA library from the conditioned-medium RNA. Then, we could screen the activity of in vitro transcribed RNA in our c/c heart bioassay.

Identification of a cDNA that Produces a Synthetic RNA that Rescues Mutant Hearts

As a first step in identifying the bioactive RNA, we constructed a cDNA library from total conditioned-medium RNA using the pcDNAII expression vector (Invitrogen) with T7 and Sp6 RNA promoters flanking the multiple cloning sites. The plasmid DNA was isolated from 202 recombinant clones originally chosen on the basis of blue-white selection. RNA was synthesized in vitro using T7 RNA polymerase and then tested for its ability to rescue the mutant hearts. Synthetic RNA (from the pool of clones 1 to 202) used at a concentration of 56 ng/10 μl rescued 60% to 70% of the mutant hearts. Because the 56 ng was the product of 202 clones, the amount of RNA derived from any one clone would average 0.28 ng/10 μl culture, which would seem to be within a physiologically significant range (see dose-response in Table 1 of Lemanski et al., 1996).

Morphological and immunocytological analyses of corrected mutant hearts revealed significant amounts of tropomyosin and organized myofibrils. Because we did not modify the synthetic RNA by capping at the 5′ end or polyadenylation at the 3′ end, it is unlikely that the bioactive RNA needs to be translated during rescue. We then tested synthetic RNAs derived from pools of plasmid DNA in several rounds of screening and ultimately determined that only the T7 transcript from a single clone, #4, rescued mutant hearts (Lemanski et al., 1996). These results strongly support the hypothesis that a specific RNA, not a protein "contaminant," promotes myofibrillogenesis in mutant hearts. If bacterial protein carried over from the in vitro transcription reactions was responsible, we would not have observed our consistent results (i.e., only pools containing Clone #4's T7 transcript were active); furthermore, we used RNA polymerases from two different suppliers (and purified by different methods) so the "impurities" in the reagents should not be responsible for correction in this case. The specificity of our result was repeated several times with different batches of RNA.

Subsequent dose-response experiments showed that only T7 (sense) RNA and not Sp6 (antisense) RNA from Clone #4 corrects mutant hearts (Lemanski et al., 1996). This confirmed our prediction (based on the assumption mentioned earlier that at least one active cDNA was present in the original pool of 202 clones) that the active RNA would be effective at a dose as low as 0.28 ng/10 μl. Correction occurred in T7 RNA at concentrations of 0.25 ng/10 μl (99% confidence interval) and as low as 0.025 ng/10 μl (95% confidence interval; see Table 1 of Lemanski et al., 1996).

Normal and mutant hearts from the experiments using synthetic RNA were immunolabeled for tropomyosin and examined by confocal laser scanning microscopy. Whereas normal hearts contain large amounts of sarcomeric tropomyosin (Figure 12.1A; see Color Plate 5), untreated mutant hearts (Figure 12.1B; see Color Plate 5) contain very little. However, *rescued* mutant hearts do contain sarcomeric arrays of tropomyosin, almost normal in appearance in some areas, when they are incubated with the bioactive RNA (Figure 12.1C; see Color Plate 5). These results provide additional proof that treatment with specific Clone #4 sense RNA stimulates myofibril formation in the mutant myocardium. Clone #4 RNA is the bioactive molecule in the normal endoderm/mesoderm conditioned medium.

In Vitro Rescue by FITC-Tagged Clone #4 RNA

In order to test a method for monitoring the intracellular location of RNA during rescue, we first labeled active Clone #4 sense RNA with FITC. After incubation in labeled RNA for specific times, hearts were then imaged by confocal microscopy. After one hour of incubation, no signal was visible, although at two hours, punctate labeling was visible near the surface of the normal heart. At four hours, punctate labeling was visible in the cytoplasm; nuclear labeling was visible after 1, 2, and 4 days of incubation. In contrast, FITC-conjugated nucleotide triphosphates (used in the labeling mixture) do not adhere to or enter hearts. The punctate labeling pattern we observed after FITC-RNA treatment is similar to images of receptor-mediated endocytosis involving clathrin-coated vesicles or caveolin-coated vesicles (caveolae) (Schnitzer et al., 1994; Tang et al., 1996). We have established that *cardiac* mutant hearts are not rescued by FITC-conjugated nucleotide triphosphates but are rescued by treatment with FITC-labeled Clone #4 sense RNA. After incubating mutant hearts in the FITC-RNA and immunostaining them for tropomyosin using a rhodamine-conjugated secondary antibody, confocal microscopic observation revealed an increase in tropomyosin arrays in those cells that also contained the FITC-RNA. Thus, it appears that the mutant heart cells take up the Clone #4 RNA, and this causes tropomyosin to be produced within the cells.

CREATION OF CHIMERIC MUTANT AXOLOTLS

The fact that c/c embryos are not identifiable until stage 35, when the normals (+/c or +/+) first begin to have beating hearts, has been a problem in analyzing the effects of Clone #4 RNA on early cardiogenesis. It is possible that the effects of Clone #4 RNA may occur as early as neurula stage 14/15 during mesoderm induction. However, we could not detect the mutant embryos until stage 35. To overcome this problem, and to be able to examine known preheartbeat mutant embryos, we have created chimeric mutant axolotls with normal (+/+ or +/c) head and heart regions with a mutant body containing mutant (c/c) reproductive organs (Figure 12.2C). Spawnings between a chimeric male (c/c) and a normal female (+/c) produce a spawning of 50% c/c, and spawnings between a chimeric male (c/c) and a chimeric female (c/c) produce a spawning of 100% c/c (Lemanski et al., 2001). Embryos from spawnings of c/c parents show no evidence of beating hearts at stages 35 to 41 and, as expected, die at stage 41. Electron microscopic observation of cardiac sections from these c/c embryos showed that mutant cardiac myocytes lack organized myofibrils. Instead, the mutant heart cells contain amorphous proteinaceous accumulations and sporadic Z-spots as well as a few thick myosin-like filaments, characteristics identical to mutant heart cells from c/c embryos obtained in heterozygous (+/c × +/c) spawnings (Figure 12.2B). Normal cardiac myocytes from the embryos (+/c or +/+) at the same stages 35 to 41 contain organized sarcomeric myofibrils (Figure 12.2A). As we mentioned previously, a deficiency of tropomyosin is a characteristic feature of mutant (c/c) axolotl heart cells. This is also a major cause for the deficiency of organized myofibrils in mutant cardiac myocytes and for the eventual death of

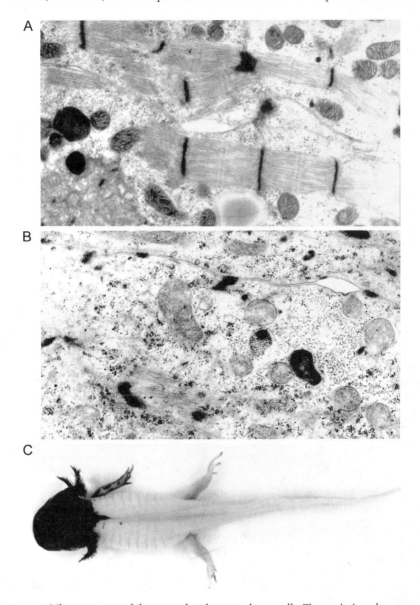

FIGURE 12.2. Ultrastructure of the normal and mutant heart cells. Transmission electron micrograph of the embryonic heart from a normal (+/+ × +/+) spawning (A) and the embryonic heart from a *c/c* chimera spawning (B) after glutaraldehyde/osmium fixation with lead citrate and uranyl staining. The normal cells show well-organized sarcomeric myofibrils. However, mutant hearts show no obvious myofibrils. Present instead at the cell peripheries are amorphous proteinaceous collections. (C) An adult axolotl chimera carrying *c/c cardiac* alleles.

mutant axolotls. We further examined the spatial distributions of tropomyosin in mutant and normal embryos with confocal microscopy. The data demonstrated that cardiac myocytes from normal embryonic axolotl hearts at stages 40 to 41 contain rich tropomyosin staining decorated by a specific antitropomyosin antibody. In contrast, the cardiac myocytes from c/c embryonic hearts show little staining for tropomyosin. These results confirm that we have created chimeric axolotls composed of a normal (+/+ or +/c) head and heart region and a mutant (c/c) body containing mutant gonads (c/c), producing the observed 100% c/c mutant offspring in spawnings between c/c chimeras. The c/c offspring have the same phenotype as that observed in naturally occurring mutant (c/c) axolotls. These chimeric axolotls provide a useful and unique way to investigate early preheartbeat embryonic development in cardiac mutant axolotls.

TARGETING THE BIOACTIVE RNA

As previously mentioned, isolated RNA from the conditioned medium was capable of rescuing mutant hearts in culture. Among them, Clone #4 RNA has been isolated and was proved to be a bioactive RNA. We have carried out further experiments to investigate Clone #4 expression in normal and mutant hearts at various stages of development and to search for the possible mutation in Clone #4 from the mutant axolotl hearts. We also searched for the full-length cDNA and encoding gene of Clone #4 RNA.

The Expression of Clone #4 RNA in Normal and Mutant Axolotl Hearts at Different Stages of Development

In order to demonstrate the relationship between Clone #4 RNA and heart development, we designed experiments to examine Clone #4 RNA expression in axolotl hearts at various developmental stages and to compare quantitatively Clone #4 RNA expression in normal and mutant axolotl hearts from various stages of development. Total RNAs were extracted from normal and mutant axolotl embryos at stages 10, 15, 20, and 30. After several steps of RNA purification, RT-PCR was performed using the purified RNAs as templates. The results revealed that Clone #4 RNA expression was barely detected at stage 10, and the expression increased significantly at stage 15. Clone #4 RNA was detectable at stage 30 (Figure 12.3). Our previous

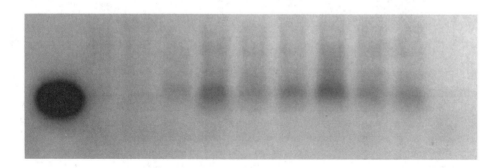

A   B   C   D   E   F   G   H   I   J   K

FIGURE 12.3. Clone #4 gene expression at different stages of embryonic development. RT-PCR amplification using primer pairs specifically designed for the Clone #4 gene shows a similar pattern for both normal and mutant axolotl embryos. Lane A: single-strand DNA of Clone #4 as positive control. Lanes B, D, F, H: normal embryos RT-PCR. Lanes C, E, G, I, J: mutant embryos RT-PCR (B, C: stage 10; D, E: stage 15; F, G: stage 20; H, I, J: stage 30). Lane K: no RT step total RNA negative control.

data showed that the expression of Clone #4 RNA in axolotl hearts after stage 40 decreased dramatically.

The expression of Clone #4 RNA in axolotl embryos was also confirmed by in situ hybridization studies. In situ hybridization was performed with antisense and sense Clone #4 RNA probes labeled with digoxigenin, which were synthesized by T7 (for antisense) and T3 (sense) RNA polymerases, respectively. A control with no probe was also included in this study. Intense staining was observed in embryonic sections with an antisense probe. The staining with the sense probe was significantly lower. No staining was observed with the no-probe control (Figure 12.4; see Color Plate 6). Interestingly, the signal was observed in the anterior endoderm and precardiac mesoderm at stages 15 to 20.

The results suggest that the initial expression of Clone #4 RNA occurs in axolotl embryos at neurulation at the same time that cardiogenesis is started. This would seem to support the idea that an RNA-mediated signaling event may occur during the onset of cardiogenesis.

Sequence Analysis of Clone #4 RNA from Normal and Mutant Axolotls

Because Clone #4 RNA is also detectable in mutant axolotl hearts from stages 15 to 30, the question then should be whether the structure of the Clone #4 RNA is the same in normal and mutant axolotls. We carried out RT-PCR using total RNAs extracted from normal or mutant axolotl embryos. The PCR products were subcloned and analyzed for nucleotide sequences. Very interestingly, a point mutation (G→T) was found in Clone #4 RNA from mutant embryos at stages 20 and 30 when compared to normal axolotl embryos at the same stages (Figure 12.7). Secondary structure analysis with the GENEBEE software package revealed a significant structure difference between Clone #4 RNA from the normal axolotls as compared to mutant (not shown). The distinctive structures are noteworthy because such structural differences may likely impose functional differences as to how the two RNAs might interact with other molecules. In order to determine the role of the point mutation in Clone #4 RNA from mutant axolotls, we plan to perform rescue experiments in mutant embryos with the mutant Clone #4 RNA.

In Vitro Mutagenesis of Clone #4 RNA

In vitro mutagenesis has been proven to be a useful method for selecting and defining the repertoire of sequences of various genetic elements. This approach has been applied to promoter sequences (also enhancers and other regulatory elements), binding sites of various proteins, and active sites of the enzymes and ribozymes. In brief, it provides a methodology for analyzing the relationship of structure to function of various nucleic acids and proteins. In our initial studies for identification of the putative active site/elements or domain of Clone #4 RNA, we have performed deletion mutagenesis using polymerase chain reaction (PCR) with a variety of primer pairs using T3 or T7 promoter sequences at the 5′ end. Amplified DNAs from each of the deletion mutants have been separated by agarose gel electrophoresis, and the desired band was extracted from the gel. Because the T7 promoter was attached directly to the 5′ end of each of the deletion mutants, it allowed us to test for in vitro RNA transcription from the PCR products following gel purification. Two deletion mutant clones have been selected: (1) 24 bp from the 3′ end (p142) and (2) 76 bp from the 3′ end (p90). Computer analysis of the structure of these two mutated RNAs (p142 and p90) revealed that p142 maintained a structure similar to that of the original p166 RNA, whereas p90 showed a significant structure difference compared to p166 RNA (Figure 12.5). Using a cationic-liposome–mediated delivery system (Loke et al., 1989; Zelphati et al., 1996), synthetic RNA from each of these deletion mutants was transfected into stage 35/36 mutant hearts. Then they were scored for the presence and relative number of organized myofibrils using confocal laser scanning microscopy after staining the hearts with antitropomyosin antibody (CH1) following our standard protocol. Bioassay data showed that p142 had the same efficiency as Clone #4 RNA in correcting the mutant hearts. However, p90 RNA had little effect in rescuing the mutant hearts (Figure 12.6; see Color Plate 7). These results strongly suggest that either nucleotide sequences of the Clone #4 RNA, its secondary structure, or both are required for the rescue activity of the synthetic RNA.

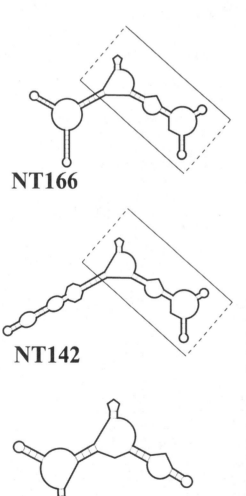

NT166

NT142

NT90

FIGURE 12.5. Secondary structure of Clone #4 RNA and its deletion mutant RNAs. The secondary structure of Clone #4 RNA (nt 166) was analyzed with the GENEBEE program (upper). The secondary structure of the RNA (nt 142) synthesized from Clone #4, which has a deletion of 24 base pairs from the 3′ end, conserves a major part of the Clone #4 RNA structure as shown in the dashed square (middle). The secondary structure of the RNA (nt 90) derived from Clone #4, which has a deletion of 76 base pairs from the 3′ end, is completely different from the original Clone #4 RNA structure (lower).

Searching for the Full Length Clone #4 Gene

As mentioned previously, we have determined that a 166-nt RNA is able to correct cardiac mutant hearts in vitro and can also be detected in embryonic tissue by RT-PCR. To search for the full length of Clone #4, we have carried out experiments both on a whole axolotl genomic library and a cDNA library from axolotls at stages 15 to 17. Primer pairs were designed using the vector flanking sequences and the known 166-bp fragment of Clone #4. The PCR products from the two-step PCR procedure (the first step using hotstart Taq beads and the second step using a high-fidelity system) were confirmed by Southern blot assay. The promising PCR bands were then subcloned into T-vectors and subsequently sequenced. Primer extension yielded a product of ~500 bp at the 5′ end. A recent 3′-RACE assay reveals a ~200 nt fragment with a poly (A) tail at the 3′ end of the Clone #4 RNA. The nucleotide sequence of the extended Clone #4 (Ext-Clone #4) was determined and was found to be unique because there were no significant homologies with other known sequences available in the gene databases (Figure 12.7). T7 sense RNA from the Ext-Clone #4 (~500-nt RNA) showed a higher efficiency in rescuing the mutant hearts than the T7 RNA from the original Clone #4 (166 nt RNA). However, the antisense RNA from the Ext-Clone #4 had no effect on mutant heart correction (Zhang et al., 1999).

Bioactive Clone #4 RNA Binds with Protein(s) in Embryonic Axolotls

Our previous results have shown that a developmentally regulated protein in the axolotl binds to Clone #4 RNA. We carried out the following experiments to further study the role of RNA protein binding and to characterize the properties of the binding protein(s).

Spatial and Temporal Distribution of the Clone #4 Binding Protein(s)

By using a Northwestern blot assay approach, we analyzed Clone #4 RNA binding protein in the protein extracts from axolotl embryos at different stages. We also examined the binding protein level in different tissues (brain, heart, liver, testes, lung, skeletal muscle, and spleen) of the adult axolotl. Our results demonstrated that Clone #4 binding proteins were initially detectable at about embryonic stage 15 and could be detected throughout the preheartbeat stages. Then, the level of Clone #4 binding proteins decreased dramatically after stage 40. No Clone #4 RNA binding activity could be found in the adult heart. However, interestingly, Clone #4 binding activity was detected in protein extracts from the adult spleen and adult skeletal muscle. The molecular weights of these proteins were identical to those detected in extracts from the early embryos.

Purification and Characterization of the Clone #4 Binding Protein(s)

We have made significant progress on the purification and identification of Clone #4 binding proteins. We have exploited the capacity of Clone #4 RNA and its binding proteins in vitro. By using UV cross-linking, we separated the RNA binding proteins from other proteins of similar molecular weights on SDS-PAGE. The gel slices containing the RNA protein complexes were detected by autoradiography. Then, the complexes were removed from the gel by electroelution. The Clone #4 RNA binding proteins were released from the RNA protein complex by RNase digestion and

CTCGATCAGC AAACTTCTTT TTGCACCATC TTCAAATGCT TCTGTGTATC   –50

TTAAGGATAG CATCACAAGG TGTATACCCC AGGATGGAAC ATTGGATCTA   –100

AGGAGGATCA CAGACTAGTC ATACCTGGAA CGCTCTGTCA GACCAGAGGA   –150

CATAAAAGAC TTTCAGGTTT GAGACGCAGA CGTTGGAATG AGTTGCCAAG   –200

TTCTGCTCAT GACAACACCT GCTTTCAGGC CCCTGTTGGA GCATCATCTG   –250

TTCAGGTAGG TGTTCATTAA TGGTCTCTAG CAGAGCCAAG ATCAGTGTGC   –300

ACAAGCACAT GATTTCACAA AGGTGATATA TGTGTAGGTC CTATGCAAAA   –350

ACACTGCAAA TTCCTAACAT CACCCAATGT TCAAAAGAA CATTTTAATT   –400

TTGTATCTCC TTATACAGCC ATCATAACAT ATTCTAGGAC TGGTATACTG   –450

TATAGACAAA CTCCCTTCCA AGGATATTTT GGGAAAGTGC TGGATAGAGG   –500

GGCAGAACAG CACCTTTCTT CTCAGGCAAT GTTAAATAGG TGCAATGTTT   –550

TCACATGTTA TGGAATATAT CTTCCAACTG ACTGACCAAG AGAAACAAT   –600

GAACCACAAT ACCGGAAACT TCATTCGTTT GTCCCTTCCA CCCACTCGAG   –650

CGTCAACTTG CCCAGGCCGC TACCCCTTGA CACACGTGT̲A̲ ̲G̲C̲A̲C̲C̲A̲C̲T̲C̲C̲   –700

A̲T̲T̲T̲T̲T̲G̲G̲A̲A̲ ̲C̲A̲C̲C̲T̲C̲C̲T̲C̲T̲ ̲A̲C̲C̲G̲T̲G̲G̲A̲T̲G̲ ̲A̲G̲A̲G̲G̲C̲G̲G̲A̲G̲ ̲C̲C̲G̲A̲T̲C̲C̲T̲T̲T̲   –750

G̲G̲A̲A̲T̲T̲T̲G̲T̲A̲ ̲C̲A̲T̲G̲T̲G̲A̲C̲C̲T̲ ̲C̲A̲A̲G̲G̲T̲T̲G̲C̲A̲ ̲C̲**G**C̲A̲T̲A̲T̲C̲C̲G̲ ̲A̲G̲C̲A̲G̲T̲T̲G̲C̲T̲   –800

G̲G̲A̲T̲T̲A̲G̲A̲G̲C̲ ̲A̲G̲G̲C̲A̲C̲T̲C̲C̲C̲ ̲T̲T̲A̲T̲C̲G̲G̲C̲T̲T̲ ̲T̲G̲G̲A̲A̲T̲G̲G̲A̲G̲ ̲A̲C̲C̲A̲G̲A̲A̲A̲G̲A̲   –850

ATACAATGTG GGACA GCTG ATATGGAGGG CAGGGCAGGG AAGTGAGAGA   –900

AAGGGCAACA ATAGAGGGCA GATAAAGGGG GGAGGGCAAG GAATAGACAG   –950

GAACTGCAGT GGGAGAAAAC GTCGAACGAG AAAAAAAAAA   –1000

FIGURE 12.7. Nucleotide sequence of Clone #4 and the extended Clone #4 (Ext-Clone #4). The underlined strand above shows the original 166 nt RNA molecule found in our laboratory. RT-PCR confirmed gene expression from 186 to 855 bases of positive strand (black). Subsequent RT-PCR results suggest that most of the genome sequence is transcribed in vivo. Sequencing of the purified 166 nt bands from RT-PCR shows a putative point mutation (G→T) in the mutant gene as indicated by the bold letter. Different clones from 3′-RACE indicate possible alternative splicing during developmental stages.

separated on SDS-PAGE. The target proteins were then transferred onto the PVDF membrane and sequenced with an N-terminal amino acid sequencer. Two sequences were obtained as potential Clone #4 RNA binding proteins. Currently, we are repeating the purification of RNA binding proteins by utilizing the method outlined previously to confirm our preliminary results.

CONCLUSIONS

We have found an RNA, derived from normal anterior endoderm and reproduced using molecular methods in bacteria, that has the ability to promote myofibrillogenesis in cardiac mutant axolotl heart cells. The usually nonbeating mutant hearts, which lack sarcomeric myofibrils, form organized myofibrils and beat when treated with the RNA. Deshpande and Siddiqui (1977) were among the first to observe that RNA promoted heart development in the chick. The RNA they described was a 7S poly A+ RNA, which hybridized to repetitive chick DNA.

For many years, RNA was considered to be confined in its role to those processes associated with and leading to protein synthesis. More recently, however, RNA has been considered to have a much broader range of biological functions both inside and, perhaps, outside of the cell. An example of how RNA may act extracellularly is monocyte angiotropin; a small RNA is an essential component of the RNA-protein-copper complex that permits monocytes to induce endothelial cells to undergo angiogenesis as part of the inflammatory response (Höckel et al., 1988; Wissler et al., 1988). There is a small regulatory RNA produced by the lin-4 gene of *Caenohofditis elegans* that appears to function as an antisense regulator of a different gene (Lee et al., 1993; Wightman et al., 1993). A 362 n+ RNA without an open ready frame was described by Retallack and Friedman (1995) in which the interaction of specific DNA binding proteins to a specific DNA was affected. That RNA can act as an enzyme to catalyze sequence-specific cleavage of RNA is well documented (Cech, 1987). It has been shown that the untranslated region of RNA derived from a gene that encodes on mRNA may play various other regulatory roles in development (Rastinejad et al., 1993; Davis and Watson, 1996; Dubnau and Struhl, 1996).

In cardiac mutant axolotls, the hearts fail to beat because they lack organized myofibrils. The hearts can be rescued by treatment with a specific synthetic RNA. The criteria for rescue include contraction of the hearts, an increase in tropomyosin staining as shown by immunofluorescent microscopy, and the presence of organized sarcomeric myofibrils as revealed by electron microscopy. Over the course of our research, we have developed and partially tested four hypotheses to account for the ability of RNA to rescue mutant hearts (Figure 12.8; see Color Plate 8).

Hypothesis A: The active RNA is a messenger RNA that enters the cell and is translated, thereby restoring to normal the level of a missing/defective protein responsible for tropomyosin synthesis and, therefore, restoring myofibrillogenesis. Because bioactive Clone #4 RNA has no homology with tropomyosin mRNA, it is not simply providing a translatable replacement for the missing tropomyosin in RNA. Furthermore, because the bioactive RNA is relatively small (166-nt long) and lacks the necessary modifications required for efficient translation, we believe that Hypothesis A is very unlikely.

Hypothesis B: The bioactive RNA acts as a ligand to a myocardial membrane receptor. Binding of the RNA activates the receptor, setting off a sequence that ultimately enhances tropomyosin accumulation. Our observation that liposome-mediated transfection of active RNA corrects mutant hearts argues strongly against a signal transduction hypothesis as the mode of action for the bioactive RNA.

Hypothesis C: The active RNA is a regulatory RNA that enters the cell and serves in gene regulation. Whatever the immediate action of the RNA, the end effect is to upregulate synthesis of tropomyosin, thereby enabling myofibril formation in the mutant myocardium. There are several possibilities for the mechanism by which the active RNA stimulates tropomyosin expression. The RNA could stimulate tropomyosin expression by interacting directly with the regulatory region of a tropomyosin gene or indirectly by enhancing the activity or synthesis of a tropomyosin gene activator or by suppressing the activity or synthesis of a tropomyosin gene inhibitor.

Hypothesis D: The RNA after entering into cells binds with specific proteins. The RNA protein complex subsequently induces myofibrillogenesis in mutant axolotl hearts. As a result, the nonbeating hearts start beating.

Whatever the final mechanism turns out to be for the action of the bioactive RNA, it is very clear from our results that this unique and novel Clone #4 RNA has the ability to promote myofibrillogenesis and rescue cardiac mutant axolotl hearts.

ACKNOWLEDGMENTS

We thank Dr. Mathew McLean for his contributions to the deletion experiments and Ms. Masako Nakatsugawa for technical assistance. This work was supported by National Institutes of Health Grants HL-58435 and HL-061246 and by an American Heart Association Grant to L.F.L.

REFERENCES

Cech, T.R. 1989. RNA as an enzyme. *Biochem. Int.* 18:7–14.

Davis, L.A. and Lemanski, L.F. 1987. Induction of myofibrillogenesis in cardiac lethal mutant axolotl hearts rescued by RNA derived from normal endoderm. *Development* 99:145–154.

Davis, S. and Watson, J.C. 1996. In vitro activation of the interferon-induced, double-stranded RNA dependent protein kinase PKR by RNA from the 3′ untranslated regions of human α-tropomyosin. *Proc. Natl. Acad. Sci. USA* 93:508–513.

Deshpande, A.K. and Siddiqui, M.A.Q. 1977. A reexamination of heart muscle differentiation in the postnodal piece of chick blastoderm mediated by exogenous RNA. *Dev. Biol.* 58:230–247.

Dubnau, J. and Struhl, G. 1996. RNA recognition and translation regulation by a homeodomain protein. *Nature* 379:694–699.

Fransen, M.E. and Lemanski, L.F. 1988. Myocardial cell relationships during morphogenesis in normal and cardiac lethal mutant axolotl, *Ambystoma mexicanum. Am. J. Anat.* 183:245–257.

Fullilove, S.L. 1970. Heart induction: distribution of active factors in newt endoderm. *J. Exp. Zool.* 175:323–326.

Höckel, M., Jung, W., Vaupel, P., Rabes, H., Khaledpour, C., and Wissler, J.H. 1988. Purified monocyte-derived angiogenic substance (angiotropin) induces controlled angiogenesis associated with regulated tissue proliferation in rabbit skin. *J. Clin. Invest.* 82:1075–1090.

Humphrey, R.R. 1968. A genetically determined absence of heart function in embryos of the Mexican axolotl (*Ambystoma mexicanum*). *Anat. Rec.* 162:475.

Humphrey, R.R. 1972. Genetic and experimental studies on a mutant gene (c) determining absence of heart action in embryos of the Mexican axolotl (*Ambystoma mexicanum*). *Dev. Biol.* 27:365–375.

Jacobson, A.G. and Duncan, J.T. 1968. Heart induction in salamanders. *J. Exp. Zool.* 167:79–104.

LaFrance, S.M., Fransen, M.E., Erginel-Unaltuna, N., Dube, D.K., Robertson, D.R., Stefanu, C., Ray, T.K., and Lemanski, L.F. 1993. RNA from anterior endoderm/mesoderm-conditioned medium corrects aberrant myofibrillogenesis in developing axolotl mutant hearts. *Cell Mol. Biol. Res.* 39:547–560.

LaFrance, S.M. and Lemanski, L.F. 1995. Immunofluorescent confocal analysis of tropomyosin in developing hearts of normal and cardiac mutant axolotls. *Int. J. Dev. Biol.* 38:695–700.

Lee, R.C., Feinbaum, R.L., and Abros, V. 1993. The *C. elegans* heterochromic gene *lin*-4 encodes small RNAs with antisense complementary to *lin*-14. *Cell* 75:843–854.

Lemanski, L.F. 1973. Morphology of developing heart in cardiac lethal mutant Mexican axolotls, *Ambystoma mexicanum. Dev. Biol.* 33:312–333.

Lemanski, L.F., Paulson, D.J., and Hill, C.S. 1979. Normal anterior endoderm corrects the heart defect in cardiac mutant salamanders (*Ambystoma mexicanum*). *Science* 204:860–862.

Lemanski, L.F., Fuldner, R.A., and Paulson, D.J. 1980. Immunofluorescence studies for myosin, alpha-actin and tropomyosin in developing hearts of normal and cardiac lethal mutant axolotls, *Ambystoma mexicanum. J. Embryol. Exp. Morphol.* 55:1–15.

Lemanski, L.F., LaFrance, S.M., Erginel-Unaltuna, N., Luque, E.A., Ward, S.M., Fransen, M.E., Mangiacapra, F.J., Nakatsugawa, M., Lemanski, S.L., Capone, R.B., Goggins, K.J., Nash, B.P., Bhatia, R., Dube, A., Gaur, A., Zajdel, R.W., Zhu, Y., Spinner, B.J., Pietras, K.M., Lemanski, S.F., Kovacs, C.P., and Dube, D.K. 1995. The cardiac mutant gene *c* in axolotls: cellular, developmental and molecular studies. (Invited Article). *Cell Mol. Biol. Res.* 41:293–305.

Lemanski, L.F., Nakatsugawa, M., Bhatia, R., Erginel-Unaltuna, N., Spinner, B., and Dube, D.K. 1996. Characterization of an RNA which promotes myofibrillogenesis in embryonic cardiac mutant axolotl hearts. *Biochem. Biophys. Res. Commun.* 229:974–981.

Lemanski, L.F., Meng, F., Lemanski, S.L., Dawson, N., Zhang, C., Foster, D., Li, Q., Nakatsugawa, M., Dube, D.K., and Huang, X.P. 2001. Creation of chimeric mutant axolotls: a model to study early embryonic heart development in Mexican axolotls. *Anatomy & Embryol.* 203:335–342.

Loke, S.L., Stein, C.A., Zhang, X.H., Mori, K., Nakanishi, M., Subasinghe, C., Cohen, J.S., and Neckers, L.M. 1989. Characterization of oligonucleotide transport into living cells. *Proc. Natl. Acad. Sci. USA* 86:3474–3478.

Muthuchamy, M., Grupp, I.L., Grupp, G., O'Toole, B.A., Kier, A.B., Boivin, G.P., Neumann, J., and Wieczorek, D.F. 1995. Molecular and physiological effects of overexpressing striated muscle β-tropomyosin in the adult murine heart. *J. Biol. Chem.* 270:30593–30603.

Nascone, N. and Mercola, M. 1995. An inductive role for the endoderm in *Xenopus* cardiogenesis. *Development* 121:515–523.

Rastinejad, F. and Blau, H.M. 1993. Genetic complementation reveals a novel role for 3′ untranslated regions in growth and differentiation. *Cell* 72:903–917.

Retallack, D.M. and Friedman, D.I. 1995. A role for a small stable RNA in modulating the activity of DNA-binding proteins. *Cell* 83:227–235.

Schnitzer, J.E., Oh, P., Pinney, E., and Allard, J. 1994. Filipin-sensitive caveolae-mediated transport in endothelium: reduced transcytosis, scavenger endocytosis, and capillary permeability of select macromolecules. *J. Cell Biol.* 127:1217–1232.

Schultheiss, T.M., Xydas, S., and Lassar, A.B. 1995. Induction of avian cardiac myogenesis by anterior endoderm. *Development* 121:4203–4214.

Smith, S.C. and Armstrong, J.F. 1990. Heart induction in wild-type and cardiac mutant axolotls (*Ambystoma mexicanum*). *J. Exp. Zool.* 254:48–54.

Starr, C., Diaz, J.G., and Lemanski, L.F. 1989. Analysis of actin and tropomyosin in hearts of cardiac mutant axolotls by two-dimensional gel electrophoresis, Western blots and immunofluorescent microscopy. *J. Morphol.* 201:1–10.

Sugi, Y. and Lough, J. 1994. Anterior endoderm is a specific effector of terminal cardiac myocyte differentiation of cells from the embryonic heart forming region. *Dev. Dyn.* 200:155–162.

Tang, Z.L., Scherer, P.E., Okamotu, T., Song, K., Chu, C., Kohtz, D.S., Nishimoto, I., Lodish, H.F., and Lisanti, M.P. 1996. Molecular cloning of carveolin-3, a novel member of the carveolin gene family expressed predominantly in muscle. *J. Biol. Chem.* 271:2255–2261.

Trivedi, R.A. and Dickson, G. 1995. Liposome-mediated gene transfer into normal and dystrophin deficient mouse myoblasts. *J. Neurochem.* 64:2230–2238.

Wightman, B., Ha, I., and Ruvkun, G. 1993. Posttranscriptional regulation of the heterochromic gene *lin*-14 by *lin*-r mediates temporal pattern formation in *C. elegans. Cell* 75:855–862.

Wissler, J.H., Kiesewtter, S., Logemann, E., Sprinzl, M., and Heilmeyer L.M.G. 1988. Monocytic angiomorphogen: RNA as constituent of a new organogenetic tissue hormone representing a copper-ribonucleo-polypeptide complex (Cu-RNP). *Biol. Chem. Hoppe-Seyler* 369:948–949.

Zelphati, O. and Szoka, F.C. 1996. Mechanism of oligonucleotide release from cationic liposomes. *Proc. Natl. Acad. Sci. USA* 93:11493–11498.

Zhang, C., Huang, X.P., Lemanski, S.L., Bhatia, R., Gaur, A., Meng, F., Dube, D.K., and Lemanski, L.F. 1999. Molecular basis of the cardiac mutation in Mexican axolotls. *Mol. Biol. Cell* 10:361a.

PART V

Cardiomyopathy and Myofibrillar Organization

CHAPTER 13

The Function of Normal and Familial Hypertrophic Cardiomyopathy-Associated Tropomyosin

Rethinasamy Prabhakar, Kathy Pieples, Ganapathy Jagatheesan, Stephanie Burge, and David F. Wieczorek

This review will focus on the role of sarcomeric tropomyosin during the assembly and function of thin filaments in striated muscle. Particular emphasis will be devoted to how mutations in familial hypertrophic cardiomyopathy influence these processes. Due to the complex and extensive nature of this topic, this review will not address the role of tropomyosin in the cytoskeleton; however, excellent articles on this topic have addressed this area (Lin et al., 1988; Hegmann et al., 1989).

TROPOMYOSIN ISOFORM EXPRESSION

There are four tropomyosin genes (α, β, TPM3, and TM4), which exhibit a very high degree of conservation in species ranging from *Drosophila* to humans. Each of the tropomyosin genes has two distinct promoter regions for generating proteins of 284 or 248 amino acids. Also, the tropomyosin genes increase the diversity of isoforms by alternative exon splicing of RNA. This complex processing occurs in the 5′, middle, and 3′ portions of the tropomyosin primary transcripts. Differential polyadenylation of messenger RNA occurs at multiple sites in the 3′ untranslated regions of the tropomyosin genes. The net effect of these complex features culminates in the production of multiple mRNA and protein isoforms from each of the genes. For example, the α-TM gene generates a minimum of ten transcripts, and the β-TM gene produces a minimum of five mRNAs (Lees-Miller and Helfman, 1991; Muthuchamy et al., 1993). The TPM3 gene has also been reported to generate numerous distinct isoforms (Dufour, 1998; Gunning et al., 1998). These distinct tropomyosin isoforms are highly regulated in their production and are expressed in nonmuscle, smooth muscle, and striated (skeletal and cardiac) muscle cells in a tissue- and development-specific manner. This tight regulatory control of expression is essential to ensure correct isoform expression during myofibrillogenesis of both striated and smooth muscles.

Tropomyosin isoform expression in the heart exhibits developmental changes from embryogenesis to adulthood. In cardiac muscle of the mouse, α-tropomyosin is the predominant isoform throughout the lifespan of the animal (Muthuchamy et al., 1993). However, β-tropomyosin transcripts comprise 20% of the total striated muscle tropomyosin in the embryonic day 11 heart. With fetal development, this percentage gradually decreases to 8% β-tropomyosin mRNA at birth. In the adult mouse, only 2% of the total tropomyosin is β-tropomyosin protein. β-tropomyosin expression remains at very low levels in the heart, unless it is subject to pressure overload. Under physiological stress, such as pressure-overload hypertrophy, reactivation of the cardiac fetal gene program occurs, which includes increases in β-tropomyosin mRNA expression (Izumo et al., 1988). Recent studies show that TPM3 is not expressed in murine cardiac tissue (Pieples and Wieczorek, 2000). In humans, α-tropomyosin is also the predominant isoform; however, β-tropomyosin and TPM3 (TM30) are expressed at lower levels (Reinach, 1995).

TROPOMYOSIN FUNCTION IN MUSCLE

Although the role of tropomyosin is well defined in striated muscle, the precise mechanism of how tropomyosin functions is still unclear. In skeletal and cardiac muscles, tropomyosin interacts with both actin and the troponin complex (troponin T, I, and C) to regulate contraction and relaxation of the sarcomere (Figure 13.1). Tropomyosin molecules form coiled-coil dimers that wrap around and stabilize filamentous actin. However, the association of tropomyosin with actin is more than just a stabilization process; it is dynamic in the sense that tropomyosin shifts its position on filamentous actin during sarcomeric contraction and relaxation in response to

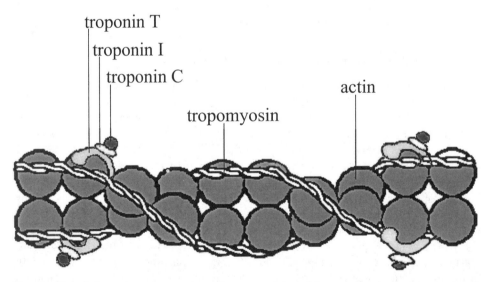

FIGURE 13.1. Schematic representation of tropomyosin in the sarcomeric thin filament and tropomyosin's interactions with both filamentous actin and the troponin complex. Tropomyosin is an α-helical coiled-coil dimer that lies in the grooves around two strands of actin monomers. Each tropomyosin molecule (comprised of a dimer) interacts with approximately seven actin monomers. Tropomyosin dimers overlap in a head-to-tail fashion along the length of the thin filament. The troponins complex (consisting of troponins T, I, and C) interacts with each tropomyosin molecule at two distinct sites in the COO⁻ terminal half of tropomyosin (amino acids 175 to 195 and 258 to 284).

changes in calcium levels in the myoplasm. It is currently unknown whether the entire tropomyosin molecule shifts as a unit or whether there is a preferential repositioning on actin by specific tropomyosin residues, followed by the remaining regions. Tropomyosin does not bind calcium directly, but rather responds to changes in calcium concentration through the troponin complex. With increases in calcium concentration, troponin C binds calcium and through its interaction with troponin T and troponin I mediates a change in the tropomyosin position on actin. Movement of tropomyosin and troponin I also facilitates myosin head interactions with actin, thereby stimulating muscle contraction.

The reaction of tropomyosin with actin and the troponin complex appears to involve at least two potential modes for control. The first modulatory effect is through steric hindrance. The steric effect proposes that tropomyosin blocks myosin cross-bridge binding to the actin thin filament during relaxing conditions (i.e., with low-calcium conditions). Activation occurs with increases in myoplasmic calcium concentration, which leads to the movement of tropomyosin and the recruitment of cross-bridge binding. The second potential mode of control is a cooperative effect, which occurs through the extension of the number of actins under the control of tropomyosin by its reaction with the troponin complex. Although one tropomyosin molecule spans seven actins, cooperativity may be maintained between neighboring functional units so that the influence of one tropomyosin molecule may influence 11 to 14 actin monomers (Geeves and Lehrer, 1994). Data suggest that a dynamic equilibrium exists among three transition states: (1) blocked—no bound calcium, and cross-bridge binding is blocked by tropomyosin; (2) closed—calcium is bound to troponin, but cross-bridge interactions are weak; and (3) open—strong cross-bridge binding that splits ATP and generates force (Solaro and Van Eyk, 1996). This three-state model supports biophysical findings on tropomyosin movement, steric hindrance, and muscle mechanics (Landis et al., 1999; Xu et al., 1999) and illustrates the bioregulatory mechanisms that control cross-bridge interactions among the sarcomeric proteins.

DOMAIN MAPPING WITHIN TROPOMYOSIN

Tropomyosin, like many proteins, appears to have regions within itself that play specific roles and interact with other protein complexes. The importance of the carboxyl end of the striated muscle tropomyosin molecule has been demonstrated through numerous studies (Pan et al., 1989; Mak and Smillie, 1991). Two critical functions are: (1) overlap in a head-to-tail fashion between adjacent tropomyosin dimers; and (2) attachment to the troponin complex through troponin T. Amino acids 258 to 284 of tropomyosin are essential for proper assembly of dimer formation and tropomyosin extension in the thin filament, presumably by facilitating head-to-tail overlap of the ends of the molecules. This region of the tropomyosin also interacts with the amino-terminal region of troponin T. The attachment of the troponin complex to this region stabilizes the TM head-to-tail assembly by forming a triple helix and may mediate calcium binding through cooperative effects along the troponin or tropomyosin helices (Hammell and Hitchcock, 1996; Lehrer et al., 1997). In addition to being an essential binding site for the troponin complex, the importance of the tropomyosin carboxyl amino acids is illustrated from in vitro experiments showing that their removal results in a loss of polymerizability and actin affinity (Butters et al., 1993; Walsh et al., 1994). We have attempted to generate transgenic mice that lack the carboxy-terminal 27 amino acids; we were unsuccessful in these attempts, possibly because the incorporation of this deletion-mutant tropomyosin may function as a "poison" peptide, or due to the inability to stably produce founder mice with this tropomyosin deletion (Jagatheesan and Wieczorek, unpublished results). Also, our previous results show that ablation of one wild-type tropomyosin allele leads to a 50% reduction in mRNA levels but no decrease in tropomyosin protein (Rethinasamy et al., 1998). This tropomyosin ablation study supports the hypothesis that mutations in one allele, which lead to decreases in tropomyosin mRNA or protein stability, would be masked by compensatory mechanisms of the second tropomyosin allele.

The importance of the highly conserved amino terminus of muscle tropomyosin has also been shown to play an essential functional role in the head-to-tail assembly of overlapping tropomyosin molecules in the sarcomere (Cho et al., 1990; Butters et al., 1993). This region is involved in binding actin and troponin. Deletion of the amino terminus results in a severe loss of the regulatory function of tropomyosin either through an inability to bind actin and/or impaired troponin binding. Thus, specific regions of tropomyosin play an integral role in its interactions with other sarcomeric proteins and their associated functions. Interestingly, a point mutation in this region which results in an amino acid charge change of the TPM3 isoform leads to a human disease, namely nemaline myopathy (Laing et al., 1995). Preliminary results from our lab show that high expression of this mutant isoform in transgenic mice is lethal due to severe cardiac abnormalities (Pieples et al., in preparation).

NATURALLY OCCURRING MUTANTS OF TROPOMYOSIN

Several animal model systems have illustrated the importance of tropomyosin in sarcomere formation, structure, and function. In the myopathic hamster, the regulatory components of the thin filaments are partially responsible for impaired cardiac function. In myofibrillar protein preparations from the hearts of these hamsters, assessment of cardiac troponin–tropomyosin demonstrates a loss in calcium sensitivity that decreases the inhibitory action on actomyosin ATPase activity (Malhotra, 1994). The axolotl, *Ambystoma mexicanum*, has served as a model for studying various aspects of cardiac muscle development and cardiogenesis (Lemanski, 1973). This model carries a mutation that results in a severe deficiency of tropomyosin that leads to an absence of both myofibril formation and contraction (Lemanski, 1979). There is no effect on the expression of the other cardiac contractile proteins in these mutants. Interestingly, addition of exogenous tropomyosin to mutant cardiac cell homogenates induces thin filament formation, suggesting that the tropomyosin deficiency in these mutant heart cells may result in a failure to stabilize filamentous actin (LaFrance et al., 1993).

Investigations on *Caenorhabditis elegans* and *Drosophila melanogaster* have demonstrated the importance of developing genetic mutant models when studying myofibrillar structure, assembly, and function (Fyrberg, 1989; Williams and Waterston, 1994). Mutations within genes encoding *Drosophila* indirect flight muscle contractile proteins often exhibit phenotypic structural defects of the sarcomere in the heterozygous state (Beall et al., 1989; Warmke et al., 1989). There is usually no compensation of expression by the remaining single functional gene. For example, mutations within the myosin light-chain 2 or tropomyosin indirect flight muscle isoforms cause disruption of the thick or thin filaments, respectively. In flies, thick and thin filaments assemble independently, but both filament types are requisite for normal sarcomere assembly and periodicity. As such, most of the defects observed in these heterozygotes are due to filament imbalances, not deficits. Furthermore, studies suggest that expression of the remaining contractile proteins in these mutants is synthesized at their normal rate but may rapidly degrade if a full complement of thick or thin proteins is not present (Beall et al., 1989; Warmke et al., 1989).

FAMILIAL HYPERTROPHIC CARDIOMYOPATHY AND TROPOMYOSIN MUTATIONS

Familial hypertrophic cardiomyopathy (FHC) is an autosomal dominant disease in humans that is caused by mutations in sarcomeric proteins. The disease is generally characterized by left ventricular hypertrophy in the absence of an increased external load, myofibrillar disarray, and fibrosis. The FHC mutations have been genetically mapped in families to occur within the myosin heavy and light chains, myosin binding protein C, actin, troponin T, and α-tropomyosin (Spirito et al., 1997; Bonne et al., 1998). Five mutations have been defined in the α-tropomyosin molecule that lead to FHC: two of the mutations occur in the troponin T binding region (Asp175Asn; Glu180Gly); two mutations lie at the amino end of the tropomyosin molecule (Ala63Val; Lys70Thr); and one mutation lies in the middle (Val95Ala) (Thierfelder et al., 1994; Nakajima-Taniguchi et al., 1995; Karibe et al., 1999) (see Figure 13.2). Interestingly, all of the troponin T mutations that are associated with FHC lie in tropomyosin binding regions, thus illustrating the significance of the troponin-T–tropomyosin interactions (see Figure 13.3).

In order to gain a greater understanding of tropomyosin's role in the sarcomere and the mechanism by which mutations in tropomyosin cause the cardiac hyper-

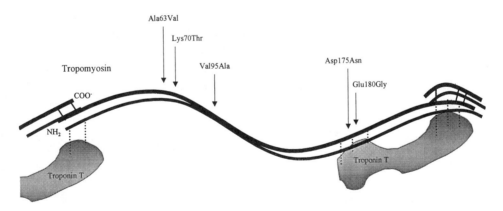

FIGURE 13.2. Schematic representation of mutations in α-tropomyosin associated with FHC (familial hypertrophic cardiomyopathy). One tropomyosin dimer, represented by the two thick lines, is shown. Adjacent tropomyosin molecules overlap at the NH₂ and COO⁻ ends. The positions of the α-tropomyosin FHC mutations with their associated amino acid changes are indicated. The relative position of the troponin T complex is shown. Solid lines indicate interacting regions between tropomyosin molecules in the overlap regions; interacting regions between tropomyosin and troponin T are indicated by dotted lines. For clarity and simplification, we have omitted actin, troponins I and C, and the second tropomyosin dimer from the drawing.

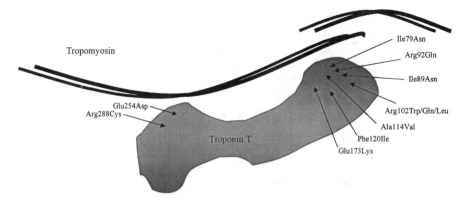

FIGURE 13.3. Schematic representation of mutations in cardiac troponin T associated with FHC (familial hypertrophic cardiomyopathy). The relative positions of the FHC mutations with their associated amino acid changes are indicated. One tropomyosin dimer is represented as pairs of thick lines. All of the troponin T point mutations associated with FHC map to regions that interact with tropomyosin. Amino acids 78 to 173 of troponin T are thought to associate with the carboxyl region of tropomyosin; amino acids 198 to 294 interact with tropomyosin in the amino acid 175 to 190 region. The position of the FHC mutations in the figure is not based upon structural data.

trophic response, we have developed animal models that express two of the tropomyosin mutations associated with FHC. We have generated transgenic mice that overexpress the Asp175Asn and the Glu180Gly mutant α-tropomyosin proteins specifically in the heart. Both of these models address how specific tropomyosin mutations influence the interaction with the troponin complex. The transgenic DNA constructs utilize the α-myosin heavy-chain promoter (α-MYHC), which restricts expression of the constructs to cardiac myocytes (see Figure 13.4). The α-tropomyosin cDNA encoding the specific amino acid change is ligated to the α-MYHC promoter at the initiation codon and extends throughout the entire coding sequence into the 3' untranslated region. Because the α-tropomyosin amino acid sequence between mice and humans is identical, the FHC mutations incorporated into the transgenes are reflective of the disease-causing mutations in humans. Either SV40 or the HGH (human growth hormone) 3' polyadenylation and termination sequences were added to the construct to ensure correct processing of the transgene. Using these constructs, multiple transgenic lines were developed and analyzed extensively. The analysis incorporated a multidimensional approach entailing biochemical, morphological, and physiological studies.

Four transgenic mouse lines were examined that expressed and incorporated the Asp175Asn mutant protein into the sarcomere (Muthuchamy et al., 1999). To determine the amount of mutant protein that was expressed, myofibrillar proteins were purified and subject to SDS-PAGE. The proteins were Western blotted and immunoreacted against a tropomyosin antibody that only recognizes striated specific isoforms. Surprisingly, although the two proteins are identical, with the exception of the single amino acid substitution, they migrate differentially on an SDS-polyacrylamide gel. The reason for this differential migration is that the amino acid substitution (Asp175Asn) corresponds to a charge change resulting in one additional negative charge in the protein. Even though SDS usually negates charge differences, tropomyosin appears to defy this situation. Results show expression of the exoge-

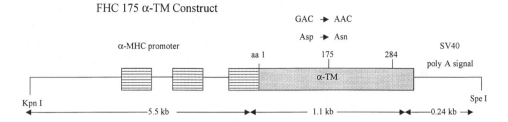

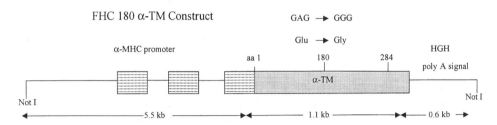

FIGURE 13.4. α-TM FHC transgenic constructs. The constructs used for microinjection are shown. Site-directed mutagenesis at codons 175 or 180 in α-tropomyosin cDNA (α-TM) was done with polymerase chain reaction methodology. The α-myosin heavy-chain (MYHC) promoter was employed to target cardiac tissue-specific expression. For the α-TM FHC 175 construct, an SV40 termination and polyadenylation cassette were ligated at the 3′ end of the α-TM cDNA to ensure correct processing of the transcripts. For the α-TM FHC 180 construct, a human growth hormone (HGH) termination and polyadenylation cassette were employed. The 5′ untranslated sequence is from the α-MYHC gene; translation initiation begins with amino acid 1 of α-tropomyosin. The 3′ untranslated region of α-TM is included in the construct, following the translation stop codon after amino acid 284. For the α-TM FHC 175 construct, KpnI and SpeI enzymes were used to release the transgene fragment; for the α-TM FHC 180 construct, NotI enzyme was employed to release the DNA fragment.

nous protein varies between 40% and 60% of the total tropomyosin that is expressed in the hearts of the transgenic mice. With usage of the α-myosin heavy-chain promoter, this exogenous expression is confined to cardiac tissue and is not expressed in other tissues/organs in these mice. Interestingly, as expression of the mutant protein increases in the heart, there is a reciprocal decrease in the endogenous α-tropomyosin levels in this tissue. The ability of different tropomyosin isoforms to regulate their expression in response to changes in endogenous or exogenous tropomyosin levels demonstrates the "cross-talk" that exists among members of this multigene family. We have previously demonstrated this phenomenon in tropomyosin expression with respect to ablation of an α-tropomyosin allele and during overexpression of various TM isoforms (Muthuchamy et al., 1995, 1998; Rethinasamy et al., 1998).

Studies addressing the physiological effect of the α-TM FHC 175 mutation reveal several interesting findings by different investigators. Work-performing heart studies on transgenic mice that carry this mutation show that there is a significant decrease in left ventricular rates of contraction and relaxation; as expected, the time to peak pressure and the time to half-relaxation are increased (Muthuchamy et al., 1999). This reduction in myocardial function is maintained during maximal stimulation with isoproterenol, a β-adrenergic agonist that stimulates muscle contraction and relaxation. When cardiac myofilament activation was measured via calcium-force measurements, results were consistent with those of the work-performing heart analyses; force developed by FHC myofilaments is significantly more sensitive to calcium than control myofilaments. These results agree with studies on skeletal myofilaments from patients with FHC, which also exhibited an increased sensitivity to calcium (Bottinelli et al., 1998). However, more recent studies conducted on isolated adult rat cardiomyocytes transfected with expression vectors containing an α-TM FHC 175 mutant construct did not exhibit alterations in calcium sensitivity of force production (Michele et al., 1999). Differences between this in vitro study and work with the mouse hearts may be due to the hemodynamic forces present in the transgenic mice, which are nonexistent in the cultured cardiomyocyte. Interestingly, studies conducted using in vitro motility assays demonstrate that the α-TM FHC 175 mutation does produce alterations in calcium-regulated thin filament movement (Bing et al., 1997). Thus, it appears that the α-TM FHC 175 mutation alters the thin filament dynamics associated with sarcomeric function.

Investigations have begun to address whether the α-TM FHC 175 mutation affects myofibril formation. Employing the axolotl heart model (Zajdel et al., 1998), Dube and coworkers have introduced the mutant α-TM FHC 175 into whole hearts in culture and in live embryos. Results demonstrate that transfection of Asp175Asn tropomyosin did not promote myofibrillogenesis in axolotl c mutant hearts, which lack endogenous α-type tropomyosin (Zajdel et al., 2000). Also, introduction of the α-TM FHC 175 cDNA into normal embryos produced localized myofibril disruptions. When the mutant TM was introduced into the pericardial cavity of live embryos, disorganized myofibrillar formation in the myocardium resulted. As such, these results indicate that in addition to the functional defect of the Asp175Asn mutation on sarcomeric function, this FHC mutant TM fails to support normal organized myofibril formation either in vitro or in vivo.

We have developed a transgenic mouse model for the FHC mutation that occurs at amino acid 180 of α-tropomyosin. Structural analyses indicate that the FHC Glu180Gly mutation in α-TM is more disruptive than the FHC Asp175Asn mutation (Michele et al., 1999). Although the pathological phenotypes from human patients with the FHC 175 or 180 mutations are similar, both in vitro and in vivo studies indicate that the Glu180Gly mutation is more severe. The precise reason for this significant difference in phenotype is unclear; however, two possible explanations are: (1) a greater disruption of the α-helical structure of tropomyosin by the amino acid substitution at position 180 versus 175; or (2) a greater perturbation in troponin T binding. Recent in vivo studies demonstrate that when the FHC 180 α-TM mutant protein is incorporated into transgenic mouse cardiomyocytes, the resulting mice develop left ventricular concentric hypertrophy and die due to heart failure within 4 to 5 months (see Figure 13.5A, see Color Plate 8, and 13.5B, see Color Plate 9) (Prabhakar et al., 2000). This pathological phenotype is dramatically different from that seen with the α-TM FHC 175 mutation, where dilated cardiomyopathy developed in only 5% of the ventricular mass. Interestingly, the FHC 180 lethal phenotype resulted with equivalent amounts of protein expression as the FHC 175 transgenic mice, namely 40% to 60% mutant tropomyosin protein. Preliminary studies indicate that this α-TM mutation causes dramatic differences in sarcomeric function, affecting both contractility and relaxation of the sarcomere. Also, myofilaments isolated from the papillary muscle show an increased sensitivity to calcium (Arteaga, Solaro, and Wieczorek, unpublished data). These transgenic mouse studies are in agreement with those conducted by others, which show an altered tropomyosin function with the FHC 180 mutation when assessed through in vitro motility assays and cardiomyocyte pCa-force measurements (Bing et al., 1997; Michele et al., 1999). It will be interesting to assess the effect of the α-TM FHC 180 mutation on myofibril formation in various animal model systems.

SUMMARY

FHC mutations can provide much information on sarcomere formation and function. Vastly different phenotypes can result from these mutations with equivalent levels of protein production. These differences may occur because of the increased severity of the glutamic acid substitution by glycine; however, detailed structural analyses will be necessary to confirm this hypothesis. In addition, modifier genes and secondary site mutations (Prado et al., 1995; Kronert et al., 1999) may also dramatically influence the resulting phenotypes associated with FHC mutations. Future investigations will need to address the mechanism(s) by which tropomyosin mutations in the troponin T binding site cause the physiological and morphological changes that occur in the myocardium. Studies of how the other α-TM FHC mutations cause this disease will also need detailed examination to fully understand the role of tropomyosin in both normal development and in pathological conditions.

REFERENCES

Beall, C., Sepanski, M., and Fyrberg. E. 1989. Genetic dissection of *Drosophila* myofibril formation: effects of actin and myosin heavy chain null alleles. *Genes Dev.* 3:131–140.

Bing, W., Redwood, C., Purcell, I., Esposito, G., Watkins, H., and Marston, S. 1997. Effects of two hypertrophic cardiomyopathy mutations in α-tropomyosin, asp175asn and glu180gly, on Ca²⁺ regulation of thin filament motility. *Biochem. Biophys. Res. Commun.* 236:760–764.

Bonne, G., Carrier, L., Richard, P., Hainque, B., and Schwartz, K. 1998. Familial hypertrophic cardiomyopathy. *Circ. Res.* 83:580–593.

Bottinelli, R., Coviello, D., Redwood, C., Pellegrino, M., Maron, B., Spirito, P., Watkins, H., and Reggiani, C. 1998. A mutant tropomyosin that causes hypertrophic cardiomyopathy is expressed in vivo and associated with an increased calcium sensitivity. *Circ. Res.* 82:106–115.

Butters, C., Willadsen, K., and Tobacman, L. 1993. Cooperative interactions between adjacent troponin–tropomyosin complexes may be transmitted through the actin filament. *J. Biol. Chem.* 268:15565–15570.

Cho, Y., Liu, J., and Hitchcock-DeGregori, S. 1990. The amino terminus of muscle tropomyosin is a major determinant for function. *J. Biol. Chem.* 265:538–545.

Dufour, C., Weinberger, R., and Gunning, P. 1998. Tropomyosin isoform diversity and neuronal morphogenesis. *Immunol. Cell Biol.* 76:424–429.

Fyrberg, E. 1989. Study of contractile and cytoskeletal proteins using *Drosophila* genetics. *Cell Motil. Cytoskel.* 14:118–127.

Geeves, M. and Lehrer, S. 1994. Dynamics of the muscle thin filament regulatory switch: the size of the cooperative unit. *Biophys. J.* 67:273–282.

Gunning, P., Hardeman, E., Jeffrey, P., and Weinberger, R. 1998. Creating intracellular structural domains: spatial segregation of actin and tropomyosin isoforms in neurons. *Bioessays* 20:892–900.

Hammell, R. and Hitchcock-DeGregori, S. 1996. Mapping the functional domains within the carboxyl terminus of α-tropomyosin encoded by the alternatively spliced ninth exon. *J. Biol. Chem.* 271:4236–4242.

Hegmann, T., Lin, J., and Lin, J. 1989. Probing the role of nonmuscle tropomyosin isoforms in intracellular granule movement by microinjection of monoclonal antibodies. *J. Cell Biol.* 109:1141–1152.

Izumo, S., Nadal-Ginard, B., and Mahdavi, V. 1988. Protooncogene induction and reprogramming of cardiac gene expression produced by pressure overload. *Proc. Natl. Acad. Sci. USA* 85:339–343.

Karibe, A., Bachinski, L., Arai, A., Tripodi, D., Roberts, R., and Fananapazir, L. 1999. Familial hypertrophic cardiomyopathy caused by a novel alpha-tropomyosin mutation (val95ala) is associated with mild cardiac hypertrophy but a high incidence of sudden death. *Circulation* 100(18):I-619a.

Kronert, W., Acebes, A., Ferrus, A., and Bernstein, S. 1999. Specific myosin heavy chain mutations suppress troponin I defects in *Drosophila* muscles. *J. Cell Biol.* 144:989–1000.

LaFrance, S., Fransen, M., Dube, D., Stefanu, C., Ray, T., and Lemanski, L. 1993. RNA from normal anterior endoderm/mesoderm-conditioned medium stimulates myofibrillogenesis in developing mutant axolotl hearts. *Cell. Mol. Biol. Res.* 39:547–560.

Laing, N., Wilton, S., Akkari, P., Dorosz, S., Boundy, K., Kneebone, C., Blumbergs, P., White, S., Watkins, H., Love, D., and Haan, E. 1995. A mutation in the α tropomyosin gene *TPM3* associated with autosomal dominant nemaline myopathy. *Nat. Genetics* 9:75–79.

Landis, C., Back, N., Homsher, E., and Tobacman, L. 1999. Effects of tropomyosin internal detections on thin filament function. *J. Biol. Chem.* 274:31279–31285.

Lees-Miller, J. and Helfman, D. 1991. The molecular basis for tropomyosin isoform diversity. *Bioessays* 13:429–437.

Lehrer, S., Golitsina, N., and Geeves, M. 1997. Actin-tropomyosin activation of myosin subfragment 1 ATPase and thin filament cooperativity. The role of tropomyosin flexibility and end-to-end interactions. *Biochemistry* 36:13449–13454.

Lemanski, L. 1973. Morphology of developing heart in cardiac lethal mutant Mexican axolotls, *Ambystoma mexicanum. Dev. Biol.* 33:312–333.

Lemanski, L. 1979. Role of tropomyosin in actin filament formation in embryonic salamander heart cells. *J. Cell Biol.* 82:227–238.

Lin, J., Hegmann, T., and Lin, J. 1988. Differential localization of tropomyosin isoforms in cultured nonmuscle cells. *J. Cell Biol.* 107:563–572.

Mak, A. and Smillie, L. 1981. Structural interpretation of the two-site binding of troponin on the muscle thin filament. *J. Mol. Biol.* 149:541–550.

Malhotra, A. 1994. Role of regulatory proteins (troponin–tropomyosin) in pathologic states. *Mol. Cell. Biochem.* 135:43–50.

Michele, D., Albayya, F., and Metzger, J. 1999. Direct, convergent hypersensitivity of calcium-activated force generation produced by hypertrophic cardiomyopathy mutant a-tropomyosins in adult cardiac myocytes. *Nat. Med.* 5:1413–1417.

Muthuchamy, M., Boivin, G., Grupp, I., and Wieczorek, D. 1998. β-tropomyosin overexpression induces severe cardiac abnormalities. *J. Mol. Cell. Cardiol.* 30:1545–1557.

Muthuchamy, M., Grupp, I., Grupp, G., O'Toole, B., Kier, A., Boivin, G., Neumann, J., and Wieczorek, D. 1995. Molecular and physiological effects of overexpressing striated muscle β-tropomyosin in the adult murine heart. *J. Biol. Chem.* 270:30593–30603.

Muthuchamy, M., Pajak, L., Howles, P., Doetschman, T., and Wieczorek, D. 1993. Developmental analysis of tropomyosin gene expression in embryonic stem cells and mouse embryos. *Mol. Cell. Biol.* 13:3311–3323.

Muthuchamy, M., Pieples, K., Rethinasamy, P., Hoit, B., Grupp, I., Boivin, G., Wolska, B., Evans, C., Solaro, R., and Wieczorek, D. 1999. Mouse model of a familial hypertrophic cardiomyopathy mutation in α-tropomyosin manifests cardiac dysfunction. *Circ. Res.* 85:47–56.

Nakajima-Taniguchi, C., Matsui, H., Nagata, S., Kishimoto, T., and Yamauchi-Takihara, K. 1995. Novel missense mutations in α-tropomyosin gene found in Japanese patients with hypertrophic cardiomyopathy. *J. Mol. Cell. Cardiol.* 27:2053–2058.

Pan, B., Gordon, A., and Luo, Z. 1989. Removal of tropomyosin overlap modifies cooperative binding of myosin S-1 to reconstituted thin filaments of rabbit striated muscle. *J. Biol. Chem.* 264:8495–8498.

Pieples, K. and Wieczorek, D.F. 2000. Tropomyosin 3 increases striated muscle isoform diversity. *Biochem.* 39:8291–8297.

Prabhakar, R., Boivin, G., Hoit, B., Arteaga, G., Grupp, I., Solaro, R., and Wieczorek, D. 2000. α-tropomyosin mutation (Glu180Gly) causes familial hypertrophic cardiomyopathy—a mouse model for this disease of the sarcomere. *Keystone Symposium: Molecular Biology of the Cardiovascular System,* p. 77a, Keystone Pub., Snow Bird, UT.

Prado, A., Canal, I., Barbas, J., Molley, J., and Ferrus, A. 1995. Functional recovery of troponin I in a *Drosophila* heldup mutant after a second site mutation. *Mol. Biol. Cell* 6:1433–1441.

Reinach, F. 1995. Nemaline myopathy mechanisms. *Nat. Genetics* 10:8.

Rethinasamy, P., Muthuchamy, M., Hewett, T., Boivin, G., Wolska, B., Evans, C., Solaro, R.J., and Wieczorek, D.F. 1998. Molecular and physiological effects of α-tropomyosin ablation in the mouse. *Circ. Res.* 82:116–123.

Solaro, R. and Van Eyk, J. 1996. Altered interactions among thin filament proteins modulate cardiac function. *J. Mol. Cell. Cardiol.* 28:217–230.

Spirito, P., Seidman, C., McKenna, W., and Maron, B. 1997. The management of hypertrophic cardiomyopathy. *N. Engl. J. Med.* 335:775–785.

Thierfelder, L., Watkins, H., MacRae, C., Lamas, R., McKenna, W., Vosberg, H., Seidman, J., and Seidman, C. 1994. α tropomyosin and cardiac troponin T mutations cause familial hypertrophic cardiomyopahty: a disease of the sarcomere. *Cell* 77:701–712.

Walsh, T., Trueblood, C., Evans, R., and Weber, A. 1984. Removal of tropomyosin overlap and the co-operative response to increasing calcium concentrations of the acto-subfragment-1 ATPase. *J. Mol. Biol.* 182:265–269.

Warmke, J., Kreuz, A., and Falkenthal, S. 1989. Co-localization to chromosome bands 99D1-3 of the *Drosophila melanogaster* myosin light chain-2 gene and a haplo-insufficient locus that affects flight behavior. *Genetics* 122:139–151.

Williams, B. and Waterston, R. 1994. Genes critical for muscle development and function in *Caenorhabditis elegans* identified through lethal mutations. *J. Cell Biol.* 124:475–490.

Xu, C., Craig, R., Tobacman, L., Horowitz, R., and Lehman, W. 1999. Tropomyosin positions in regulated thin filaments revealed by cryoelectron microscopy. *Biophys. J.* 77:985–992.

Zajdel, R., McLean, M., Lemanski, S., Muthuchamy, M., Wieczorek, D., Lemanski, L., and Dube, D. 1998. Ectopic expression of tropomyosin promotes myofibrillogenesis in mutant axolotl hearts. *Dev. Dyn.* 213:412–420.

Zajdel R., McLean, M., Isitmangil, G., Lemanski, L., Wieczorek, D., and Dube, D. 2000. Alteration of cardiac myofibrillogenesis by liposome-mediated delivery of exogenous proteins and nucleic acids into whole embryonic hearts. *Anatomy and Embryology* 201:217–228.

CHAPTER 14

Cardiomyopathies and Myofibril Abnormalities

Jeffrey A. Towbin and Neil E. Bowles

Cardiomyopathies are primary disorders of the myocardium in which systolic and/or diastolic dysfunction occurs (Towbin, 1993, 1999). These cardiac muscle diseases are classified based on phenotypic features (Towbin, 1993, 1999; Richardson et al., 1996) and include: (1) dilated cardiomyopathy (DCM), (2) hypertrophic cardiomyopathy (HCM), (3) restrictive cardiomyopathy (RCM), and (4) arrhythmogenic right ventricular dysplasia/cardiomyopathy (ARVD/ARVC). Since the early 1990s, molecular insight into these disorders has emerged and, based on the genotype–phenotype correlations, we have suggested a unifying hypothesis called the "final common pathway" hypothesis (Towbin et al., 1999a,b; Bowles et al., 2000). In this chapter, we will describe the clinical features of these myocardial disorders, outline the current understanding of the molecular basis of these diseases, and discuss the final common pathways of each of these tragic disorders, including the effect on the myofibril in each case.

DILATED CARDIOMYOPATHY (DCM)

Dilated cardiomyopathy (DCM) is characterized by left ventricular dilation and systolic dysfunction, which, in many instances, leads to clinical congestive heart failure (CHF) (Manolio et al., 1992; Towbin, 1999b). In some patients, the right ventricle is also affected and diastolic dysfunction may be secondarily noted (Towbin, 1999). Due to left ventricular dilation and stretching of the mitral ring, varying degrees of mitral regurgitation (MR) are common (Figure 14.1). In addition, fibrosis and myocyte hypertrophy occur, and cardiac arrhythmias, particularly ventricular tachycardia (VT), are relatively common. These findings are translated into the clinical picture of a patient with dyspnea on exertion and exercise intolerance, abdominal pain with hepatomegaly, abdominal ascites, peripheral edema, jugular venous distention, easy fatigue, chest pain and palpitations, and in some patients circulatory collapse and sudden death (Wiles et al., 1991). Chronic CHF is a common outcome and, in some cases, cardiac transplantation becomes necessary.

Diagnostic Laboratory Evaluation

In cases in which clinical symptoms suggest a possible diagnosis of DCM, a variety of tests are pursued, including chest x-ray, electrocardiography (ECG), and an

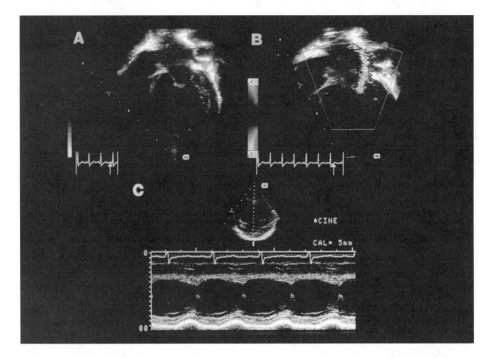

FIGURE 14.1. Dilated cardiomyopathy with dilated left ventricle seen in panel A, mitral regurgitation in panel B, and poor LV function in panel C.

echocardiogram. In familial forms of DCM, these tests are used as screening studies of relatives despite the lack of symptoms of DCM in these relatives.

Chest x-ray: The chest radiograph is typically helpful in identifying cardiomegaly (enlarged heart) and increased pulmonary vascular markings consistent with pulmonary edema. In some patients, the left atrium is quite large due to mitral regurgitation and causes atelectasis of the left lung due to compression. Pleural effusions may also occur.

Electrocardiogram (ECG): ECGs are usually abnormal in DCM with sinus tachycardia, ventricular hypertrophy, and left atrial enlargement common. Rhythm abnormalities such as VT may also occur.

Echocardiogram: In patients with DCM, a dilated and dysfunctional left ventricle with or without mitral regurgitation and left atrial enlargement is standard. A pericardial effusion may be seen in some individuals as well.

Histology

Myocyte hypertrophy, fibrosis, loss of myofibrils, and in some cases inflammatory infiltrates are notable on light microscopy. Electron microscopy further details these abnormalities as well as demonstrating loss of contractile elements, and sarcolemmal and mitochondrial analyses have demonstrated reduction of such structural and contractile proteins as vinculin, titin, dystrophin, dystrophin-associated proteins, myosin, and actin, among others (Schaper et al., 1991).

Genetics

Approximately 30% of patients (or more) with DCM have a genetically transmitted disease (Michels et al., 1992; Keeling et al., 1995; Grunig et al., 1998; Baig et al., 1998). The most common inheritance pattern is autosomal dominant transmission, but X-linked inheritance is also prominent. Autosomal recessive and mitochondrial transmission of DCM are well known, albeit uncommon (Towbin, 1993).

The genetic etiologies of patients with inherited forms of DCM have started to unravel over the past several years (Graham and Owens, 1999). The first genes to be identified were those involved in X-linked forms of DCM; more recently, the genes for autosomal dominant DCM have started to be identified.

X-Linked DCM

Two X-linked forms of DCM have been described: X-linked cardiomyopathy (XLCM) and Barth syndrome.

X-linked cardiomyopathy (XLCM): First described by Berko and Swift (1987) as DCM occurring in males in the teen years and early twenties with rapid progression from CHF to death or transplantation, these patients are distinguished by elevated serum creatine kinase muscle isoforms (CK-MM). Female carriers tend to develop mild to moderate DCM in the fifth decade and the disease is slowly progressive. Towbin and colleagues (Towbin, 1993, 1995) were the first to identify the disease-causing gene and characterize the functional defect. In this report, the dystrophin gene was shown to be responsible for the clinical abnormalities, and protein analysis by immunoblotting demonstrated severe reduction or absence of dystrophin protein in the hearts of these patients (Figure 14.2). These findings were later confirmed by Muntoni et al. (1993) when a mutation in the muscle promoter and exon 1 of dystrophin was identified in another family with XLCM. Subsequently, multiple mutations have been identified in dystrophin (Figure 14.3) in patients with XLCM (Nigro et al., 1994; Milasin et al., 1996; Ortiz-Lopez et al., 1997; Ferlini et al., 1998; Yoshida et al., 1998; Franz et al., 1995, 2000).

Dystrophin is a cytoskeletal protein which provides structural support to the myocyte by creating a lattice-like network to the sarcolemma (Hoffman et al., 1988, 1990). In addition, dystrophin plays a major role in linking the sarcomeric contractile apparatus to the sarcolemma and extracellular matrix (Campbell, 1995; Meng et al., 1996; Cox and Kunkel, 1997; Kaprielian et al., 2000). Furthermore, dystrophin is involved in cell signaling, particularly through its interactions with nitric oxide synthase (Chang et al., 1996). The dystrophin gene is responsible for Duchenne and Becker muscular dystrophy (DMD/BMD) when mutated as well (Koenig et al., 1987). These skeletal myopathies are present early in life (DMD is diagnosed before age 12 years, while BMD is seen in teenage males older than 16 years of age), and the vast majority of patients develop DCM before the 25th birthday (Hunsaker et al., 1982; Melacini et al., 1996). In most patients, CK-MM is elevated, similar to that seen in XLCM; in addition, manifesting female carriers develop disease late in life, similar to XLCM. Furthermore, immunohistochemical analysis demonstrates reduced levels (or absence) of dystrophin, similar to that seen in the hearts of patients with XLCM.

Murine models of dystrophin deficiency demonstrate abnormalities of muscle physiology based on membrane structural support abnormalities. In addition to the dysfunction of dystrophin, mutations in dystrophin secondarily affect proteins that

WESTERN BLOT: XLCM VS. N-TERMINAL ANTI-DYSTROPHIN ANTIBODY

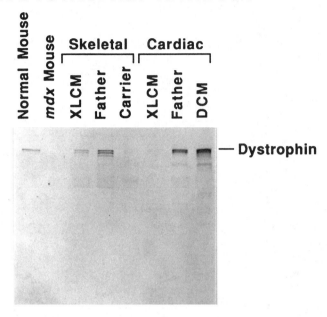

FIGURE 14.2. Western blot using dystrophin antibody to the N-terminus of dystrophin shows reduced or no cardiac dystrophin and mild reduction in skeletal muscle dystrophin in patients with X-linked cardiomyopathy (XLCM). The mdx mouse is a dystrophin-deficient mouse.

DYSTROPHIN MUTATIONS CAUSING XLCM

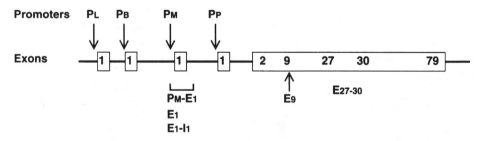

FIGURE 14.3. Positions in the dystrophin gene where mutations causing X-linked cardiomyopathy (XLCM) have been identified. The gene has 79 exons and multiple promoters.

interact with dystrophin. At the amino-terminus (N-terminus), dystrophin binds to the sarcomeric protein actin, a member of the thin filament of the contractile apparatus. At the carboxy-terminus (C-terminus), dystrophin interacts with α-dystroglycan, a dystrophin-associated membrane-bound protein involved in the function of the dystrophin-associated protein complex (DAPC), which includes β-dystroglycan, the sarcoglycan subcomplex (α-, β-, γ-, δ-, and ε-sarcoglycan), syntrophins, and dystrobrevins (Ohlendieck, 1996) (Figure 14.4). In turn, this complex interacts with α2-laminin and the extracellular matrix (Helbling-Leclerc et al., 1995). Like dystrophin, mutations in these genes lead to muscular dystrophies with or without cardiomyopathy, supporting the contention that this group of proteins is important to the normal function of the myocytes of the heart of skeletal muscles (Klietsch et al., 1993; Campbell, 1995; Ozawa et al., 1995). In both cases, mechanical stress (Petrof et al., 1993) appears to play a significant role in the age-onset-dependent dysfunction of these muscles. The information gained from the studies on XLCM, DMD, and BMD led us to hypothesize that DCM is a disease of the cytoskeleton/sarcolemma (Towbin, 1998).

Barth Syndrome: Initially described as X-linked cardioskeletal myopathy with abnormal mitochondria and neutropenia by Neustein et al. (1979) and Barth et al. (1983), this disorder typically presents in male infants as CHF associated with neutropenia (cyclic) and 3-methylglutaconic aciduria (Kelley et al., 1991). Mitochondrial dysfunction is noted on EM and electron transport chain biochemical analysis. Echocardiographically, these infants typically have left ventricular dysfunction with left ventricular dilation, endocardial fibroelastosis, or a dilated hypertrophic left

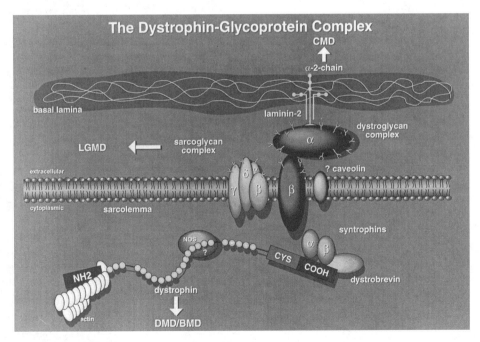

FIGURE 14.4. Dystrophin–glycoprotein complex. Mutations in the subcomplexes lead to various muscular dystrophies. DMD = Duchenne muscular dystrophy; BMD = Becker muscular dystrophy; LGMD = limb girdle muscular dystrophy; BMD = Becker muscular dystrophy; CMD = congenital muscular dystrophy.

ventricle. In some cases, these infants succumb due to CHF/sudden death or sepsis due to leukocyte dysfunction. The majority of these children survive infancy and do well clinically, although DCM usually persists. In some cases, cardiac transplantation has been performed. Histopathologic evaluation typically demonstrates the features of DCM, although endocardial fibroelastosis may be prominent and the mitochondria are abnormal in shape and abundance.

The genetic basis of Barth syndrome was first described by Bione et al. (1996), who cloned the disease-causing gene G4.5. This gene encodes a novel protein called tafazzin, whose function is not currently known. However, mutations in G4.5 result in a wide clinical spectrum (D'Adamo et al., 1997; Johnston et al., 1997), which includes apparent classic DCM, hypertrophic DCM, endocardial fibroelastosis (EFE), or left ventricular noncompaction (LVNC) (Bleyl et al., 1997). In the latter case, the left ventricular noncompaction is characterized by deep trabeculations giving the appearance of a "spongiform" myocardium (Figure 14.5) (Chin et al., 1990). The mechanisms responsible for this clinical heterogeneity are not currently known.

Autosomal Dominant DCM

The most common form of inherited DCM, these patients present as classic "pure" DCM or DCM associated with conduction system disease (CDDC). In the latter case, patients usually present in the twenties with mild conduction system disease, which can progress to complete heart block over decades. DCM usually presents late in the course but is out of proportion to the degree of conduction system disease. The echocardiographic and histologic findings in both subgroups are classic for DCM, although the conduction system may be fibrotic in patients with CDDC.

Genetic heterogeneity exists for autosomal dominant DCM, with seven genetic loci mapped for pure DCM and five loci for CDDC. In the case of pure DCM, the genetic loci identified thus far include three mapped by our group (1q32, 5q33, 10q21-23) (Durand et al., 1995; Bowles et al., 1996; Tsubata et al., 2000), as well as those mapped to chromosomes 2q31, 2q35, 9q13-22, and 15q14 by others (Krajinovic et al., 1995; Olson et al., 1998; Li et al., 1999; Siu et al., 1999). Three of these genes are now known: actin (chromosome 15q14-linked) (Olson et al., 1998), desmin (chromosome 2q35) (Li et al., 1999), and δ-sarcoglycan (chromosome 5q33) (Tsubata et al., 2000).

Cardiac actin is a sarcomeric protein that is a member of the sarcomeric thin filament interacting with tropomyosin and the troponin complex. As previously noted, actin plays a significant role in linking the sarcomere to the sarcolemma via its binding to the N-terminus of dystrophin. Interestingly, the mutations in actin that resulted in DCM as described by Olson et al. (1998) appear to be directly involved in the binding of dystrophin, whereas mutations in the sarcomeric end of actin result in hypertrophic cardiomyopathy (Mogensen et al., 1999). The DCM-causing mutations are believed to result in DCM by causing force transmission abnormalities.

Desmin is a cytoskeletal protein that forms intermediate filaments specific for muscle. This muscle-specific 53-kd subunit of class III intermediate filaments forms connections between the nuclear and plasma membranes of cardiac, skeletal, and smooth muscle. Desmin is found at the Z-lines and intercalated disk of muscle, and its role in muscle function appears to involve attachment or stabilization of the sarcomere. Mutations in this gene appear to cause abnormalities of force and signal transmission similar to that believed to occur with actin mutations (Li et al., 1999).

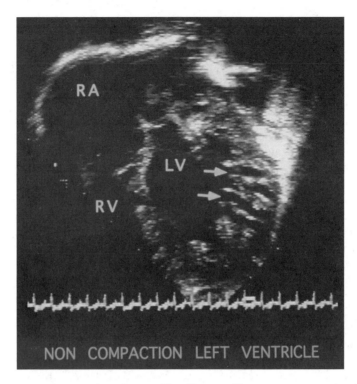

FIGURE 14.5. LV noncompaction. Arrows identify spongiform noncompacted regions of LV.

Desmin mutations in murine models produce a similar phenotype (Li et al., 1996; Milner et al., 1996).

The most recently identified DCM-causing gene was reported by our group (Tsubata et al., 2000) to be δ-sarcoglycan, a member of the sarcoglycan subcomplex of the DAPC. This gene encodes for a protein involved in stabilization of the myocyte sarcolemma as well as signal transduction. Mutations identified in familial and sporadic cases resulted in reduction of the protein within the myocardium.

As seen in pure autosomal dominant DCM, genetic heterogeneity also exists for CDDC. To date, CDDC genes have been mapped to chromosomes 1p1–1q1 (Kass et al., 1994), 2q14-21 (Jung et al., 1999), 3p25-22 (Olson and Keating, 1996), and 6q23 (Messina et al., 1997). Schonberger et al. (2000) mapped the gene for DCM with sensorineural deafness to 6q23 as well. The only gene thus far identified was reported by Fatkin et al. (1999) and Brodsky et al. (2000) to be lamin A/C on chromosome 1q21, which encodes a nuclear envelope intermediate filament protein. Similar to emesis, the gene responsible for X-linked Emery-Dreifuss muscular dystrophy has similar clinical features (Bione et al., 1994). The mechanisms responsible for the development of DCM and conduction system abnormalities are currently unknown.

Interestingly, all of the genes identified for inherited DCM are also known to cause skeletal myopathy. In the case of dystrophin, mutations cause DMD and BMD (Koenig et al., 1987), whereas sarcoglycan mutations cause limb girdle muscular dystrophy (LGMD2F) (Campbell, 1995; Jung et al., 1996; Melacini et al., 1999; Nigro et al., 1996a; Ozawa et al., 1998). Lamin A/C has been shown to cause autosomal dominant Emery–Dreifuss muscular dystrophy (EDMD) (Bonne et al., 1999; Muchir et al., 2000; Raffaela Di Barletta et al., 2000), while actin mutations are associated with nemaline myopathy (Nowak et al., 1999). Desmin and G4.5 mutations also have associated skeletal myopathy (Bione et al., 1996; Goldfarb et al., 1998; Dalakas et al., 2000), suggesting that cardiac and skeletal muscle functions are interrelated and that possibly the skeletal muscle fatigue seen in patients with DCM with and without CHF may be due to primary skeletal muscle disease and not related to the cardiac dysfunction. It also suggests that the function of these muscles has a final common pathway and that cardiologists and neurologists should consider evaluation of both sets of muscles (Van der Kooi et al., 1996, 1997).

Further support for this concept comes from studies of both humans and animal models. Maeda et al. (1997) identified absence of the metavinculin transcript in the heart of a patient with idiopathic DCM and confirmed the metavinculin abnormality by immunoblot, which demonstrated the absence of metavinculin protein in the myocardium. Metavinculin plays a role in attaching the sarcomere to the cardiomyocyte membrane by complexing with nonsarcomeric actin microfilaments complexed with other cytoskeletal proteins such as talin, vinculin and α-actinin, which are linked to cadherin or to the integrin receptor. Mutations in δ-sarcoglycan in hamsters result in cardiomyopathy (Nigro et al., 1996, 1997; Sakamoto et al., 1997, 1999), while mutations in all sarcoglycan subcomplex genes in mice cause skeletal and cardiac muscle disease (Araishi et al., 1999; Coral-Vazquez et al., 1999; Hack et al., 1999; Straub et al., 1998). Mutations in other DAPC genes as well as dystrophin in murine models also consistently demonstrate abnormalities of skeletal and cardiac muscle function (Fadic et al., 1996; Lim et al., 1995). Arber et al. (1997) also produced a mouse deficient in muscle LIM protein (MLP), a structural protein that links the actin cytoskeleton to the contractile apparatus. The resultant mice develop severe DCM, CHF, and disruption of cardiac myocyte cytoskeletal architecture. Human abnormalities in MLP have been seen in DCM patients as well (Katz, 2000; Zolk et al., 2000). Finally, Badorff et al. (1999) have shown that the DCM that develops after viral myocarditis has a mechanism similar to the inherited forms. Using coxsackievirus B3 (CVB3) infection of mice, the authors showed that the CVB3 genome encodes for a protease (enteroviral protease 2A) that cleaves dystrophin at the third hinge region of dystrophin, resulting in force transmission abnormalities and DCM. Interestingly, a similar dystrophin mutation, which affects the first hinge region of dystrophin in patients with XLCM, was previously reported by our laboratory (Ortiz-Lopez et al., 1997), demonstrating a consistent mechanism of DCM development abnormalities of the cytoskeleton/sarcolemma.

HYPERTROPHIC CARDIOMYOPATHY (HCM)

Hypertrophic cardiomyopathy (HCM) is a complex cardiac disease with unique pathophysiological characteristics and a wide spectrum of morphologic, functional, and clinical features (Maron, 1997; Spirito et al., 1997). In these patients, the interventricular septum and left ventricular posterior wall are thickened (usually asymmetric septal hypertrophy), and systolic hypercontractile function is noted along with diastolic dysfunction (Figure 14.6). Systolic anterior motion of the mitral valve and left ventricular outflow tract obstruction may also occur. Although HCM has been considered a relatively uncommon cardiac disease by many, the prevalence of echocardiographically defined HCM in a large cohort of apparently healthy young adults selected from a community-based general population was reported to be as high as 0.2% (Maron et al., 1995). Although sporadic cases are common,

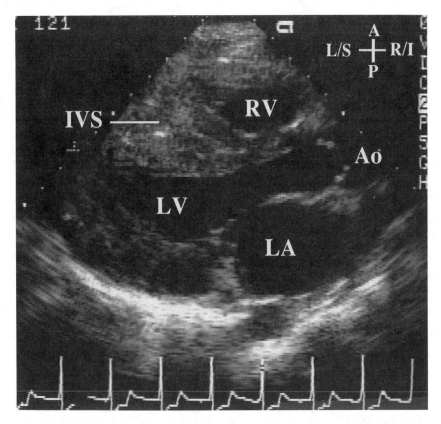

FIGURE 14.6. Hypertrophic cardiomyopathy. Note the thick LV posterior wall and interventricular septum (IVS), as well as the small LV chamber size.

familial disease (FHC) with autosomal dominant inheritance predominates (Maron, 1997). Hypertrophic cardiomyopathy appears to have a triphasic pattern of presentation (Maron et al., 1986; Maron, 1997). The most common presentation occurs in children above the age of ten years to those individuals younger than twenty-five years. However, it has been found that late-onset disease also occurs. Finally, young children also may be diagnosed with HCM, and these children usually present by the first year of life. In young children, the underlying disease process is thought to differ from that of older children and adults, with different etiologies apparent. Because HCM is a major cause of sudden death in young healthy individuals and athletes (Maron et al., 1996), it has a major impact on any population. The purpose of this review is to describe the current understanding of the genetic basis of this disease.

MAPPING OF FAMILIAL HYPERTROPHIC CARDIOMYOPATHY GENES

The first gene for familial hypertrophic cardiomyopathy (FHC) was mapped to chromosome 14q11.2–q12 using genome-wide linkage analysis in a large Canadian family (Jarcho et al., 1989). Soon afterward, FHC locus heterogeneity was reported

(Solomon et al., 1990; Schwartz et al., 1992) and subsequently confirmed by Thierfelder et al. (1993) and Watkins et al. (1993) when they mapped the second FHC locus to chromosome 1q3 and the third locus to chromosome 15q2 (Watkins et al., 1993). Carrier et al. (1993) mapped the fourth FHC locus to chromosome 11p11.2. Five other loci were subsequently reported, located on chromosomes 7q3 (MacRae et al., 1995), 3p21.2–3p21.3 (Poetter et al., 1996), 12q23–q24.3 (Kimura et al., 1997), 15q14 (Mogensen et al., 1999), and 2q31 (Satho et al., 1999). Several other families are not linked to any known FHC loci, indicating the existence of additional FHC-causing genes. These genes have been identified in patients presenting late in child-hood or adulthood and not in neonates, infants, and young children.

GENE IDENTIFICATION IN FAMILIAL HYPERTROPHIC CARDIOMYOPATHY

All the disease genes encode proteins that are part of the sarcomere (i.e., a sarcomy-opathy), which is a complex structure with an exact stoichiometry and multiple sites of protein–protein interactions (Rayment et al., 1993; Thierfelder et al., 1994; Bonne et al., 1998). The encoded proteins include three myofilament proteins, the β-myosin heavy chain (β-MyHC) (Geisterfer-Lowrance et al., 1990), the ventricular myosin essential light chain 1 (MLC-1s/v), and the ventricular myosin regulatory light chain 2 (MLC-2s/v) (Poetter et al., 1996); four thin filament proteins: cardiac actin (Mogensen et al., 1999), cardiac troponin T (cTnT) (Thierfelder et al., 1994), cardiac troponin I (cTnI) (Kimura et al., 1997), and α-tropomyosin (α-TM) (Thierfelder et al., 1994); and finally one myosin binding protein, the cardiac myosin binding protein C (cMyBP-C) (Figure 14.7) (Bonne et al., 1995; Watkins et al., 1995a,b; Satoh et al., 1999). Titin has been described as well (Satoh et al., 1999). Each of these proteins is encoded by multigene families that exhibit tissue-specific, developmental, and physiologically regulated patterns of expression.

Thick Filament Proteins

Myosin Subunits

Myosin is the molecular motor that transduces energy from the hydrolysis of ATP into directed movement and that, by doing so, drives sarcomere shortening and muscle contraction. Cardiac myosin consists of two heavy chains (MyHC) and two pairs of light chains (MLC), referred to as essential (or alkali) light chains (MLC-1) and regulatory (or phosphorylatable) light chains (MLC-2), respectively (Schiaffino and Reggiani, 1996). The myosin molecule is highly asymmetric, consisting of two globular heads joined to a long rod-like tail. The light chains are arranged in tandem in the head–tail junction. Their function is not fully understood. Neither myosin light-chain type is required for the adenosine triphosphatase (ATPase) activity of the myosin head, but they probably modulate it in the presence of actin and contribute to the rigidity of the neck, which is hypothesized to function as a lever arm for generating an effective power stroke. Mutations have been found in the heavy chains and in the two types of ventricular light chains (Bonne et al., 1998).

Concerning the heavy chains, the β isoform (β-MyHC) is the major isoform of the human ventricle and of slow-twitch skeletal fibers. It is encoded by *MYH7*. At least 50 mutations have been found in unrelated families with FHC, and three hot spots

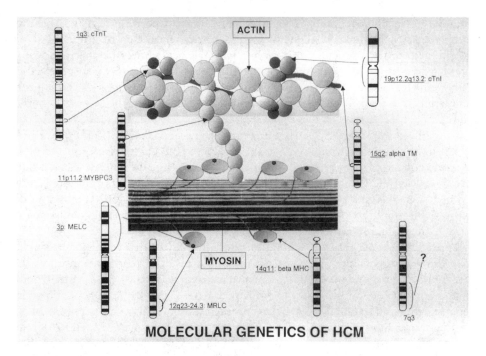

FIGURE 14.7. Molecular genetics of hypertrophic cardiomyopathy (HCM). The genetic loci, genes, and encoded proteins (all sarcomeric proteins) are shown.

for mutations have been identified, including codons 403, 719, and 741. All but three of these mutations are missense mutations located in the head or in the head–rod junction of the molecule (Bonne et al., 1998). The three exceptions are two 3-bp deletions that do not disrupt the reading frame (codon 10, codon 930) and a 2.4-kb deletion in the 3′ r egion. In the kindred with the latter mutation, only the proband developed clinical evidence of HCM, and in this case it occurred at a very late age of onset (59 years).

Animal models have been reported by Vikstrom et al. (1996) and Geisterfer-Lowrance et al. (1996). The mice produced by these groups developed HCM-like disease. Heterozygotes have been produced with the Arg403Gln mutation and have been shown to have normal survival, whereas homozygotes had premature death, mimicking the human mutations. Fatkin et al. (1999) demonstrated neonatal cardiomyopathy in homozygotes.

As for the light chains, the isoforms expressed in the ventricular myocardium and in the slow-twitch muscles are the so-called ventricular myosin regulatory light chains (MLC-2s/v) encoded by *MYL2*, and the ventricular myosin essential light chain (MLC-1s/v) encoded by *MYL3*. Both belong to the superfamily of EF-hand proteins. Two missense mutations have been reported in *MYL3* and five in *MYL2* (Poetter et al., 1996).

Myosin Binding Protein C (MyBP-C)

Myosin binding protein C (MyBP-C) is part of the thick filaments of the sarcomere, being located at the level of the transverse stripes, 43 nm apart, seen by electron microscopy in the sarcomere A-band. Its function is uncertain, but it is believed to play both structural and regulatory roles. Partial extraction of cMyBP-C from rat skinned cardiac myocytes and rabbit skeletal muscle fibers has been shown to alter Ca^{2+}-sensitive tension (Hofmann et al., 1991). In addition, that phosphorylation of cMyBP-C was shown to alter myosin cross-bridges in native thick filaments, suggesting that cMyBP-C can modify force production in activated cardiac muscles.

The cardiac isoform of myosin binding protein C is encoded by the *MYBPC3* gene, which has three distinct regions that are specific to the cardiac isoform: the NH_2-terminal domain C0 Ig-I containing 101 residues, the MyBP-C motif (a 105-residue stretch linking the C1 and C2 Ig-I domains), and a 28-residue loop inserted in the C5 Ig-I domain (Gautel et al., 1995). It has been shown that cMyBP-C is specifically expressed in the heart during human and murine development (Fougerousse et al., 1998; Gautel et al., 1998).

Approximately 30 *MYBPC3* mutations have been identified in unrelated families with FHC. In the majority of cases, the mutations result in aberrant transcripts that are predicted to encode COOH-terminal truncated cardiac MyBP-C polypeptides lacking at least the myosin binding domain (Carrier et al., 1997). Of the remaining reported mutations, mutated or deleted proteins occur without disruption of the reading frame; most of these are missense mutations (exons 6, 17, 21, and 23), but a splice donor site mutation (intron 27) and an 18-residue duplication (exon 33) have also been reported (Bonne et al., 1995). In addition, three mutations are predicted to produce either a mutated protein or a truncated one: two are missense mutations in exons 15 and 17 and one is a branch point mutation in intron 23. It has been suggested that most mutations result in a dominant-negative action of the mutant protein (McConnell et al., 1999).

A mouse model has been reported by McConnell et al. (1999) in which mice expressing altered forms of MYBPC3 were produced. The engineered mutations encoded truncated forms of MyBP-C in which the cardiac myosin heavy-chain binding and titin binding domain was replaced with novel amino acid residues. Homozygous mice exhibited neonatal onset of a progressive dilated cardiomyopathy with myocyte hypertrophy, myofibrillar disarray, fibrosis, and dystrophic calcification on histopathology. Left ventricular dilation with reduced contractility was noted on echocardiography in the neonatal period, but myocardial hypertrophy and diastolic dysfunction developed later in the life of these animals. Previously, Yang et al. (Yang, 1998) reported a mouse model of HCM due to mutant MyBP-C, and this clinical phenotype was more similar to the human disorder. Here, however, the transgenic animals had very high levels of mutant gene (driven by the α-MyHC promoter) and diminished wild-type gene expression to very low levels.

Thin Filament Proteins

The thin filament contains actin, the troponin complex, and tropomyosin. The troponin complex and tropomyosin constitute the Ca^{2+}-sensitive switch that regulates the contraction of cardiac muscle fibers. Mutations were found in α-TM and in two of the subunits of the troponin complex: cTnI, the inhibitory subunit, and cTnT, the tropomyosin binding subunit. Actin mutations have also been reported (Mogensen et al., 1999).

α-tropomyosin (α-TM) is encoded by *TPM1*. The cardiac isoform is expressed both in the ventricular myocardium and in fast-twitch skeletal muscles (Lees-Miller and Helfman, 1991). It shares the overall structure of other tropomyosins that are rod-like proteins that possess a simple dimeric α-coiled-coil structure in parallel orientation along their entire length (Lees-Miller and Helfman, 1991). Four missense mutations were initially described in unrelated FHC families. Two of them, A63V and K70T, are located in exon 2b within the consensus pattern of sequence repeats of α-TM and could alter tropomyosin binding to actin. Mutations D175N and E180G are located within constitutive exon 5 in a region near the C190 and near the calcium-dependent TnT binding domain. This appears to be a relatively rare cause of HCM.

Cardiac troponin T (cTnT) is encoded by *TNNT2*. In human cardiac muscle, multiple isoforms of cTnT have been described, which are expressed in the fetal, adult, and diseased hearts and result from alternative splicing of the single gene *TNNT2* (Mesnard et al., 1995). The precise physiological relevance of these isoforms is currently poorly understood, but the organization of the human gene has been partially established (Forissier et al., 1996; Farza et al., 1998), thus allowing precise identification of the position of the mutations within exons, including those alternatively spliced during development. Eleven mutations (10 missense, 1 deletion) have been found in unrelated FHC families, three of which are located in a hot spot (codon 102).

Cardiac troponin I (cTnI) is encoded by *TNNI3*. The cTnI isoform is expressed only in cardiac muscles (Hunkeler et al., 1991). Cooperative binding of cTnI-actin-tropomyosin is a unique property of the cardiac variant (Solaro and Van Eyk, 1996). To date, six mutations have been identified, all single base substitutions.

α-cardiac actin (ACTC) was identified as a cause of FHC. Mogensen et al. (1999) studied a family with heterogeneous phenotypes, ranging from asymptomatic with mild hypertrophy to pronounced septal hypertrophy and left ventricular outflow tract obstruction. Using linkage analysis and mutation screening, the gene was mapped to chromosome 15q14 with a lod score of 3.6, and a missense mutation (G → T in position 253, exon 5) resulted in an Ala295Ser amino acid substitution. The mutation is localized at the surface of actin in proximity to a putative myosin binding site. This mutation, which causes FHC, differs from the two mutations reported to cause dilated cardiomyopathy (DCM) (Olson et al., 1998), which were localized in the immobilized end of actin that cross-binds to the anchor polypeptides in the Z-bands. It appears that mutations near the dystrophin binding region result in DCM, whereas mutations affecting the sarcomeric end result in an HCM phenotype.

GENOTYPE–PHENOTYPE RELATIONS IN FAMILIAL HYPERTROPHIC CARDIOMYOPATHY

The pattern and extent of left ventricular hypertrophy in patients with HCM vary greatly even in first-degree relatives, and a high incidence of sudden deaths is reported in selected families. An important issue therefore is to determine whether the genotype heterogeneity observed in FHC accounts for the phenotypic diversity of the disease. However, the results must be seen as preliminary because the available data relate to only a few hundred individuals. Several concepts nevertheless begin to emerge, at least for mutations in the *MYH7*, *TNNT2*, and *MYBPC3* genes. For *MYH7*, it is clear that the prognosis for patients with different mutations varies considerably (Spirito et al., 1997). For example, the R403Q mutation appears to be associated with markedly reduced survival, whereas some others, such as V606M, appear more benign (Watkins et al., 1992). The disease caused by *TNNT2* mutations is usually associated with a 20% incidence of nonpenetrance, a relatively mild (and sometimes subclinical) hypertrophy, but a high incidence of sudden death, which can occur even in the absence of significant clinical left ventricular hypertrophy (Watkins et al., 1995b; Moolman et al., 1997). However, one family with a *TNNT2* mutation reportedly had complete penetrance, and echocardiographic data showed a wide range of hypertrophy and there was no sudden cardiac death (Farza et al., 1998). Thus, more data are needed before final conclusions are drawn. Mutations in *MYBPC3* seem to be characterized by specific clinical features, with a mild phenotype in young subjects, a delayed age at the onset of symptoms, and a favorable prognosis before the age of 40 years (Bonne et al., 1995; Watkins et al., 1995a; Charron et al., 1998; Nimura et al., 1998).

Genetic studies have also revealed the presence of clinically healthy individuals carrying a mutant allele (no matter which gene is affected), which is associated in first-degree relatives with a typical phenotype of the disease. Several mechanisms could account for the large variability of the phenotypic expression of the mutations: the role of environmental differences and acquired traits (e.g., differences in lifestyle, risk factors, and exercise), and the existence of modifier genes and/or polymorphisms that could modulate the phenotypic expression of the disease. The only significant results obtained so far concern the influence of the angiotensin I converting enzyme insertion/deletion (ACE I/D) polymorphism. Association studies showed that, compared to a control population, the D allele is more common in patients with hypertrophic cardiomyopathy and in patients with a high incidence of sudden cardiac death (Marian et al., 1993; Yonega et al., 1995). It has been shown that the association between the D allele and hypertrophy is observed in the case of *MYH7* R403 codon mutations but not with *MYBPC3* mutation carriers (Tesson et al., 1997), raising the concept of multiple genetic modifiers in FHC.

NEONATAL HYPERTROPHIC CARDIOMYOPATHY

Clinically, this appears to be the same disorder as that seen in older children and adults (Towbin and Lipshultz, 1999). Echocardiograms and electrocardiograms tend to mimic the findings described for adults with HCM. However, the underlying etiologies appear to differ from the sarcomeric gene-related disorders of adults. In the neonate and infant with HCM, abnormalities of mitochondrial function (Wallace, 1992) and metabolic disorders tend to predominate. In addition, genetic syndromes are common in this age group as well. Mott et al. (2000) have demonstrated that Noonan syndrome, mitochondrial myopathies, and metabolic disorders such as glycogen storage diseases predominate along with abnormalities of fatty acid oxidation. Although sarcomeric gene mutations could be involved as well, essentially no patients have been described to date in the literature. Instead, it is likely that the sarcomere is secondarily affected by the mitochondrial and metabolic abnormalities, which probably create an energy production–utilization mismatch.

Finally, neonates with a dilated form of HCM with combined systolic and diastolic dysfunction are well described (Towbin and Lipshultz, 1999). In some cases, deep endomyocardial trabeculations are noted, consistent with left ventricular noncompaction (Figure 14.5). Males with this abnormality have been shown, in some cases, to carry mutations in G4.5, an X-linked gene (Xq28), which encodes a novel protein called tafazzin. Although the function of this protein is not known, clinically these children have abnormal mitochondria on electron microscopy and abnormal energy production. Other children with early presentation of dilated HCM have been found to have mitochondrial abnormalities as well, and this occurs in both sexes.

RESTRICTIVE CARDIOMYOPATHY (RCM)

This disorder is characterized by extreme atrial dilation in the face of normal-sized ventricles with normal systolic function and diastolic dysfunction. In some cases, ventricular hypertrophy occurs. The intracardiac pressures are equivalent in all chambers, and pulmonary hypertension is common. Clinically, these patients tend to present with syncope and sudden death, although in many patients CHF occurs. Ventricular arrhythmias and ischemia commonly are seen as well. Cardiac transplantation is the treatment of choice in young patients. Histologically, myocyte hypertrophy and fibrosis dominate the picture.

Diagnostic Laboratory Evaluation

As with other forms of cardiomyopathy, the standard tests for diagnosis include chest radiography, ECG, and echocardiography. Cardiomegaly is seen on chest x-rays, and biatrial enlargement associated with ST-segment abnormalities is classically seen on ECG. On echocardiography, biatrial enlargement with normal ventricular function is seen (Figure 14.8). Holter monitor evaluation (24-hour ambulatory ECG) may be useful in the diagnosis of ischemia and ventricular arrhythmias.

Genetics

Autosomal dominant inheritance has been reported for some cases of RCM, although most commonly this disorder is sporadic. To date, no genetic mutations have been consistently identified in inherited RCM. However, in elderly patients, mutations in the transthyretin (prealbumin) gene have been identified.

Desmin, the same gene that causes DCM, has been seen in the skeletal muscle and heart as desmin protein inclusions in some patients with RCM with and without conduction system disease. Arbustini et al. (1998) have demonstrated desmin abnormalities in these patients, and this has been supported by Thornell et al. (1997), who identified desmin mutations in patients with RCM, and by Capetenaki and colleagues, who developed a desmin-deficient murine model in which a restrictive phenotype occurred.

The mechanism responsible for RCM is not understood, but based on the abnormalities in desmin it is likely that intermediate filaments and the interaction between sarcomeric structures result in the development of restrictive physiology with or without hypertrophy.

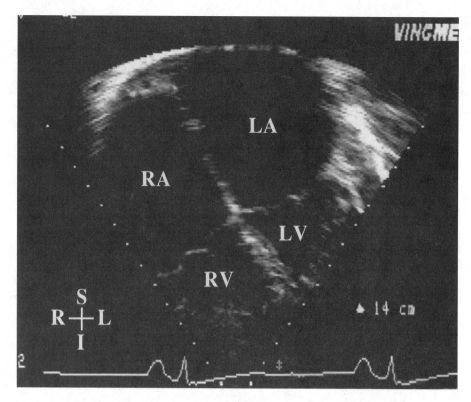

FIGURE 14.8. Restrictive cardiomyopathy. Note the very large atria (RA, LA) and normal ventricles.

ARRHYTHMOGENIC RIGHT VENTRICULAR DYSPLASIA/ CARDIOMYOPATHY (ARVD/ARVC)

ARVD/ARVC is a primary disorder of the right ventricle in which fibrofatty infiltration of the right ventricular wall occurs, with dilation and systolic dysfunction of the right ventricle resulting (Marcus et al., 1982; Thiene et al., 1988; Corrado et al., 2000). Ventricular arrhythmias commonly occur, and sudden death, particularly in athletes, is an important and tragic outcome in many patients (Thiene et al., 1988).

Diagnostic Laboratory Evaluation

Patients with ARVD/ARVC typically present with syncope and ventricular arrhythmias with a left bundle branch (LBB) pattern on ECG. Holter monitoring typically identifies runs of VT of the LBB type as well (Corrado et al., 1997). The chest x-ray may be normal or have mild cardiomegaly, while the echocardiogram may demonstrate right ventricular dilation and dysfunction. In some cases, magnetic resonance imaging (MRI) demonstrates right ventricular outflow tract and abnormal fatty infiltration. Right ventricular endomyocardial biopsy may identify fibrofatty infiltration, confirming the diagnosis (Thiene et al., 1988).

Genetics

A significant percentage of patients with ARVD/ARVC have autosomal dominant transmission of disease (Nava et al., 1988). In one subgroup of patients with ARVD/ARVC associated with wooly hair, palmoplantar keratosis, and autosomal recessive inheritance, known as Naxos disease because of its identification in individuals from Naxos, Greece, the disorder is complex (Coonar et al., 1998).

ARVD/ARVC is another genetically heterogeneous disorder with genetic loci identified for pure ARVD/ARVC (chromosomes 14q24 (Rampazzo et al., 1994), 1q42 (Rampazzo et al., 1995), 14q12–22 (Severini et al., 1996), 2q35 (Rampazzo et al., 1997), 10p12–14 (Li et al., 2000), 3p23 (Ahmad et al., 1998)) and one locus for Naxos disease (chromosome 17q21) (Coonar et al., 1998) and myofibrillar myopathy with ARVD/ARVC (chromosome 10q) (Melberg et al., 1999). The only gene identified thus far is for Naxos disease, where mutations in plakoglobin, an intermediate filament desmosomal protein, has been reported (McKoy et al., 2000). Although this protein type could be involved in cases of pure ARVD/ARVC, it is not clear whether the pure form of this disease and the complex phenotype of Naxos disease are interchangeable. Another possibility is that ARVD/ARVC is a primary rhythm disorder with secondary cardiomyopathy. To date, all inherited ventricular arrhythmias have been found to occur due to mutations in ion channel-encoding genes (Vatta et al., 2000). In long QT syndrome, five genes have been identified, including two potassium channel α-subunits (KVLQT1, HERG), two potassium channel β-subunits (minK, MiRP1), and the cardiac sodium channel α-subunit (SCN5A). Brugada syndrome, a disorder in which ventricular fibrillation occurs, is also due to ion-channel–encoding gene mutations (SCN5A), suggesting that ventricular arrhythmias are ion channelopathies (Towbin, 2000; Vatta et al., 2000). Therefore, it is possible that ARVD/ARVC could be caused by ion channel abnormalities, particularly calcium channel genes.

FINAL COMMON PATHWAY HYPOTHESIS

Clearly, FHC of adults is a disease of the sarcomere. Similarly, patients with other cardiac disorders, such as familial dilated cardiomyopathy (FDCM) and familial ventricular arrhythmias (i.e., long QT syndrome and Brugada syndrome), have been shown to have mutations in genes encoding a consistent family of proteins. In familial ventricular arrhythmias, ion channel gene mutations (i.e., ion channelopathy) have been found in all cases thus far reported. In FDCM, cytoskeletal protein-encoding genes have been speculated to be causative (i.e., cytoskeletal myopathy) (Towbin, 1998; Towbin et al., 1999). Hence, the final common pathways of these disorders include ion channels and cytoskeletal proteins, similar to the sarcomyopathy in HCM (Figure 14.5). Although it is not yet certain what the underlying pathways and targets are for RCM and ARVD, hints have been forthcoming. Desmin and other intermediate filament proteins appear to be at play in RCM, while cell–cell adhesion proteins (i.e., desmosomes, adherens junction proteins) or possibly ion channels are involved in ARVD. In addition, it appears that cascade pathways are involved directly in some cases (i.e., mitochondrial abnormalities in HCM, DCM), while secondary influences are likely to result in the wide clinical spectrum seen in patients with similar mutations. In HCM, mitochondrial and metabolic influences are probably important (Figure 14.6). Additionally, molecular interactions with such molecules as calcineurin, sex hormones, and growth factors, among others, are probably involved in development of clinical signs, symptoms, and age of presentation. In the future, these factors are expected to be uncovered, allowing for development of new therapeutic strategies.

Relevance

The relevance of the hypothesis is its ability to classify disease entities on a molecular and mechanistic basis. This reclassification of disorders on the basis of molecular abnormalities such as "dystrophinopathies," "ion channelopathies," "sarcomyopathies," or "cytoskeletopathies" could lead to more focused approaches to gene discovery and future therapeutic interventions. For instance, on the basis of the understanding of the molecular aspects of long QT syndrome, we considered the possibility that all ventricular arrhythmias are the result of ion channel abnormalities. On the basis of the hypothesis, we studied the possibility that the cardiac sodium channel gene (SCN5A) was mutated in patients with the idiopathic ventricular fibrillation disorder called Brugada syndrome, identifying mutations in three separate unrelated families. Use of this hypothesis for disorders such as inherited DCM is likely to more narrowly focus efforts at gene identification. In the near future, when the human genome project is completed, this will allow for investigations to more rapidly identify disease-responsible genes. Once the genes are known, and the mechanisms causing the clinical phenotype and natural history are known, improved pharmacologic therapies based on the actual disease mechanism can be produced and utilized. At that time, the impact of molecular genetics on clinical practice and patient care will become fully evident.

REFERENCES

Ahmad, F., Li, D., Karibe, A., Gonzalez, O., Tapscott, T., Hill, R., Weilbaecher, D., Blackie, P., Furey, M., Gardner, M., Bachinski, L.L., and Roberts, R. 1998. Localization of a gene responsible for arrhythmogenic right ventricular dysplasia to chromosome 3p23. *Circulation* 98:2791–2795.

Araishi, K., Sasaoka, T., Imamura, M., Noguchi, S., Hama, H., Wakabayashi, E., Yoshida, M., Hori, T., and Ozawa, E. 1999. Loss of the sarcoglycan complex and sarcospan leads to muscular dystrophy in beta-sarcoglycan-deficient mice. *Hum. Mol. Genetics* 8:1589–1598.

Arber, S., Hunter, J.J., Ross, J., Jr., Hongo, M., Sansig, G., Borg, J., Perriard, J.C., Chien, K.R., and Caroni, P. 1997. MLP-deficient mice exhibit a disruption of cardiac cytoarchitectural organization, dilated cardiomyopathy, and heart failure. *Cell* 88:393–403.

Arbustini, E., Morbini, P., Grasso, M., Fasani, R., Verga, L., Bellini, O., Dal Bello, B., Campana, C., Piccolo, G., Febo, O., Opasich, C., Gavazzi, A., and Ferrans, V.J. 1998. Restrictive cardiomyopathy, atrioventricular block and mild to subclinical myopathy in patients with desmin-immunoreactive material deposits. *J. Am. Coll. Cardiol.* 31:645–653.

Badorff, C., Lee, G.H., Lamphear, B.J., Martone, M.E., Campbell, K.P., Rhoads, R.E., and Knowlton, K.U. 1999. Enteroviral protease 2A cleaves dystrophin: Evidence of cytoskeletal disruption in an acquired cardiomyopathy. *Nat. Med.* 5:320–326.

Baig, M.K., Goldman, J.H., Caforio, A.L.P., Coonar, A.S., Keeling, P.J., and McKenna, W.J. 1998. Familial dilated cardiomyopathy: Cardiac abnormalities are common in asymptomatic relatives and may represent early disease. *J. Am. Coll. Cardiol.* 31:195–201.

Barth, P.G., Scholte, H.R., Berden, J.A., Van der Klei-Van Moorsel, J.M., Luyt-Houwen, I.E., Van 't Veer-Korthof, E.T., Van der Harten, J.J., and Sobotka-Plojhar, M.A. 1983. An X-linked mitochondrial disease affecting cardiac muscle, skeletal muscle and neutrophil leucocytes. *J. Neurol. Sci.* 62:327–355.

Berko, B.A. and Swift, M. 1987. X-linked dilated cardiomyopathy. *N. Engl. J. Med.* 316:1186–1191.

Bione, S., D'Adamo, P., Maestrini, E., Gedeon, A.K., Bolhuis, P.A., and Toniolo, D. 1996. A novel X-linked gene, G4.5, is responsible for Barth syndrome. *Nat. Genetics* 12:385–389.

Bione, S., Maestrini, E., Rivella, S., Mancini, M., Regis, S., Romeo, G., and Toniolo, D. 1994. Identification of a novel X-linked gene responsible for Emery–Dreifuss muscular dystrophy. *Nat. Genetics* 8:323–327.

Bleyl, S.B., Mumford, B.R., Thompson, V., Carey, J.C., Pysher, T.J., Chin, T.K., and Ward, K. 1997. Neonatal, lethal noncompaction of the left ventricular myocardium is allelic with Barth syndrome. *Am. J. Hum. Genetics* 61:868–872.

Bonne, G., Carrier, L., Bercovici, J., Cruaud, C., Richard, P., Hainque, B., Gautel, M., Labeit, S., James, M., and Beckmann, J. 1995. Cardiac myosin binding protein-C gene splice acceptor site mutation is associated with familial hypertrophic cardiomyopathy. *Nat. Genetics* 11:438–440.

Bonne, G., Carrier, L., Richard, P., Hainque, B., and Schwartz, K. 1998. Familial hypertrophic cardiomyopathy from mutations to functional defects. *Circ. Res.* 83:380–593.

Bonne, G., DiBarletta, M.R., Varnous, S., Becane, H.M., Hammouda, E.H., Merlini, L., Muntoni, F., Greenberg, C.R., Gary, F., Urtizberea, J.A., and Schwartz, K. 1999. Mutations in the gene encoding lamin A/C cause autosomal dominant Emery-Dreifuss muscular dystrophy. *Nat. Genetics* 21:285–288.

Bowles, K.R., Gajarski, R., Porter, P., Goytia, V., Bachinski, L., Roberts, R., Pignatelli, R., and Towbin, J.A. 1996. Gene mapping of familial autosomal dominant dilated cardiomyopathy to chromosome 10q21–23. *J. Clin. Invest.* 98:1355–1360.

Bowles, N.E., Bowles, K.R., and Towbin, J.A. 2000. The "final common pathway" hypothesis and inherited cardiovascular disease: the role of cytoskeletal proteins in dilated cardiomyopathy. *Herz* 25:168–175.

Brodsky, G.L., Muntoni, F., Miocic, S., Sinagra, G., Sewry, C., and Mestroni, L. 2000. Lamin A/C gene mutation associated with dilated cardiomyopathy with variable skeletal muscle involvement. *Circulation* 101:473–476.

Campbell, K.P. 1995. Three muscular dystrophies: Loss of cytoskeleton-extracellular matrix linkage. *Cell* 80:675–679.

Carrier, L., Bonne, G., Bahrend, E., Yu, B., Richard, P., Niel, F., Hainque, B., Cruaud, C., Gary, F., Labeit, S., Bouhour, J.B., Dubourg, O., Desnos, M., Hagege, A.A., Trent, R.J., Komajda, M., Fiszman, M., and Schwartz, K. 1997. Organization and sequence of human cardiac myosin binding protein C gene (*MYBPC3*) and identification of mutations predicted to produce truncated proteins in familial hypertrophic cardiomyopathy. *Circ. Res.* 80:427–434.

Carrier, L., Hengstenberg, C., Beckmann, J.S., Guicheney, P., Dufour, C., Bercovici, J., Dausse, E., Berebbi-Bertrand, I., Wisnewsky, C., and Pulvenis, D. 1993. Mapping of a novel gene for familial hypertrophic cardiomyopathy to chromosome 11. *Nat. Genetics* 4:311–313.

Chang, W.J., Iannaccone, S.T., Lau, K.S., Masters, B.S., McCabe, T.J., McMillan, K., Padre, R.C., Spencer, M.J., Tidball, J.G., and Stull, J.T. 1996. Neuronal nitric oxide synthase and dystrophin-deficient muscular dystrophy. *Proc. Natl. Acad. Sci. USA* 93:9142–9147.

Charron, P., Dubourg, O., Desnos, M., Bennaceur, M., Carrier, L., Camproux, A.C., Isnard, R., Hagege, A., Langlard, J.M., Bonne, G., Richard, P., Hainque, B., Bouhour, J.B., Schwartz, K., and Komajda, M. 1998. Clinical features and prognostic implications of familial hypertrophic cardiomyopathy related to cardiac myosin binding protein C gene. *Circulation* 97:2230–2236.

Chin, T.K., Perloff, J.K., and Williams, R.G. 1990. Isolated noncompaction of left ventricular myocardium. A study of eight cases. *Circulation* 82:507–513.

Coonar, A.S., Protonotarios, N., Tsatsopoulou, A., Needham, E.W.A., Houlston, R.S., Cliff, S., Otter, M.I., Murday, V.A., Mattu, R.K., and McKenna, W.J. 1998. Gene for arrhythmogenic right ventricular cardiomyopathy with diffuse nonepidermolytic palmoplantar keratoderma and wooly hair (Naxos disease) maps to 17q21. *Circulation* 97:2049–2058.

Coral-Vazquez, R., Cohn, R.D., Moore, S.A., Hill, J.A., Weiss, R.M., Davisson, R.L., Straub, V., Barresi, R., Bansal, D., Hrstka, R.F., Williamson, R., and Campbell, K.P. 1999. Disruption of the sarcoglycan-sarcospan complex in vascular smooth muscle: A novel mechanism for cardiomyopathy and muscular dystrophy. *Cell* 98:465–474.

Corrado, D., Basso, C., Thiene, G., McKenna, W.J., Davies, M.J., Fontaliran, F., Nava, A., Silvestri, F., Blomstrom-Lundquist, C., Wlodarska, E.K., Fontaine, G., and Camerini, F. 1997. Spectrum of clinicopathologic manifestations of arrhythmogenic right ventricular cardiomyopathy/dysplasia: A multicenter study. *J. Am. Coll. Cardiol.* 30:1512–1520.

Corrado, D., Fontaine, G., Marcus, F.I., McKenna, W.J., Nava, A., Thiene, G., and Wichter, T. 2000. Arrhythmogenic right ventricular dysplasia/cardiomyopathy: Need for an international registry. *J. Cardiovasc. Electrophysiol.* 11:827–832.

Cox, G.F. and Kunkel, L.M. 1997. Dystrophies and heart disease. *Curr. Opin. Cardiol.* 12:329–343.

D'Adamo, P., Fassone, L., Gedeon, A., Janssen, E.A., Bione, S., Bolhuis, P.A., Barth, P.G., Wilson, M., Haan, E., Orstavik, K.H., Patton, M.A., Green, A.J., Zammarchi, E., Donati, M.A., and Toniolo, D. 1997. The X-linked gene G4.5 is responsible for different infantile dilated cardiomyopathies. *Am. J. Hum. Genetics* 61:862–867.

Dalakas, M.C., Park, K.-Y., Semino-Mora, C., Lee, H.S., Sivakumar, K., and Goldfarb, L.G. 2000. Desmin myopathy: a skeletal myopathy with cardiomyopathy caused by mutations in the desmin gene. *N. Engl. J. Med.* 342:770–780.

Durand, J.B., Bachinski, L.L., Bieling, L., Czernuszewicz, G.Z., Abchee, A.B., Yu, Q.T., Tapscott, T., Hill, R., Ifegwu, J., and Marian, A.J. 1995. Localization of a gene responsible for familial dilated cardiomyopathy to chromosome 1q32. *Circulation* 92:3387–3389.

Emery, A.E.H. 1987. X-linked muscular dystrophy with early contractures and cardiomyopathy (Emery–Dreifuss type). *Clin. Genetics* 32:360–367.

Fadic, R., Sunada, Y., Walclawik, A.J., Buck, S., Lewandoski, P.J., Campbell, K.P., and Lotz, B.P. 1996. Brief report: Deficiency of a dystrophin-associated glycoprotein (adhalin) in a patient with muscular dystrophy and cardiomyopathy. *N. Engl. J. Med.* 334:362–366.

Farza, H., Townsend, P.J., Carrier, L., Carrier, L., Barton, P.J., Mesnard, L., Bahrend, E., Forissier, J.F., Fiszman, M., Yacoub, M.H., and Schwartz, K. 1998. Genomic organization, alternative splicing and polymorphisms of the human cardiac troponin T gene. *J. Mol. Cell. Cardiol.* 30:1247–1253.

Fatkin, D., Christe, M.E., Aristizabal, O., McConnell, B.K., Srinivasan, S., Schoen, F.J., Seidman, C.E., Turnball, D.H., and Seidman, J.G. 1999. Neonatal cardiomyopathy in mice homozygous for the Arg403Gln mutant in the α-cardiac myosin heavy chain gene. *J. Clin. Invest.* 103:147–153.

Ferlini, A., Galie, N., Merlini, L., Sewry, C., Branzi, A., and Muntoni, E. 1998. A novel Alu-like element rearranged in the dystrophin gene causes a splicing mutation in a family with X-linked dilated cardiomyopathy. *Am. J. Hum. Genetics* 63:436–460.

Forissier, J.F., Carrier, L., Farza, H., Bonne, G., Bercovici, J., Richard, P., Hainque, B., Townsend, P.J., Yacoub, M.H., Faure, S., Dubourg, O., Millaire, A., Hagege, A.A., Desnos, M., Komajda, M., and Schwartz, K. 1996. Codon 102 of the cardiac troponin T gene is a putative hot spot for mutations in familial hypertrophic cardiomyopathy. *Circulation* 94:3069–3073.

Fougerousse, F., Delezoide, A.L., Fiszman, M.Y., Schwartz, K., Beckman, J.S., and Carrier, L. 1998. Cardiac myosin binding protein C gene is specifically expressed in heart during murine and human development. *Circ. Res.* 82:130–133.

Franz, W.M., Cremer, M., and Hermann, R. 1995. X-linked dilated cardiomyopathy: Novel mutation of the dystrophin gene. *Ann. NY Acad. Sci.* 751:470–491.

Franz, W.-M., Muller, M., Muller, A.J., Herrmann, R., Rothmann, T., Cremer, M., Cohn, R.D., Voit, T., and Katus, H.A. 2000. Association of nonsense mutation of dystrophin gene with disruption of sarcoglycan complex in X-linked dilated cardiomyopathy. *Lancet* 355:1781–1785.

Friedman, R.A., Moak, J.P., and Garson, A., Jr. 1991. Clinical course of idiopathic dilated cardiomyopathy in children. *J. Am. Coll. Cardiol.* 18:152–156.

Gautel, M., Fürst, D.O., Cocco, A., and Schiaffino, S. 1998. Isoform transitions of the myosin-binding protein C family in developing human and mouse muscles. Lack of isoform transcomplementation in cardiac muscle. *Circ. Res.* 82:124–129.

Gautel, M., Zuffardi, O., Freiburg, A., and Labeit, S. 1995. Phosphorylation switches specific for the cardiac isoform of myosin binding protein C: a modulator of cardiac contraction? *EMBO J.* 14:1952–1960.

Geisterfer-Lowrance, A.A., Christie, M., and Conner, D.A. 1996. A mouse model of familial hypertrophic cardiomyopathy. *Science* 272:731–734.

Geisterfer-Lowrance, A.A., Kass, S., Tanigawa, G., Vosberg, H.P., McKenna, W., Seidman, C.E., and Seidman, J.G. 1990. A molecular basis for familial hypertrophic cardiomyopathy β-cardiac myosin heavy chain gene missense mutation. *Cell* 62:999–1006.

Goldfarb, L.G., Park, K.-Y., Cervenakova, L., Gorokhova, S., Lee, H.-S., Vasconcelos, O., Nagle, J.W., Semino-Mora, C., Sivakumar, K., and Dalakas, M.C. 1998. Missense mutations in desmin associated with familial cardiac and skeletal myopathy. *Nat. Genetics* 19:402–403.

Graham, R.M. and Owens, W.A. 1999. Pathogenesis of inherited forms of dilated cardiomyopathy. *N. Engl. J. Med.* 341:1759–1762.

Grunig, E., Tasman, J.A., Kucherer, H., Franz, W., Kubler, W., and Katus, H.A. 1998. Frequency and phenotypes of familial dilated cardiomyopathy. *J. Am. Coll. Cardiol.* 31:186–194.

Helbling-Leclerc, A., Zhang, X., Topaloglu, H., Cruand, C., Tesson, F., Weissenbach, J., Tome, F.M., Schwartz, K., Fardeau, M., and Traggvason, K. 1995. Mutations in the laminin α2-chain (LAMA2) cause merosin-deficient congenital muscular dystrophy. *Nat. Genetics* 11:216–218.

Hoffman, E.P., Brown, R.H., and Kunkel, L.M. 1987. Dystrophin: The protein product of the Duchenne muscular dystrophy locus. *Cell* 51:919–928.

Hofmann, P.A., Hartzell, H.C., and Moss, R. 1991. Alterations in Ca^{2+} sensitive tension due to partial extraction of C-protein from rat skinned cardiac myocytes and rabbit skeletal muscle fibers. *J. Gen. Physiol.* 97:1141–1163.

Hunkeler, N.M., Kullman, J., and Murphy, A.M. 1991. Troponin I isoform expression in human heart. *Circ. Res.* 69:1409–1414.

Hunsaker, R.H., Fulkerson, P.K., Barry, F.J., Lewis, R.P., Leier, C.V., and Unverferth, D.V. 1982. Cardiac function in Duchenne's muscular dystrophy: Result of 10-year follow-up study and noninvasive test. *Am. J. Med.* 73:235–238.

Jarcho, J.A., McKenna, W., Pare, J.A.P., Solomon, S.D., Holcombe, R.F., Dickie, S., Levi, T., Donis-Keller, H., and Seidman, J.G. 1989. Mapping a gene for familial hypertrophic cardiomyopathy to chromosome 14q1. *N. Engl. J. Med.* 321:1372–1378.

Johnston, J., Kelley, R.I., Feigenbaum, A., Cox, G.F., Iyer, G.S., Funanage, V.L., and Proujansky, R. 1997. Mutation characterization and genotype–phenotype correlation in Barth syndrome. *Am. J. Hum. Genetics* 61:1053–1058.

Jung, D., Duclos, F., Apostal, B., Straub, V., Lee, J.C., Allamand, V., Venzke, D.P., Sunada, Y., Moomaw, C.R., Leveille, C.J., Slaughter, C.A., Crawford, T.O., McPherson, J.D., and Campbell, K.P. 1996. Characterization of delta-sarcoglycan, a novel component of the oligomeric sarcoglycan complex involved in limb-girdle muscular dystrophy. *J. Biol. Chem.* 271:32321–32329.

Jung, M., Poepping, I., Perrot, A., Ellmer, A.E., Wienker, T.F., Dietz, R., Reis, A., and Osterziel, K.J. 1999. Investigation of a family with autosomal dominant dilated cardiomyopathy defines a novel locus on chromosome 2q14–q22. *Am. J. Hum. Genetics* 65:1068–1077.

Kaprielian, R.R., Stevenson, S., Rothery, S.M., Cullen, M.J., and Severs, N.J. 2000. Distinct patterns of dystrophin organization in myocyte sarcolemma and transverse tubules of normal and diseased human myocardium. *Circulation* 101:2586–2594.

Kass, S., MacRae, C., Graber, H.L., Sparks, E.A., McNamara, D., Boudoulas, H., Basson, C.T., Baker III, P.B., Cody, R.J., and Fishman, M.C. 1994. A gene defect that causes conduction system disease and dilated cardiomyopathy maps to chromosome 1p1–1q1. *Nat. Genetics* 7:546–551.

Katz, A.M. 2000. Cytoskeletal abnormalities in the failing heart. Out on a LIM? *Circulation* 101:2672–2673.

Keeling, P.J., Gang, Y., Smith, G., Seo, H., Bent, S.E., Murday, V., Caforio, A.L., and McKenna, W.J. 1995. Familial dilated cardiomyopathy in the United Kingdom. *Br. Heart J.* 73:417–421.

Kelley, R.I., Cheatham, J.P., Clark, B.J., et al. 1991. X-linked dilated cardiomyopathy with neutropenia, growth retardation, and 3-methylglutaconic aciduria. *J. Pediatr.* 119:738–747.

Kimura, A., Harada, H., Park, J.E., Nishi, H., Satoh, M., Takahashi, M., Hiroi, S., Sasaoka, T., Ohbuchi, N., Nakamura, T., Koyanagi, T., and Hwang, T.H. 1997. Mutations in the cardiac troponin I gene associated with hypertrophic cardiomyopathy. *Nat. Genetics* 16:379–382.

Klietsch, R., Ervasti, J.M., Arnold, W., Campbell, K.P., and Jorgensen, A.O. 1993. Dystrophin-glycoprotein complex and laminin colocalize to the sarcolemma and transverse tubules of cardiac muscle. *Circ. Res.* 72:349–360.

Koenig, M., Hoffman, E.P., Bertelson, C.J., et al. 1987. Complete cloning of the Duchenne muscular dystrophy (DMD) cDNA and preliminary genomic organization of the DMD gene in normal and affected individuals. *Cell* 50:509–517.

Koenig, M. and Kunkel, L.M. 1990. Detailed analysis of the repeat domain of dystrophin reveals four potential hinge segments that may confer flexibility. *J. Biol. Chem.* 265:4560–4566.

Koenig, M., Monaco, A.P., and Kunkel, L.M. 1988. The complete sequence of dystrophin predicts a rod-shaped cytoskeletal protein. *Cell* 53:219–226.

Krajinovic, M., Pinamonti, B., Sinagra G., Vatta, M., Severini, G.M., Milasin, J., Falaschi, A., Camerini, F., Giacca, M., and Mestroni, L. 1995. Linkage of familial dilated cardiomyopathy to chromosome 9. *Am. J. Hum. Genetics* 57:846–852.

Lees-Miller, J.P. and Helfman, D.M. 1991. The molecular basis for tropomyosin isoform diversity. *Bioessays* 13:429–437.

Li, D., Tapscott, T., Gonzalez, O., Burch, P.E., Quinones, M.A., Zoghbi, W.A., Hill, R., Bachinski, L.L., Mann, D.L., and Roberts, R. 1999. Desmin mutation responsible for idiopathic dilated cardiomyopathy. *Circulation* 100:461–464.

Li, D., Ahmad, F., Gardner, M.J., Weilbaecher, D., Hill, R., Karibe, A., Gonzalez, O., Tapscott, T., Sharratt, G.P., Bachinski, L.L., and Roberts, R. 2000. The locus of a novel gene responsible for arrhythmogenic right ventricular dysplasia characterized by early onset and high penetrance maps to chromosome 10p12–p14. *Am. J. Hum. Genetics* 66:148–156.

Li, Z., Colucci-Guyon, E., Pincon-Raymond, M., Mericskay, M., Pournin, S., Paulin, D., and Babinet, C. 1996. Cardiovascular lesions and skeletal myopathy in mice lacking desmin. *Dev. Biol.* 175:362–366.

Lim, L.E., Duclos, F., Broux, O., Bourg, N., Sunada, Y., Allamand, V., Meyer, J., Richard, J., Moomaw, C., and Slaughter, C. 1995. Beta-sarcoglycan: Characterization and role in limb-girdle muscular dystrophy linked to 4q12. *Nat. Genetics* 11:257–265.

MacRae, C.A., Ghaisas, N., Kass, S., Donnelly, S., Basson, C.T., Watkins, H.C., Anan, R., Thierfelder, L.H., McGarry, K., and Rowland, E. 1995. Familial hypertrophic cardiomyopathy with Wolff–Parkinson–White Syndrome maps to a locus on chromosome 7q3. *J. Clin. Invest.* 96:1216–1220.

Maeda, M., Holder, E., Lowes, B., Valent, S., and Bies, R.D. 1997. Dilated cardiomyopathy associated with deficiency of the cytoskeletal protein metavinculin. *Circulation* 95:17–20.

Manolio, T.A., Baughman, K.L., Rodeheffer, R., Pearson, T.A., Bristow, J.D., Michels, V.V., Abelmann, W.H., and Harlan, W.R. 1992. Prevalence and etiology of idiopathic dilated cardiomyopathy (summary of a National Heart, Lung and Blood Institute Workshop). *Am. J. Cardiol.* 69:1458–1466.

Marcus, F.I., Fontaine, G., Guiraudon, G., Frank, R., Laurenceau, J.L., Malergue, C., and Grosgogeat, Y. 1982. Right ventricular dysplasia. A report of 24 adult cases. *Circulation* 65: 384–398.

Marian, A.J., Yu, Q.-T., Workman, R., Greve, G., and Roberts, R. 1993. Angiotensin-converting enzyme polymorphism in hypertrophic cardiomyopathy and sudden cardiac death. *Lancet* 342:1085–1086.

Maron, B.J., Gardin, J.M., Flack, J.M., Gidding, S.S., Kurosaki, T.T., and Bild, D.E. 1995. Prevalence of hypertrophic cardiomyopathy in a general population of young adults: echocardiographic analysis of 4111 subjects in the CARDIA study. *Circulation* 92:785–789.

Maron, B.J., Shirani, J., Pollac, L.C., Mathenge, R., Roberts, W.C., and Mueller, F.O. 1996. Sudden death in young competitive athletes. Clinical demographic and pathological profiles. *JAMA* 27:199–204.

Maron, B.J., Spirito, P., Wesley, Y.E., and Arce, J. 1986. Development and progression of left ventricular hypertrophy in children with hypertrophic cardiomyopathy. *N. Engl. J. Med.* 315:610–614.

Maron, B.J. 1997. Hypertrophic cardiomyopathy. *Lancet* 350:127–133.

McConnell, B.K., Jones, K.A., Fatkin, D., Arroyo, L.H., Lee, R.T., Aristizabal, O., Turnbull, D.H., Georgakopoulos, D., Kass, D., Bond, M., Nimura, H., Schoen, F.J., Conner, D., Fischman, D.A., Seidman, C.E., and Seidman, J.G. 1999. Dilated cardiomyopathy in homozygous myosin-binding protein-C mutant mice. *J. Clin. Invest.* 104:1235–1244.

McKoy, G., Protonotarios, N., Cosby, A., Tsatsopoulou, A., Anastasakis, A., Coonar, A., Norman, M., Baboonian, C., Jeffery, S., and McKenna, W.J. 2000. Identification of a deletion in plakoglobin in arrhythmogenic right ventricular cardiomyopathy with palmoplantar keratoderma and woolly hair (Naxos disease). *Lancet* 355:2119–2124.

Melacini, P., Fanin, M., Danieli, G.A., Villanova, C., Martinello, F., Miorin, M., Freda, M.P., Miorelli, M., Mostacciuolo, M.L., Fasoli, G., Angelini, C., and Dalla Volta, S. 1996. Myocardial involvement is very frequent among patients affected with subclinical Becker's muscular dystrophy. *Circulation* 94:3168–3175.

Melacini, P., Fanin, M., Duggan, D.J., Freda, M.P., Berardinelli, A., Danieli, G.A., Barchitta, A., Hoffman, E.P., Dalla Volta, S., and Angelini, C. 1999. Heart involvement in muscular dystrophies due to sarcoglycan gene mutations. *Muscle Nerve* 22:473–479.

Melberg, A., Oldfors, A., Blomstrom-Lundquist, C., Stalberg, E., Carlsson, B., Larsson, E., Lidell, C., Eeg-Olofsson, K.E., Wikstrom, G., Henriksson, K.G., and Dahl, N. 1999. Autosomal dominant myofibrillar myopathy with arrhythmogenic right ventricular cardiomyopathy linked to chromosome 10q. *Ann. Neurol.* 46:684–692.

Meng, H., Leddy, J.J., Frank, J., Holland, P., and Tuana, B.S. 1996. The association of cardiac dystrophin with myofibrils/z-discs regions in cardiac muscle suggests a novel role in the contractile apparatus. *J. Biol. Chem.* 271:12364–12371.

Mesnard, L., Logeart, D., Taviaux, S., Diriong, S., Mercadier, J.J., and Samson, F. 1995. Human cardiac troponin T. Cloning and expression of new isoforms in the normal and failing heart. *Circ. Res.* 76:687–692.

Messina, D.N., Speer, M.C., Pericak-Vance, M.A., and McNally, E.M. 1997. Linkage of familial dilated cardiomyopathy with conduction defect and muscular dystrophy to chromosome 6q23. *Am. J. Hum. Genetics* 61:909–917.

Michels, V.V., Moll, P.P., Miller, F.A., Tajik, A.J., Chu, J.S., Dirscoll, D.J., Burnett, J.C., Rodeheffer, R.J., Chesebro, J.H., and Tazelaar, H.D. 1992. The frequency of familial dilated cardiomyopathy in a series of patients with idiopathic dilated cardiomyopathy. *N. Engl. J. Med.* 326:77–82.

Milasin, J., Muntoni, F., Severini, G.M., Bartoloni, L., Vatta, M., Krajinovic, M., Mateddu, A., Angelini, C., Camerini, F., Falaschi, A., Mestroni, L., and Giacca, M. 1996. A point mutation in the 5′ splice site of the dystrophin gene first intron responsible for X-linked dilated cardiomyopathy. *Hum. Mol. Genetics* 5:73–79.

Milner, D.J., Weitzer, G., Tran, D., Bradley, A., and Capetanaki, Y. 1996. Disruption of muscle architecture and myocardial degeneration in mice lacking desmin. *J. Cell Biol.* 134: 1255–1270.

Mogensen, J., Klausen, I.C., Pederson, A.K., Egeblad, H., Bross, P., Kruse, T.A., Gregersen, N., Hansen, P.S., Baandrup, U., and Borglum, A.D. 1999. α-cardiac actin is a novel disease gene in familial hypertrophic cardiomyopathy. *J. Clin. Invest.* 103:R39–R43.

Moolman, J.C., Corfield, V.A., Posen, B., Ngumbela, K., Seidman, C., Brink, P.A., and Watkins, H. 1997. Sudden death due to troponin T mutations. *J. Am. Coll. Cardiol.* 29:549–555.

Mott, A.R., Pac, F., Denfield, S.W., Price, J.K., Belmont, J., Craigen, W., Lipshultz, S.E., and Towbin, J.A. 2000. Hypertrophic cardiomyopathy in children. *J. Am. Coll. Cardiol.* In press.

Muchir, A., Bonne, G., van der Kooi, A.J., van Meegen, M., Baas, F., Bolhuis, P.A., de Visser, M., and Schwartz, K. 2000. Identification of mutations in the gene encoding lamin A/C in autosomal dominant limb girdle muscular dystrophy with atrioventricular conduction disturbance (LGMD1B). *Hum. Mol. Genetics* 9:1453–1459.

Muntoni, F., Cau, M., Ganau, A., Congiu R., Arvedi, G., Mateddu, A., Marrosu, M.G., Cianchetti, C., Realdi, G., Cao, A., and Melis, M.A. 1993. Brief report: Deletion of the dystrophin muscle-promoter region associated with X-linked dilated cardiomyopathy. *N. Engl. J. Med.* 329:921–925.

Nava, A., Thiene, G., Canciani, B., Scognamiglio, R., Daliento, L., Buja, G.F., Martini, B., Stritoni, P., and Fasoli, G. 1988. Familial occurrence of right ventricular dysplasia: A study involving nine families. *J. Am. Coll. Cardiol.* 12:1222–1228.

Neustein, H.D., Lurie, P.R., Dahms, B., and Takahashi, M. 1979. An X-linked recessive cardiomyopathy with abnormal mitochondria. *Pediatrics* 64:24–29.

Nigro, G., Politano, L., Nigro, V., Petretta, V.R., and Comi, L.I. 1994. Mutation of dystrophin gene and cardiomyopathy. *Neuromusc. Disord.* 4:371–379.

Nigro, V., de Sa Moreira, E., Piluso, G., Vainzof, M., Belsito, A., Politano, L., Puca, A.A., Passos-Bueno, M.R., and Zatz, M. 1996a. Autosomal recessive limb-girdle muscular dystrophy, LGMD2F, is caused by a mutation in the delta-sarcoglycan gene. *Nat. Genetics* 14:195–198.

Nigro, V., Piluso, G., Belsito, A., Politano, L., Puca, A.A., Papparella, S., Rossi, E., Viglietto, G., Esposito, M.G., Abbondanza, C., Medici, N., Molinari, A.M., Nigro, G., and Puca, G.A. 1996b. Identification of a novel sarcoglycan gene at 5q33 encoding a sarcolemmal 35 kDa glycoprotein. *Hum. Mol. Genetics* 5:1179–1186.

Nigro, V., Okazaki, Y., Belsito, A., Piluso, G., Matsuda, Y., Politano, L., Nigro, G., Ventura, C., Abbondanza, C., Molinari, A.M., Acampora, D., Nishimura, M., Hayashizaki, Y., and Puca, G.A. 1997. Identification of the Syrian hamster cardiomyopathy gene. *Hum. Mol. Genetics* 6:601–607.

Nimura, H., Bachinski, L.L., Sangwatanaroh, S., Watkins, H., Chudley, A.E., McKenna, W., Kristinsoon, A., Roberts, R., Sole, M., Maron, B.J., Seidman, J.G., and Seidman, C.E. 1998. Mutations in the gene for cardiac myosin-binding protein C and late-onset familial hypertrophic cardiomyopathy. *N. Engl. J. Med.* 33:1248–1257.

Nowak, K.J., Wattanasirichaigoon, D., Goebel, H.H., Wilce, M., Pelin, K., Donner, K., Jacob, R.L., Hubner, C., Oexle, K., Anderson, J.R., Verity, C.M., and North, K.N. 1999. Mutations in the skeletal muscle alpha-actin gene in patients with actin myopathy and namaline myopathy. *Nat. Genetics* 23:208–212.

Ohlendieck, K. 1996. Towards an understanding of the dystrophin-glycoprotein complex: Linkage between the extracellular matrix and the membrane cytoskeleton in muscle fibers. *Eur. J. Cell Biol.* 69:1–10.

Olson, T.M. and Keating, M.T. 1996. Mapping a cardiomyopathy locus to chromosome 3p22–p25. *J. Clin. Invest.* 97:528–532.

Olson, T.M., Michels, V.V., Thibodeau, S.N., Tai, Y.S., and Keating, M.T. 1998. Actin mutations in dilated cardiomyopathy, a heritable form of heart failure. *Science* 280:750–752.

Ortiz-Lopez, R., Li, H., Su, J., Goytia, V., and Towbin, J.A. 1997. Evidence for a dystrophin missense mutation as a cause of X-linked dilated cardiomyopathy. *Circulation* 95:2434–2440.

Ozawa, E., Noguchi, S., Mizuno, Y., Hagiwara, Y., and Yoshida, M. 1998. From dystrophinopathy to sarcoglycanopathy: evolution of a concept of muscular dystrophy. *Muscle Nerve* 21:421–438.

Ozawa, E., Yoshida, M., Suzuki, A., Mizuno, Y., Hagiwara, Y., and Noguchi, S. 1995. Dystrophin-associated proteins in muscular dystrophy. *Hum. Mol. Genetics* 4:1711–1716.

Petrof, B.J., Shrager, J.B., Stedman, H.H., Kelly, A.M., and Sweeney, H.L. 1993. Dystrophin protects the sarcolemma from stresses developed during muscle contraction. *Proc. Natl. Acad. Sci. USA* 90:3710–3714.

Poetter, K., Jiang, H., Hassanzadeh S., Master, S.R., Chang, A., Dalakas, M.C., Rayment, I., Sellers, J.R., Fananapazir, L., and Epstein, N.D. 1996. Mutation in either of the essential regulatory light chains of myosin are associated with a rare myopathy in human heart and skeletal muscle. *Nat. Genetics* 13:63–69.

Raffaele Di Barletta, M., Ricci, E., Galluzzi, G., Tonali, P., Mora, M., Morandi, L., Romorini, A., Voit, T., Orstavik, K.H., Merlini, L., and Trevisan, C. 2000. Different mutates in the LMNA gene cause autosomal dominant and autosomal recessive Emery–Dreifuss muscular dystroopy. *Am. J. Hum. Genetics* 66:1407–1412.

Rampazzo, A., Nava, A., Danieli, G.A., Buja, G.F., Daliento, L., Fasoli, G., Scognamiglio, R., Corrado, D., and Thiene, G. 1994. The gene for arrhythmogenic right ventricular cardiomyopathy maps to chromosome 14q23–q24. *Hum. Mol. Genetics* 3:959–962.

Rampazzo, A., Nava, A., Erne, P., Eberhard, M., Vian, E., Slomp, P., Tiso, N., Thiene, G., and Danieli, G.A. 1995. A new locus for arrhythmogenic right ventricular cardiomyopathy (ARVD2) maps to chromosome 1q42–q43. *Hum. Mol. Genetics* 4:2151–2154.

Rampazzo, A., Nava, A., Miorin, M., Fonderico, P., Pope, B., Tiso, N., Livolsi, B., Lerman, B., Thiene, G., and Danieli, G.A. 1997. A new locus for arrhythmogenic right ventricular cardiomyopathy (ARVD4) maps to chromosome 2q32. *Genomics* 45:259–263.

Rayment, I., Holden, H.M., Whittaker, M., Yohn, C.B., Lorenz, M., Holmes, K.C., and Milligan, R.A. 1993. Structure of the actin-myosin complex and its implications for muscle contraction. *Science* 261:58–65.

Richardson, P., McKenna, W.J., Bristow, M., Maisch, B., Mautner, B., O'Connell, J., Olsen, E., Thiene, G., Goodwin, J., Gyarfas, I., Martin, I., and Nordet, P. 1996. Report of the 1995 World Health Organization/International Society and Federation of Cardiology Task Force on the definition and classification of cardiomyopathies. *Circulation* 93:841–842.

Sakamoto, A., Abe, M., and Masaki, T. 1999. Delineation of genomic deletion in cardiomyopathic hamster. *FEBS Lett.* 447:124–128.

Sakamoto, A., Ono, K., Abe, M., Jasmin, G., Eki, T., Murakami, Y., Masaki, T., Toyo-oka, T., and Hanaoka, F. 1997. Both hypertrophic and dilated cardiomyopathies are caused by mutation of the same gene, delta-sarcoglycan, in hamster: an animal model of disrupted dystrophin-associated glycoprotein complex. *Proc. Natl. Acad. Sci. USA* 94:13873–13878.

Satoh, M., Takahashi, M., Sakamoto, T., Hiroe, M., Marumo, I., and Kimura, A. 1999. Structural analysis of the titin gene in hypertrophic cardiomyopathy: Identification of a novel disease gene. *Biochem. Biophys. Res. Commun.* 262:411–417.

Schaper, J., Froede, R., Hein, S., Buck, A., Hashizume, H., Speiser, B., Friedl, A., and Bleese, N. 1991. Impairment of the myocardial ultrastructure and changes of the cytoskeleton in dilated cardiomyopathy. *Circulation* 83:503–514.

Schiaffino, S. and Reggiani, C. 1996. Molecular diversity of myofibrillar proteins: gene regulation and functional significance. *Physiol. Rev.* 76:371–423.

Schonberger, J., Levy, H., Grunig, E., Sangwatanoroj, S., Fatkin, D., MacRae, C., Stacker, H., Halpin, C., Eavey, R., Philbin, E.F., Katus, H., Seidman, J.G., and Seidman, C.E. 2000. Dilated cardiomyopathy and sensorineural hearing loss. A heritable syndrome that maps to 6q23–24. *Circulation* 101:1812–1818.

Severini, G.A., Krajinovic, M., Pinamonti, B., Sinagra, G., Fioretti, P., Brunazzi, M.C., Falaschi, A., Camerini, F., Giacca, M., and Mestroni, L. 1996. A new locus for arrhythmogenic right ventricular dysplasia on the long arm of chromosome 14. *Genomics* 31:193–200.

Siu, B.L., Nimura, H., Osborne, J.A., Fatkin, D., MacRae, C., Solomon, S., and Benson, D.W. 1999. Familial dilated cardiomyopathy locus maps to chromosome 2q31. *Circulation* 99: 1022–1026.

Solaro, R.J. and Van Eyk, J. 1996. Altered interactions among thin filament proteins modulate cardiac function. *J. Mol. Cell. Cardiol.* 28:217–230.

Solomon, S.D., Jarcho, J.A., McKenna, W.J., Geisterfer-Lowrance, A., Germain, R., Salerni, R., Seidman, J.G., and Seidman, C.E. 1990. Familial hypertrophic cardiomyopathy is a genetically heterogeneous disease. *J. Clin. Invest.* 86:993–999.

Spirito, P., Seidman, C.E., McKenna, W.J., and Maron, B.J. 1997. The management of hypertrophic cardiomyopathy. *N. Engl. J. Med.* 336:775–785.

Straub, V., Duclos, F., Venzke, D.P., Lee, J.C., Cutshall, S., Leveille, C.J., and Campbell, K.P. 1998. Molecular pathogenesis of muscle degeneration in the delta-sarcoglycan deficient hamster. *Am. J. Pathol.* 153:1623–1630.

Tesson, F., Dufour, C., Moolman, J.C., Carrier, L., Al-Mahdawi, S., Chojnowska, L., Dubourg, O., Soubrier, E., Brink, J., Komajda, M., Guicheney, P., Schwartz, K., and Feingold, J. 1997. The influence of the angiotensin I converting enzyme genotype in familial hypertrophic cardiomyopathy varies with the disease gene mutation. *J. Mol. Cell. Cardiol.* 29:831–838.

Thiene, G., Nava, A., Corrado, D., Rossi, L., and Pennelli, N. 1988. Right ventricular cardiomyopathy and sudden death in young people. *N. Engl. J. Med.* 318:129–133.

Thierfelder, L., MacRae, C., Watkins, H., Tomfohrde, J., Williams, M., McKenna, W., Bohm, K., Noeske, G., Schlepper, M., and Bowcock, A. 1993. A familial hypertrophic cardiomyopathy locus maps to chromosome 15q2. *Proc. Natl. Acad. Sci. USA* 90:6270–6274.

Thierfelder, L., Watkins, H., MacRae, C., Lamas, R., McKenna, W., Vosberg, H.P., Seidman, J.G., and Seidman, C.E. 1994. α-tropomyosin and cardiac troponin T mutations cause familial hypertrophic cardiomyopathy: a disease of the sarcomere. *Cell* 77:701–712.

Thornell, L., Carlsson, L., Li, Z., Mericskay, M., and Paulin, D. 1997. Null mutation in the desmin gene gives rise to a cardiomyopathy. *J. Mol. Cell. Cardiol.* 29:2107–2124.

Towbin, J. 1993. Molecular genetic aspects of cardiomyopathy. *Biochem. Med. Metab. Biol.* 49:285–320.

Towbin, J.A. 1995. *Biochemical and Molecular Characterization of X-Linked Dilated Cardiomyopathy (XLCM): Developmental Mechanisms of Heart Disease*, Eds. E.B. Clark, R.R. Markwald, and A. Takao, Futura Publishing Co., Inc., New York, pp. 121–132.

Towbin, J.A. 1998. The role of cytoskeletal proteins in cardiomyopathies. *Curr. Opin. Cell Biol.* 10:131–139.

Towbin, J.A. 1999. Pediatric myocardial disease. *Pediatr. Clin. North. Am.* 46:289–312.

Towbin, J.A. 2000. Cardiac arrhythmias: The genetic connection. *J. Cardiovasc. Electrophysiol.* 11:601–602.

Towbin, J.A., Bowles, K.R., and Bowles, N.E. 1999a. Etiologies of cardiomyopathy and heart failure. *Nat. Med.* 5:266–267.

Towbin, J.A. and Lipshultz, S.E. 1999b. Genetics of neonatal cardiomyopathy. *Curr. Opin. Cardiol.* 14:250–262.

Tsubata, S., Bowles, K.R., Vatta, M., Zintz, C., Titus, J., Muhonen, L., Bowles, N.E., and Towbin, J.A. 2000. Mutations in the human delta-sarcoglycan gene infamilial and sporadic dilated cardiomyopathy. *J. Clin. Invest.* 106:655–662.

Van der Kooi, A.J., Ledderhof, T.M., de Voogt, W.G., Res, C.J., Bouwsma, G., Troost, D., Busch, H.F., Becker, A.E., and de Visser, M. 1996. A newly recognized autosomal dominant limb girdle muscular dystrophy with cardiac involvement. *Ann. Neurol.* 39:636–642.

Van der Kooi, A.J., van Meeger, M., Ledderhof, T.M., McNally, E.M., de Visser, M., and Bolhuis, P.A. 1997. Genetic heterogeneity of a newly recognized autosomal dominant limb girdle muscular dystrophy with cardiac involvement (LGMD1B). *Am. J. Hum. Genetics* 60:891–895.

Vatta, M., Li, H., and Towbin, J.A. 2000. Molecular biology of arrhythmic syndromes. *Curr. Opin. Cardiol.* 15:12–22.

Vikstrom, K.L., Factor, S.M., and Leinwand, L.A. 1996. Mice expressing mutant myosin heavy chains are a model for familial hypertrophic cardiomyopathy. *Mol. Med.* 2:556–567.

Wallace, D.C. 1992. Mitochondrial genetics: A paradigm for aging and degenerative diseases? *Science* 256:628–632.

Watkins, H., Conner, D., Thierfelder, L., Jarcho, J.A., MacRae, C., McKenna, W.J., and Matson, B.J. 1995a. Mutations in the cardiac myosin binding protein-C gene on chromosome 11 cause familial hypertrophic cardiomyopathy. *Nat. Genetics* 11:434–437.

Watkins, H., MacRae, C., Thierfelder, L., Chou, Y.H., Frenneaux, M., McKenna, W., Seidman, J.G., and Seidman, C.E. 1993. A disease locus for familial hypertrophic cardiomyopathy maps to chromosome 1q3. *Nat. Genetics* 3:333–337.

Watkins, H., McKenna, W.J., Thierfelder, L., Suk, H.J., Anan, R., O'Donoghue, A., Spirito, P., Matsumori, A., Moravec, C.S., and Seidman, J.G. 1995b. Mutations in the genes for cardiac troponin T and α-tropomyosin in hypertrophic cardiomyopathy. *N. Engl. J. Med.* 332: 1058–1064.

Watkins, H., Rosenzweig, T., Hwang, D.S., Levi, T., McKenna, W., Seidman, C.E., and Seidman, J.G. 1992. Characteristic and prognostic implications of myosin missense mutations in familial hypertrophic cardiomyopathy. *N. Engl. J. Med.* 326:1106–1114.

Wiles, H.B., McArthur, P.D., Taylor, A.B., Gillette, P.C., Fyfe, D.A., Matthews, J.P., and Shelton, L.W. 1991. Prognostic features of children with idiopathic dilated cardiomyopathy. *Am. J. Cardiol.* 68:1372–1376.

Yang, Q. 1998. A mouse model of myosin binding protein C human familial hypertrophic cardiomyopathy. *J. Clin. Invest.* 102:1292–1300.

Yonega, K., Okamoto, H., Machida, M., Onozuka, H., Noguchi, M., Mikami, T., Kawaguchi, H., Murakami, M., Uede, T., and Kitabatake, A. 1995. Angiotensin-converting enzyme gene polymorphism in Japanese patients with hypertrophic cardiomyopathy. *Am. Heart J.* 130:1089–1093.

Yoshida, K., Nakamura, A., Yazak, M., Ikeda, S., and Takeda, S. 1998. Insertional mutation by transposable element, L1, in the DMD gene results in X-linked dilated cardiomyopathy. *Hum. Mol. Genetics* 7:1129–1132.

Zolk, O., Caroni, P., and Bohm, M. 2000. Decreased expression of the cardiac LIM domain protein MLP in chronic human heart failure. *Circulation* 101:2674–2677.

Index